Archaeology and the Methodology of Science

ARCHAEOLOGY AND THE METHODOLOGY OF SCIENCE

By Jane H. Kelley and Marsha P. Hanen

University of New Mexico Press
Albuquerque

For Dave and Bob

Library of Congress Cataloging-in-Publication Data

Kelley, Jane Holden, 1928–
    Archaeology and the methodology of science.

    Bibliography: p.
    Includes index.
    1. Archaeology—Philosophy.   2. Archaeology—
Methodology.   I. Hanen, Marsha P., 1936–
II. Title.
CC72.K45   1988   930.1   87-19228
ISBN 0-8263-1030-3

Permissions acknowledgments may be found on page 419.

/ Contents

/ Preface

This book has had an unusually long period of gestation. Originally conceived as a short paper in response to some of the early extravagances of the "New Archaeology," it grew as the literature grew, changed direction and emphasis several times as the archaeological response to philosophical theory became more settled, and ultimately emerged as a kind of summary of developments over the past fifteen years, together with some attempt to provide a framework for future theoretical discussion. The book is intended primarily for an archaeological audience—as a useful source book for archaeologists at all stages of their careers and as primary or secondary reading in method and theory courses in archaeology. It is an attempt at an understanding of contemporary philosophy of science in a very general form in order to illuminate issues in archaeological theory and method. It is thus not so much an exercise in directly applying philosophy to archaeology as it is an examination of ways in which philosophical insights might help us to understand, to clarify, and to improve theorizing in archaeology. As well as its primary direction to an audience of archaeologists, persons working in related social sciences such as anthropology or history and philosophers interested in the social sciences should find the book useful.

Approaches to the connections between philosophy and archaeology have included primarily philosophical works which apply philosophical concepts to archaeology, and work more in the tradition of archaeological method and theory, in which the primary emphasis is on ways in which an understanding of conceptual foundations can help in the practice of archaeology.

Our own orientation exemplifies the third most obvious stance one could take on these matters—that of a conscious and deliberate interdisciplinary approach. Not surprisingly, this approach, like the work itself, has evolved over a number of years. The project began with the puzzlement of an archaeologist about the nature and validity of the philosophical arguments to which the early New Archaeologist writers appealed. What evolved was a collaboration which, to some extent, mirrored the development of the general collaboration between philosophy and archaeology. Thus, what began in 1971 as a short critical paper exposing some misconceptions about philosophical concepts that had entered the archaeological literature, soon developed into a monograph attempting to clarify for an archaeological audience the standard concepts of philosophy of science—explanation, confirmation, law, theory, and so on. Along the way, we had come to see that the essentially adversarial stance wherein philosophers criticized the apparently wholesale adoption into archaeology of a particular philosophical approach and the misunderstanding on the part of archaeologists of certain central concepts such as induction led only to defensiveness and questioning of the relevance of philosophy to archaeology.

Fortunately (in retrospect), the work continued, partly because new material that required consideration kept appearing; and, as we were both busy with other concerns, that consideration took time. As we proceeded, two things became evident: that the positivist view of the philosophy of science on which most of the New Archaeology had been based was being found inadequate and was being replaced by a much broader perspective, and at the same time, that archaeologists were beginning to ask theoretical questions more relevant to their actual doing of archaeology. The nature of the contact between the two disciplines had changed from that of the early 1970s, in which the New Archaeologists had been quite rightly concerned at the paucity of systematic attention to method in the "old" archaeology, and had turned to philosophy in order to supply this lack. This had given rise to two problems— archaeologists were working with unfamiliar material and thus unable to evaluate it properly or to distinguish what was helpful from what was not, and philosophers were playing the role of sometimes patronizing mentors, offering a general theory frequently without careful examination of specifically archaeological problems. They often seemed not to realize, or to forget, that their own views were context dependent in important ways. The new contact, where it survived the disillusionment of the earlier adversarial approaches, was much more cooperative in spirit.

Our own collaboration has passed through all of the stages that have characterized the broader contact between the two disciplines. The process of coming to understand something of a discipline in which one wasn't specifically trained and communicating about it is difficult and time consuming; and beyond this, to learn to leave behind one's disciplinary biases and try to see a problem from a new perspective—one that is characteristic of neither discipline but rather employs and integrates insights from both—is even more difficult. For one thing, virtually everything in our academic training and surroundings works against such an approach. We live in the midst of ever greater specialization and the cult of the expert. Almost everything we do academically is within a fairly narrowly characterized discipline; the departmental structure of universities fosters an attitude of intellectual territoriality which denies the possibility of anyone's competence to speak outside his or her area of specialization. It is this attitude that we have tried to counteract in our work, and it is this that probably represents the greatest difference between our approach and other current ones, in spite of the similarities in our substantive views. We are painfully aware that we have done no more than make a beginning with this approach, and we hope others will find that beginning useful enough to pursue it further.

Specifically, the structure of the book is as follows. Chapter 1 provides a general introduction to the philosophical background relevant to method and theory in archaeology, including some fairly detailed discussion of the historical background both in archaeology and in philosophy and of the use of philosophical concepts in archaeology. Chapter 2 attempts to clarify a number of confusions that have arisen and were never properly sorted out, especially concerning testing, discovery, and justification and the use of inductive and deductive inference in archaeology. In Chapter 3 we discuss two views of science—the Kuhnian and the Popperian—which have received considerable attention from both scientists and philosophers, and we try to provide an assessment of the relevance of these views for archaeology. The treatment of Kuhn's work leads, in Chapter 4, to a discussion of the social context of archaeology, including cultural and institutional influences, technology, and communication in archaeology. This area is generating increasing interest within the discipline, and we have tried here to provide only a very preliminary overview of the issues that will need further research. The first four chapters of the book thus treat the general context—historical, philosophical, and social—within which any approach to theorizing in archaeology must take place. Characteristically, we attempt to integrate, as far as possible, the archaeological material into that context.

Chapters 5 and 6 treat the two standard issues of explanation and inference in archaeology. In both cases, in addition to the usual material, we discuss more recent developments than is customary in the archaeological literature. In particular, we provide an analysis of current debates about scientific realism and an evaluation of their relevance for archaeology, as well as a discussion of the relations among confirmation, explanation, and acceptance of hypotheses. Chapter 7 offers a number of case studies to show the use of various forms of inference in archaeology. This is intended to provide an indication both of the variety of forms that acceptable inference in archaeology can take and of the ways in which the rather abstract philosophical discriptions of the structure of scientific inference find concrete application. It is our view that there is as much to be learned from an examination of the contextual features of particular cases of archaeological inference as from an attempt to abstract from those features to expose the logical structure of the inference.

In the last chapter we make explicit some of what has been implicit about theorizing in archaeology throughout the book. Using another example, we show how much such theorizing can be viewed as a form of inference to the best explanation, and we draw further conclusions about the importance of analogical reasoning in archaeology. Again, we argue that the acceptance of these forms of reasoning does not imply a commitment to scientific realism, and that recent fascination in archaeological circles with that view can no more be expected to solve the outstanding problems of archaeological theorizing than did the earlier attachment to a positivist perspective. What is required is the integration of insights, wherever they originate, to help us draw a coherent picture of both the matter and the method of archaeology; but this by no means entails the adoption of whatever "ism" may be currently fashionable in philosophy. Indeed, such espousal may be counterproductive in the long run, for it may focus attention away from the sort of analysis of examples of successful inference in archaeology that we have attempted in Chapter 7; and this, we believe, is really where our understanding of theorizing must begin.

Any work of this sort, written over a period of years, necessarily exemplifies a perspective emerging from the context in which it was written. If it were to be written at a different time, the perspective would be different. This book is no exception. The perspective of this book grew from a series of developments in both archaeology and philosophy, having to do with a stress on the importance of science, of method, of precision, of clarity and of objectivity. The topics considered, and the ways in which they are considered in the book, reflect these concerns.

If we were to undertake to write a book on this subject now, its focus would be somewhat different. It would, for example, emphasize more than we have done a sense of history, and the historical context for developments in archaeology. It would also attempt to be even more interdisciplinary and more holistic, with greater stress on the interrelationships among topics and approaches. Most importantly, perhaps, it would attempt to incorporate much more thoroughly than we have managed here the insight available from the growing body of feminist scholarship on these areas. We have done some of this in Chapter 4, but there is much more to be done. In particular, there must be a rethinking of what is problematic in the first place, what requires explanation and what is interesting and worthy of research. As well, we would have more to say on the unfruitfulness of most traditional dualisms—a view that we adopt in the book but would now argue even more strongly; and we would attempt a more thoroughgoing revision of our understanding of the nature of objectivity—a concept which has for too long been tied to a rather inflexible and primarily male notion of rationality.

These are all areas that need to be developed. The beginnings are here, but the working out is for another time.

/ Acknowledgments

For more than a decade, we have profited from the encouragement and criticisms of numerous friends and colleagues. Our first vote of thanks goes to Janet Keeping, who, at a very critical intermediate stage, performed miracles as a go-between for us as researcher, writer, and bibliophile. We are also deeply indebted to Alison Wylie for her ongoing interest in our manuscript and her many and invariably useful critical comments, insights, and suggestions. Linda Cordell, Nicholas David, Jonathan Driver, Robert Janes, David Kelley, John Robertson, Robert Weyant, Peter White, Michael Wilson, Margaret Glass, and anonymous reviewers have read some or all of the manuscript in one or more of its incarnations and offered helpful criticisms.

A grant from the University of Calgary for research assistance and manuscript production is gratefully acknowledged.

A number of typists have worked on various versions of the manuscript and we are grateful to all of them. For the last draft of the manuscript the primary typists were Ornella Cavaliere, Lois Morris, and Donna Russell. Coleen Barry tidied up bibliographic references.

Archaeology and the Methodology of Science

1 / Philosophy and Archaeology

I n the early 1960s, traditional archaeology was challenged on the ground that it had failed to provide any substantive understanding of the cultural past. Major criticisms of older archaeology were that it had remained heavily descriptive, that it accepted too much as unknowable, and that when interpretations were offered, they were likely to be generalizations extrapolated from a particular data base in what was called a narrow inductivist mode. "New Archaeology" sought to clarify the discipline's basic goals of explanation and theory building and to expand the vision of just how much about the past was knowable or accessible through archaeological investigation. Methodological improvement was seen as the most important means for exploring and achieving these goals. For methodological guidance, archaeologists quite rightly looked to philosophy of science. Improved methodology would be scientific methodology. Indeed, the idea that archaeology should be or become a science has been a fundamental part of subsequent developments.

In the appeal to philosophy of science during the last two decades, there has been growing sophistication in the archaeological awareness of the nature of what philosophy of science can and cannot offer in the way of help to a discipline seeking to become more scientific. A growing number of archaeologists are exploring philosophy of science, and the complexity of the field is becoming increasingly appreciated.

One reason that archaeologists turned initially to philosophy of science was that the central concerns in the archaeological debates

1

had also been central concerns in the philosophical debates—namely concepts of and assumptions about the nature of science. Issues such as the theory-laden nature of facts, objectivity in science, the nature of explanation, ways of evaluating support for inferences (confirmation), appropriate and inappropriate forms of reasoning, and numerous other issues concerned with the conduct of science as a rational enterprise had been explored in detail in philosophy of science, and so this was a natural place for archaeologists to begin their own explorations.

The appeal to philosophy of science was, initially, incomplete and somewhat shallow, however. For one thing, the work of a few philosophers (notably Carl Hempel, Ernest Nagel, and Karl Popper) tended to dominate the archaeological perspective, and this in itself gave a rather narrow perspective to the discussions. For another thing, the archaeological appeal to philosophy occurred at a time when philosophy of science was itself in a transitional state, but this was not evident at the time to archaeologists. A consequence was that archaeologists were not at first aware that what they were borrowing from philosophy of science was under serious question and in the process of being replaced by philosophical views thought to be more adequate. Archaeologists interested in foundations and method were deeply committed to trying to make archaeology as scientific and logically rigorous as possible, a task similar to the one philosophers of the school known as logical positivism had set for themselves and their own discipline in the 1920s and 1930s.

As is so often the case when concepts are borrowed, the introduction of philosophical concepts into archaeology has been selective and incomplete. Some fundamental confusions have crept into archaeological understanding of philosophical concepts (*see* Chapter 2). Debates in philosophy of science surrounding concepts archaeologists have adopted have been incompletely reviewed by archaeologists even though some of these debates contain much that is relevant to archaeological concerns. The result at first was that archaeologists were operating with philosophical ideas that were themselves becoming less widely accepted. In addition, with the exception of the work of Thomas Kuhn, archaeologists have not for the most part followed developments in philosophy of science since the heyday of logical positivism. Thus the conception of what philosophy could offer to archaeology became fossilized in a view of philosophy that was increasingly being viewed as incorrect. This was unfortunate, because more recent philosophy of science has focused on such areas as the nature of theories and sociological aspects of science that might well have been more appropriate to archaeological concerns.

Initially, the pattern for the introduction of philosophical concepts into archaeology was for a few archaeologists to investigate the philosophical literature and then for the majority of archaeologists to quote those philosophically minded archaeologists rather than exploring philosophy of science for themselves. Of course, it would have been extremely difficult for great numbers of archaeologists to explore another field in depth, and indeed one could hardly expect this to happen, given the obvious constraints of working within disciplinary perspectives. Still, one of the results of the rather incomplete exploration that took place was that the richness of the philosophical debates and the range of philosophical positions has not adequately reached archaeologists. Since this richness is not only interesting in itself but also instructive for an understanding of the foundations of science, we take it as one of our tasks in this book to place before archaeologists a broader range of philosophical positions than have previously been available in the archaeological literature. We regard it as important also to review some of the confusions that arose in the archaeological literature as a result of the focus on philosophy as a way of understanding the discipline's foundations, and to try to develop a clear perspective on the ways in which philosophical insights can be helpful to archaeology.

Archaeological interest in philosophy of science

It would be a mistake to think that concern with the theoretical foundations of archaeology began in the 1960s, but not until that time was concern with such issues very widespread in the discipline. With the publication of Binford's article "Archaeology as Anthropology" in 1962, however, archaeologists began to show increasing interest in the philosophy of science, in particular in the concept of explanation, and particularly in the writings of Hempel, beginning with the seminal article "Studies in the Logic of Explanation" by Hempel and Oppenheim (1948); more generally, *see* Hempel (1965 and 1966), Nagel (1961), and, to a lesser extent, Popper

(1961; first published 1935). The seriousness with which archaeologists quite quickly came to view a number of central issues in the philosophy of science was reflected not only in the major publications in methodology in archaeology, but also in introductory textbooks, such as the one by Fagan (1972).

Early in his career, Clyde Kluckhohn (1940), influenced by the work of A. N. Whitehead (1953; first published 1926) among others, appealed to archaeologists to make explicit the conceptual frameworks within which they were operating, to avoid "obsessive wallowing in detail in and for itself" (1940:42), and to devote a reasonable proportion of their attention to methodology, theory, and the questioning of fundamental assumptions. Kluckhohn influenced many archaeologists, but none more than Walter Taylor, who developed Kluckhohn's criticisms of archaeological method and theory more fully. Taylor (1948) drew heavily on the work of philosophers of history such as Teggart (1916; 1918; 1919; 1925) and Mandelbaum (1967, second edition); of philosophers of science such as Whitehead (1953) and Cohen and Nagel (1934); and of Znaniecki (1968; first published 1934) and Lesser (1939), who wrote on methodology in the social sciences. Taylor (1948:157) was concerned, as Kluckhohn had been, that archaeologists make their underlying assumptions explicit and that they put forward testable hypotheses. He was to some extent interested in the nature of confirmation as well (Ibid.:157–159).

Raymond Thompson (1958) is another archaeologist who was already concerned with the theoretical foundations of archaeology before the appearance of the New Archaeology. Thompson was strongly influenced by Dewey's views on the nature of scientific inference (see especially Dewey 1938) and he held that the literature on inductive logic was what archaeologists should explore: "for the deductive methods of formal logic are not appropriate to the interpretation of the empirical data of a discipline like archaeology" (Thompson 1958b:1). Thompson's work includes discussions of, for example, the importance of context in hypothesis formation and of the testing of hypotheses and the use of analogy in archaeology. But in general Thompson did not develop these views in any great detail. Because he claimed that deduction has no place in archaeology, his work has received little attention from most archaeologists currently interested in the philosophy of science except as a foil for differing views.

Robert Butler (1965) was among the first archaeologists to discuss deductive models of explanation and confirmation, and he was particularly interested in Karl Popper's (1961) work on these subjects. The interest on the part of archaeologists in philosophy seems to have developed independently of Butler, however.

The primary interest among New Archaeologists has been in the D-N (deductive-nomological) model of explanation most often associated with the name of Carl Hempel (1965). Though the D-N model of explanation did not originate with Hempel, the version of this account of explanation that became current both in philosophy and in archaeology was Hempel's. Incidentally, though, it is interesting that Popper, a contemporary of Hempel's and in many crucial respects a philosopher of a very different theoretical persuasion from Hempel (*see* Chapter 3), was also an advocate of the deductive-nomological model of explanation. We shall try to show that Hempel's work has been imperfectly understood and must be viewed in the broader historical context of which it was a part.

Some archaeologists, although committed to the idea that archaeologists should look outside their own discipline for a scientific basis for their work, nevertheless thought Hempel's model inappropriate, and substituted another analysis. Thus, Tuggle, Townsend, and Riley (1972) drew on Systems Theory whereas Johnson (1972) turned to a method of Michael Scriven's he referred to as detectivism. Some thought that Systems Theory might be able to provide the needed foundation (*see* F. Plog 1975), but it was never made very clear just what role it was to play. The idea that Systems Theory was somehow a model of explanation was cogently challenged by Merrilee Salmon (1978) in an article in *American Antiquity* and again with Wesley Salmon (1979) in an article in the *American Anthropologist*. In the latter article, Salmon and Salmon sum up their position by saying: "Vague claims, such as 'to explain some event is to show its place in a system' do not count as providing a model of explanation" (Ibid., p. 65).

As these few examples show, archaeologists have been prone to turn to a favored philosopher for the articulation of and support for particular views of the foundations of their discipline. What they have not always seen is that the point is not to amass a suitable number of appeals to authority, for philosophers and other theoreticians are not providing neutral data which anyone who cared to take the time and trouble could replicate. Analyses put forward by philosophers are always controversial and can be established only by argument. Archaeologists who wish to adopt one among a number of disputed philosophical views ought therefore to be able to evaluate claims for and against the superiority of that view over its rivals. In order to do this it would be useful for such archaeologists to appreciate the context within which the philosophical analyses they are trying to defend or refute were developed. Since Hempel's work has been both of primary interest to the New Archaeologists and of seminal importance to twentieth-century philosophy of science, an understanding of the milieu in which Hempel worked is essential.

Philosophy of science and logical positivism

Philosophers of science have traditionally devoted much more attention to the physical than to the social sciences. Indeed, many philosophers of science were originally trained in the physical sciences and only later in their careers turned their attention to philosophical questions. In the first decades of the twentieth century this pattern was particularly evident with the emergence of the philosophical movement known as logical positivism (a movement that constituted the dominant force in philosophy of science and whose members were primarily mathematicians and physicists). The close relationship between philosophy of science and the physical sciences may be attributable partly to the fact that the social sciences developed much later as sciences than the physical sciences. One result of this relationship has been that the philosophy of physics, for example, is a well-developed discipline, while the philosophy of sociology is in its infancy; but it is important to see that this is primarily a matter of the sort of maturity that involves sophistication rather than age. Jarvie, for example, points out that it may not even be correct to speak of the social sciences as younger than the physical sciences: "since their problems and therefore speculations on their problems has to be as old as speculation of any kind. To explain their poor performance by their youth may be to put the cart before the horse. They appear not to be mature, but since they have longevity we cannot infer their youth" (1979:77). But what is clear is that the traditional marks of science—replicability, mathematical precision, objectivity, and the like have been slower to appear in the social sciences than in the physical. A further result of the traditional emphasis on the physical sciences is that the examples used in philosophy of science are usually from the physical sciences. These examples are, therefore, seldom as useful as they otherwise could be for the social scientist.

Of the social sciences, psychology has received the most attention from philosophers of science. Psychoanalytic theory, for example, has given rise to much philosophical debate. In turn, psychologists have been more concerned than have other social scientists about the philosophical foundations of their discipline. Philosophical interest in the other social sciences has developed, of course, but it has developed piecemeal. Articles in the philosophy of psychology have been appearing since at least the 1930s, but not until after the publication of Von Neumann and Morgenstern's *Theory of Games and Economic Behaviour* in 1944 did economics, for example, receive any sustained attention from philosophers of science. While archaeologists have for many years been to some extent concerned with methodology, it was not until the 1960s and the emergence of the New Archaeology that their concerns with the methodology and theoretical basis of archaeology became explicitly philosophical (*see* Binford 1962; 1964; and 1965).

The philosophical movement that dominated philosophy of science throughout the first half of the twentieth century is usually called logical positivism or logical empiricism. This movement had its beginnings in a reaction against nineteenth-century idealism and a desire to place philosophy on a sound scientific footing— goals similar to those that motivated the turning to philosophy that took place in archaeology in the 1960s. The logical positivists thought of themselves as continuing a nineteenth-century Viennese empirical tradition traceable to British empiricism and especially the philosophy of David Hume, to the scientific philosophy of Mach, Helmholtz, Poincaré, Duhem, and Einstein, and to the logical writings of Bertrand Russell.[1] In the early 1920s a group that became known as the "Vienna Circle" established itself around Moritz Schlick, with members such as Carnap, Feigl, Gödel, Neurath, Waismann, and others. Wittgenstein and Popper, though not members of the circle, had regular discussions with its members; and in Berlin a similar group arose, including Reichenbach and Hempel as members. Their aims were varied: some positivists wished not merely to destroy traditional metaphysics of the German idealist variety but even to reject all philosophical questions in favor of constructing a unified science (indeed their journal, *Erkenntnis*, later became known as the *Journal of Unified Science*); but others, like Schlick, more plausibly saw logical positivism itself as a philosophy, albeit a new and better grounded one, much as the New Archaeologists saw their contribution as developing a better grounded archaeology.

Since the overriding concern of the logical positivists was to make philosophy more scientific, they sought a *formal* criterion for sorting sense, or what they saw as science, from nonsense, or what they saw as nonscience, in philosophy. Absolutely central to logical positivism in general, and this undertaking in particular, was the view that "science is, fundamentally, the description of experience" (Hempel 1965:52). Despite decades of effort, however, the logical positivists never succeeded in formulating a satisfactory criterion for differentiating what is grounded in experience from what is not.

The positivists claimed that the scientific was the "cognitively significant" and that the cognitively significant was that which had meaning. Thus, they characterized their search as one for a criterion of meaningfulness—something that would allow them to reject as meaningless much of traditional metaphysics. Their first approach to this end focused on the meaningfulness of terms. In this view, called operationism, meaning was seen as "a characteristic of concepts or of the terms representing them—namely, as their susceptibility to operational definition" (Hempel 1968:101). The context in which operationism was first explored was physics, though the view was soon imported into psychology where, in some quarters at least, only concepts capable of being given operational definitions were thought suitable for study by "scientific" psychology (cf. Stevens 1935a, 1935b, 1936, and 1945).

The fundamental tenets of operationism were (1) that "every meaningful scientific term must . . . permit of an operational definition" (Hempel 1968:102) and (2) that an operational definition is "a rule to the effect that the term is to apply to a particular case if the performance of specified operations in that case yields a certain characteristic result" (Ibid.:101–102). Thus, as an illustration: "The term 'harder than' might be operationally defined by the rule that a piece of mineral, x, is to be called harder than another piece of mineral, y, if the operation of drawing a sharp point of x across the surface of y results in a scratch mark on the latter" (Ibid.:102).

Operationism may have appeared moderately satisfactory when applied to terms such as "harder than" but it seemed to break down completely when applied to terms such as "speed" or "distance." One difficulty was that the concept of operational definition was so characterized that the successful performances of distinct operations yielded separate operational definitions. Thus, speed measured by radar would receive a different operational definition from speed measured by a stopwatch and similarly for distance measured with a metre stick or by triangulation. One can argue that these are in fact different, but for the concept of operationism, it was disconcerting to have definitions proliferate in such ways. There seemed to be no limit to the number of definitions for things or processes that, one might also argue, were similar or identical. One might assume that, whether measured by radar or a stopwatch, the speed of a car is the same, i.e., how fast it is moving; but on a strict construal of operationism we have no way to be sure of this. One could postulate that the same thing was being measured in the two cases, but this would be to make assumptions not warranted under strict operationism.

Proponents of operationism came to see that to rely on operational definitions alone was to place too severe a restriction on scientific language and thus on theory, but to loosen these restrictions was to turn away from operationism. This, in fact, is what happened, and philosophers began to look elsewhere for a criterion of meaningfulness, though versions of operationism are still relied on in many areas in social science even if the concept itself is theoretically unsound. Thus psychologists do not consider it respectable to talk about "learning" or "motivation" unless these terms can be, as they often put it, operationalized; and sociologists seek operational definitions of such terms as "social institution." Much of the motivation for this is the same one that moved the positivists: a desire for precision and clarity.

Operationism—the concept and the associated problems—has direct relevance to archaeology. In some senses, assemblage-specific typologies or diagnostic types taken to represent cultures can be seen as representing operational concepts of cultures. At least, individual cultures were sometimes treated as if they were just configurations of archaeologically detectable attributes. Exactly the same criticisms were levelled against such concepts of culture as have been directed against operationally defined terms—i.e., that such concepts of culture were unstable, changing as new data shifted the range of attributes used to define them. Similarly, types (especially in type/variety), cultures, and other operationally defined phenomena tended to proliferate beyond reason when defined exclusively in terms of—indeed even *as*—covariants of select classes of formal attributes. Such operational concepts appear to exhaust the data they subsume. Classifications which operate internally within a collection or assemblage will not correspond to wider classification systems unless the program incorporates such information or the analyst makes the connection. This internal ordering is characteristic of numerical taxonomy. Barrais and Freeman (1958) did a statistically based typology on Aztalan pottery in which rim forms previously used to sort two ceramic types were collapsed into a single category. When they undertook a similar analysis for another site previously regarded as having Aztalan pottery, the types created statistically were site specific. Such operational procedures are useful for certain kinds of problems requiring fuller disclosure of intra-assemblage variability. For other purposes, however, archaeologists seem to want terms of reference that will accept new data, terms that remain fairly stable, and terms that go beyond the archaeologically discernible attributes.

The difficulty, then, is that many terms useful to science do not seem to be easily operationalizable, and this was a difficulty recognized very early. One possible explanation of what had gone wrong with operationism was that the chosen unit of meaningfulness was too small; we should perhaps not demand that each term in a scientific system be certifiable as meaningful, but rather that the sentences we utter or statements we make should have cognitive significance. Proponents of this view developed the approach that came to be called verificationism because of its attempt to formulate a criterion for meaningfulness based on the notion of empirical verifiability. One of the early formulations of the Verification Criterion of Meaning was the following: "The meaning of a proposition is the method of its verification" (Schlick 1949:148), though this principle was often attributed by members of the Vienna Circle themselves to Wittgenstein. Of course, if a proposition was not verifiable there was no method for its verification, so it followed from this formulation that it was meaningless.

The criterion was also interpreted as saying that only propositions that were verifiable in experience, i.e., through observation, were meaningful. It is not really surprising that it should have been so construed, since it is essentially worries about observability that lie behind empiricism, of which logical positivism is a species. Indeed, as was mentioned earlier, another name for logical positivism is logical empiricism, and Schlick liked to refer to what others called "logical positivism" as "Consistent Empiricism" (Ibid.:149).

The positivists attributed meaning to analytic propositions such as "All bachelors are male" whose truth follows simply from the meaning of its constituent terms. In addition, observation sentences such as "This book is blue" were meaningful, and so were all sentences that could be deduced from a finite set of observation sentences. This formulation of the Verification Criterion suffered from a number of problems, the most disturbing of which to the positivists was that it excluded too much—it was too strong. Not only was that which they believed to be meaningless shown by the criterion to be meaningless, but many other propositions they believed to be paradigmatically meaningful, e.g., scientific laws, were also placed in the category of the meaningless. Since one of the uses to which we put such laws is prediction, universal generalizations that are of any interest to scientists are necessarily not amenable to conclusive verification. A statement of the form "all A's are B's" would be so verified only if all A's had been examined and found to be B's, but then it would be useless for purposes of prediction since there would exist no *un*examined A's.

Any historical discipline is an interesting case for examining problems of verification, because nothing of great significance is directly verifiable by observation. The problem of what is verifiable about the past is one of the core issues of the archaeological debates of the last two decades. From some perspectives, it is the key issue.

In one form, this problem is crystallized in the two contrasting points of view presented by Marion A. Smith (1955), whose passage about the limitations inherent in archaeological data is often cited by critics, and Binford's (1968) more encompassing view that past systems are knowable. Klejn (1977) notes that some archaeologists worry that they cannot verify anything—that archaeology is rather like speculative fiction. Archaeological concerns with verifiability are also manifest in more diffuse ways, perhaps, in the general methodological awareness and in the attention paid to problems of objectivity and ethnocentrism. In a more focused form, the concern with verifiability is highlighted in the debates over the use of analogy (cf. Wylie 1982a). If analogy is imperfect, but essential, the quandary becomes, as DeBoer and Lathrap put it (1979:103), the familiar one "of choosing between a significant pursuit based on a faulty method or one which is methodologically sound but trivial in purpose." The difficulty, as always, is one of making significant empirical claims that go beyond the data and that are also true, or at least that are very likely to be true. The propositions that are most likely to be true are probably the ones that are most closely connected to what we can observe, but there is great difficulty in stating this in a way not open to objections. Naturally enough, archaeologists, like other scientists and like philosophers of science, have been deeply concerned about the principle of verifiability, and have independently tackled the problem.

A clarification is perhaps in order here. The logical positivists were not suggesting that only propositions which actually could be verified by experience were meaningful. They distinguished carefully between verifiability-in-fact and verifiability-in-principle. Propositions verifiable in fact were those which actually could be verified in experience. For example, a verificationist would say that "some kivas are not round" is verifiable in fact since its verification, through the observation of a not-round kiva, is technically possible.

On the other hand, propositions verifiable in principle were those whose verification was logically but not necessarily technically possible. For example, in the 1880s, in respect to any particular archaeological kiva, the proposition "ground maize was used in this kiva" could not be verified in the absence of preserved cornmeal because the technology necessary for its verification (palynology, for example) was not then, although it is now, available. It was possible even then, however, to specify what *would* count as that proposition's verification, and thus it was verifiable in principle. Since what is technically possible must be logically possible as well, all propositions verifiable in fact are verifiable in principle. The reverse, of course, is not true.

It was, then, verifiability in principle that the logical positivists required for meaningfulness, and the famous case of the other side of the moon discussed by Moritz Schlick (1949:157–58) shows why that had to be the case. Schlick argued that none of us would be willing to accept a view according to which it would be nonsense to speak of the other side of the moon because it is not normally observable by us. We could answer questions about what the other side of the moon is like by describing what someone located somewhere "behind" the moon would observe, and this even at a time when we did not know whether it was possible to travel around the moon. In 1936, when Schlick discussed this example, human beings had not yet observed the far side of the moon, but this was, as he makes clear, completely irrelevant to the meaningfulness of the question, "What is the other side of the moon like?" What counted was that states of affairs could be described which, if observed, would answer the question. There is actually no agreement in the literature about what should be the appropriate standard of possibility in these cases. Schlick used logical possibility, but others such as Reichenbach preferred physical possibility, where what is physically possible is what is in accord with the known laws of nature, whether this is technically possible or not. On this conception, verification of statements about the other side of the moon (in 1936) would be possible, but not, for example, verification of statements that require that someone be able to run faster than the speed of light, even if such a thing can be imagined without contradiction. It seems clear that technical possibility is too narrow for appropriate application of the Verifiability Criterion, but for our purposes it makes no difference whether we then choose physical or logical possibility.

Since the Verification Criterion of Meaning was too strong, what the positivists needed to replace it was a criterion both weak enough to accord meaningfulness to the propositions they believed were meaningful and precise enough clearly to distinguish those propositions from the ones they believed were nonsense. An attempt to formulate such a criterion was made by A. J. Ayer. He formulated what he called the "weak verifiability" requirement as follows:

> The principle is that a statement is verifiable, and consequently meaningful, if some observation-statement—a sentence which records the outcome of an observation or sense experience—can be deduced from it in conjunction with certain other premises, without being deducible from those other premises alone (Ayer 1946:11).

The criticism that followed upon this reformulation of Ayer's criterion was that it was too weak—that, instead of ruling out any statements as meaningless, it judged all statements to be meaning-ful and thus served no purpose at all. Ayer himself eventually conceded that while the new criterion was liberal enough to allow scientific laws meaningfulness it also

> allows meaning to any statement whatsoever. . . . Thus, the state-ments "the Absolute is lazy" and "if the Absolute is lazy, this is white" jointly entail the observation-statement "this is white," and since "this is white" does not follow from either of these premises, taken by itself, both of them satisfy my criterion of meaning. Fur-thermore, this would hold good for any other piece of nonsense that one cared to put, as an example, in place of "the Absolute is lazy," provided only that it had the grammatical form of an indicative sentence. (Ibid.:11–12).

Noting that this state of affairs is unacceptable, Ayer went on to revise the criterion of verifiability to meet this objection. There is no need to consider it in detail here as, although it was considerably more complicated than the original criterion, it was refuted by a more complicated but similar objection (cf. Church 1949). This new criterion, too, it was shown, allowed meaningfulness to any non-sense whatsoever, and was therefore unacceptable. At this stage, then, it remains an open question whether the criterion can be restated so as to distinguish between meaningful and meaningless propositions. It seems doubtful that it can, for it would appear that, for each new formulation to take account of past objections, a new objection can be constructed. Actually, the various formulations and their respective flaws may well be systematically related to one another, with perhaps all the flaws caused by a common problem (cf. W. Salmon 1966a:354–376). If so, the essential difficulty proba-bly lies in the assumption upon which the formulations are based. The positivists held that a proposition is meaningful if and only if it is either true or false. Another tenet of the logical system they accepted was the idea that deductive arguments were composed of propositions that were either true or false. In rendering sentences to be tested for meaningfulness as propositions in deductive argu-ments, they were implicitly assuming them to be either true or false and so (though they did not realize it), meaningful (Ibid.:359).

The criticisms to which the various versions of the verifiability criterion were subjected were of two different kinds. From "inside" the philosophy of science came, first, the charge that the criterion was virtually unsatisfiable and, second, upon its reformulation in Ayer's several forms, that it was trivially satisfiable. These criticisms came to be generally accepted as showing not only that the criterion had failed but that it had failed on its own assumptions. This result led philosophers of science to abandon verifiability as the criterion of meaningfulness.

It is perhaps slightly misleading to speak of verifiability *tout court* here, for Ayer's formulations were really of a criterion of incomplete verifiability or confirmability. That is to say, what was being required in order to attribute significance to a sentence was that it have some tie to observation; but the tie could be somewhat looser than the requirement that the sentence in question be deducible from observation sentences. When these criteria of incomplete verifiability or confirmability turned out to be too easily satisfied, and thus to fail to draw a line between what was thought to be cognitively significant and what wasn't, philosophers turned to other means of making the demarcation. There were efforts to formulate an empiricist language into which all approved terms could be translated or introduced in some otherwise acceptable way. But it soon became clear that terms such as "neutrino" or "unconscious desire"—or, for that matter, "Adena Culture" or "Maya Collapse"—could not be so translated, and thus the sentences in which they occurred would have to be labelled meaningless.

These and similar unpalatable results of trying to draw a precise line of demarcation led philosophers to look for empirical significance in whole theoretical systems rather than in terms or sentences; and to fasten not so much on questions of significance as on an assessment of the clarity and precision with which the theories are formulated, their explanatory and predictive power, their formal simplicity, and the extent to which they have been confirmed by experiential evidence (Hempel 1965:117). Clearly, this approach places us back in the realm of widely accepted but more or less vaguely understood approaches to scientific respectability, leaving us without any precise standard that would allow us to judge in a mechanical way the scientific status of a theory.[2]

From "outside" the philosophy of science came another sort of criticism. It was argued by some that even if an adequately formulated criterion of cognitive significance was a sufficient condition of meaningfulness, it was not a necessary condition. These philosophers insisted that there are clearly meaningful propositions—for example, many that express religious beliefs—which are not open even to confirmation, let alone verification, by sense experience. To argue as the positivists did that such propositions are meaningless because they are not confirmable by sense experience was, these objectors claimed, to beg the question (*see* Ewing 1973; Joad 1950; Copleston 1956). Others felt that it was to answer the question and to show that indeed they were meaningless, while still others maintained that religious propositions were indeed confirmable by means of experience, even if not by sense experience as it is traditionally conceived. We seem to have here, as with operationism, another case in which, even though philosophers have abandoned the search for a precisely formulated principle of verifiability, the notion—however incompletely understood—is still employed in other areas. The fact that we do not seem able to formulate an unassailable principle does not turn nonsense into sense, nor does it mean that we should make no effort to distinguish what is verifiable or confirmable by empirical evidence from what is not (*see* Chapter 6).

Again, the parallels between the dilemmas of the logical positivists and those of archaeology are striking. Archaeologists are concerned with problems of cognitive significance, with separating science from nonscience, with rationality, and with the eternal quandary of addressing significant problems rather than trivial ones. Thus, archaeologists debate the merits of various approaches such as the cognitive structural and materialist (cf. Leone 1982), they wonder whether interesting laws will be trivial (cf. Flannery 1973), and they wonder if the most accessible research areas are really the most significant (DeBoer and Lathrap 1979).

At the same time as the positivists were searching for a means of distinguishing science from nonscience, another approach to the problem of formally identifying the propositions of science was underway. Karl Popper was also concerned with demarcating science from nonscience (*see* Chapter 3). While Popper's work also grew out of the Vienna Circle, he did not share the logical positivists' view that anything which was not science was meaningless and therefore he never characterized his project as one of distinguishing sense from nonsense. Rather, he thought it important to try to distinguish empirical science from other bodies of assertions that might be confused with it: metaphysics, such traditional pseudo sciences as astrology and phrenology, and the more imposing pseudo sciences of the present age, such as the Marxist theory of history and Freudian psychoanalysis (Quinton 1967:398).

Popper thought that to "identify this distinction with that between sense and nonsense" was "to make an arbitrary verbal stipulation" (Ibid.:398).

Popper suggested that what characterized scientific propositions was their falsifiability. While universal laws, the search for which is presumably one of the main goals of science, are not conclusively verifiable—as we have already noted, they are subject to empirical falsification by negative instances. One example of an A which is not a B will falsify "All A's are B's," whereas its verification would seem to be beyond our powers altogether. This is because, no matter how many A's we have observed that have been B's, if there are still some unexamined A's, it is always possible that one or more of them will turn out not to be B's. It is, then, this "logical asymmetry in the relation of general statements to observations [which] underlies Popper's view that falsifiability by observation is the criterion of the empirical and scientific character of a theory" (Ibid.:399).

The history of the positivists' attempt to formulate a criterion of cognitive significance is of interest not only because it illustrates the most fundamental problem the positivists tackled but because the criticisms the various formulations of the criterion encountered are strikingly similar to the criticisms that were later levelled at Hempel's deductive-nomological model of explanation. In turn, these debates echo the concerns of archaeologists. We have earlier suggested that the difficulties with the criterion of cognitive significance may be systematically related to the assumption upon which the positivists worked: that a proposition was meaningful if and only if it could be formulated in a deductive system. We will suggest in later chapters that perhaps all attempts, including Hempel's, to find formal models of, for example, explanation that are genuinely applicable in the sciences are doomed to failure. As with the positivists' search for the criterion of cognitive significance, the assumption of formalizability may be disastrously inappropriate.

Recent developments

Logical positivism dominated the philosophy of science from early in the twentieth century until at least the late 1940s–1950s. Throughout the 1920s and 1930s, there was constant tinkering with the many concepts presented, and in general, each modification elicited new reasons why the concepts presented were not generally applicable.

Carl Hempel's formulations emerged toward the end of the heyday of logical positivism. Originally trained in mathematics and physics, Hempel was a member of the Berlin-based Society for Empirical Philosophy which, as mentioned earlier, was closely associated with the Vienna Circle. Hempel eventually rejected some of the central tenets of positivism, but in both his educational background and his approach to philosophy he is representative of that school of thought.

Hempel's writings were, perhaps, the most comprehensive and clearly formulated of the entire logical positivist movement. They were so clear that they exposed both the best and the worst aspects of all that had gone before. Without a doubt Hempel's work is an exceptionally fine summing up of the decades of tinkering, and certainly his work is among the best statements of ideals of rational scientific procedures (*see* Chapters 5 and 6). Even his definitive statement was not without some fundamental problems, however, and a new generation of criticisms soon developed. It is worth noting that in the intervening years since 1948, Hempel himself has modified his position to the point that he accepts more nonlogical aspects of science as playing a significant role in the scientific endeavor than he did earlier.

By the 1950s, then, it was clear that logical positivism had a number of recalcitrant problems that had persisted despite several decades of serious and devoted efforts to resolve them. It seemed to philosophers of science that reliance on formal axiomatic representations of scientific theories, sharp divisions between fact and theory, and attempts to develop precise, formal accounts of such concepts as explanation and confirmation did not really capture the essence of science or scientific method. Rom Harré, for instance, says:

> Recent philosophy, though in many ways an improvement on the philosophy of the past, has been particularly prone to myths, to

charismatically proclaimed standpoints from which the extremely complex facts of human intellectual activity have seemed suddenly to be capable of definitive, explicative analysis in relatively simple terms. Contemporary philosophy has been dominated by a particular style of logic, exemplified in a group of systems forged in the study of the foundations of mathematics. The main analytical tool in the philosophy of science has been the propositional and predicate logic developed from Frege's and Russell's studies of mathematical reasoning. Intimately associated with the use of mathematical logic has been a predilection for a metaphysical system derived from Mach, and ultimately from Hume, formed in the mould of classical positivism (1970:3).

Harré claims that the whole positivist structure leads us away from perfectly sensible intuitions about what constitutes rational thinking toward dogmas that absorb us but do not really help us to understand the problems with which we started:

> For instance, the insight that it is sometimes a sufficient condition for a theory to be said to explain a fact, that the statements expressing the fact be deducible from the statements expressing the theory, together with statements of conditions, has been transformed into the dogma that explanation consists in nothing but the setting up of a deductive link from what is to be explained to what explains it. We watch with consternation the consequent contortions of the dogmatist as he endeavors to stretch his dogma to accommodate the cases which do not fit; cases for instance when the explanation is not given in deductive form at all, but say in a form which essentially involves an analogy (Ibid.:4; *see also* Chapters 5 and 6).

Harré is here expressing one version of a view that has attained considerable currency. Indeed, even by the 1960s, the view that further tinkering with logical models was unprofitable had gained a certain following. A number of philosophers of science had begun to say, as does Harré, that explanation is by no means always just a matter of formal relationships between a specific instance and a generalized pattern, and some philosophers began to turn to other directions of inquiry, looking anew at some of the fundamental assumptions of philosophy of science. The new directions included a more focused exploration of the way science actually operates, with explicit attention to the historical, sociological, economic, and political factors that affect the doing of science.

A broad spectrum of positions developed rather rapidly. One source of a very influential new perspective was the work of Thomas Kuhn, which we discuss at some length in Chapters 3 and 4. Some writers, such as Feyerabend and the members of the Edinburgh School of sociologists of science (Barnes 1972a, 1972b, 1973, 1974; Bloor 1971, 1973), not only dismissed logical models of scientific explanation but also insisted that there is no such thing as a neutral conception of rational scientific method. Indeed, Feyerabend argues (as we discuss more fully elsewhere) that "anything goes." Both Feyerabend and those occupying the extreme or so-called strong sociological position see scientific rationality as a matter of what is acceptable to the scientists of the day, and thus as conditioned by political and economic factors as much as anything. Others, like Newton-Smith (1981), want to retain a rational core of science while admitting that nonrational aspects play a significant role in science.

We, like Newton-Smith, take the position that there is a rational core to science and that, although conceptualizing that core in a formal way presents very real problems (as we try to make clear in the next four chapters), there is no justification for supposing that there is no way of demarcating the rational from the nonrational or the irrational. At the same time, there is certainly no doubt that sociological, economic, and political factors exert a great influence on the doing of science. While both of these sets of influences are real and significant, it seems useful to try to treat them separately.

Within philosophy of science, the recent trend has been toward assigning more weight to the substantive aspects of different scientific disciplines and thus toward taking seriously some of the differences among the sciences. A corollary of this is that Science with a capital S is viewed less as a monolithic entity, a Golden Hall of Science, than previously. Acceptance of the view that the substantive aspects of different disciplines create different kinds of sciences has another corollary—namely, that a uniform scientific method and uniform rules of logic, even if they could be formulated, might be less helpful than previously believed. This is not to say that there are no reasonable rules of inference or method at all, but that the actual methodology developed in individual disciplines will reflect the particular problems associated with research in a discipline, and that these differences may be as interesting as the similarities. We look at the "sociology" of archaeology in Chapter 4, arguing there that these factors do in fact exercise an enormous influence on the doing of archaeology.

One can argue that by insisting on a rational core of science and by assigning importance to the substantive aspects of science, we are espousing a form of empiricism (since the empiricists placed great emphasis on observable facts), and we claim, in part, that the observable facts of different disciplines affect the methodology of each discipline. Indeed, we would further claim that the difficulties with logical positivism are not difficulties with empiricism but with a particularly rigid version of it. Recent more flexible versions, such as the "constructive empiricism" of Bas van Fraassen (1980), are more promising and have real possibilities for providing an understanding of scientific theories that is not tied to the rigid formalism of the positivists.

Philosophy of science and science

The logical positivists turned to the foundations of science in order to make philosophy more precise and to free it from what were thought to be the wild speculations of nineteenth-century metaphysics. Since that time scientists in one discipline after another have been turning to the philosophy of science, which, until the 1950s, was heavily dominated by the work of the logical positivists, in order to put their investigations on a sounder scientific footing. It might look as though in the end scientists have nothing to learn from philosophers of science if the core of that branch of philosophy was taken directly from science—but, of course, it was not. The philosophy of science is not, and does not purport to be, merely descriptive of the scientific process. It is, in addition, normative in the sense that it is meant also to provide norms for science—to prescribe, in part, what scientific method, explanation, or confirmation ought to be as well as to describe what each of these actually is. There is a serious problem concerning the precise extent to which philosophy of science is either descriptive or normative and the way these two aspects are to be integrated; but that it is to some extent both seems to be a tenet of most major writers in the field. The term "normative," of course, has acquired other, largely derogatory connotations for archaeologists. "Normative" to archaeologists today often means descriptive, nonrepresentational, or characterizing culture as a uniform, learned entity. This is frequently contrasted with systemic or processual views of culture which stress representativeness, variability, and an altogether more dynamic and adaptive concept of culture. Unfortunately, contemporary archaeological usage of the term "normative" appears to lump together a series of concepts that might be more usefully differentiated, and, since there is virtually no overlap between the archaeological and the philosophical uses of the term, considerable confusion has been generated by its use in writings meant to be intelligible to both audiences.

Philosophers of science attempt to elucidate the patterns of reasoning adequate for the justification of knowledge claims and the concepts essential to the conduct of science. The concepts of justification as well as of adequacy are necessarily normative as opposed to being merely descriptive of how people actually claim justification and adequacy for their findings; but there must be some relationship between the process of scientific investigation and the philosophy of science if (1) the philosophy of science is to be at all relevant to science, and (2) it is to be at all appropriately called "the philosophy of *science*." Philosophy of science is often said to be a rational reconstruction, but, if so, it is a rational reconstruction of science rather than of something else, and, as such, must capture reasonably closely what scientists actually do. Hempel and others have wanted, in part, to describe in a systematic way what scientists actually do when, for example, they try to explain various phenomena; that is, they have wanted to give an account of what the logical structure of explanations offered by scientists actually is. But this is not the whole story, as they have also wanted to offer a normative account of what the logic of adequate explanations in science ought to be, whether any existing explanations completely conform to this pattern or not.

The interaction between philosophy of science and science must be an on-going, two-way exchange. Philosophy of science is rooted in science. Because of philosophy's location one step removed from actual science, it commands a view of science that is both useful and somewhat remote. By addressing the common problems of all science, philosophers can offer useful guidance to the sciences that gave rise to philosophy of science. As philosophical inquiries extend beyond the physical sciences, as additional factors other than logical ones are assigned a role in doing and understanding science, and as philosophy takes greater cognizance of historical and actual cases, philosophy of science will have more to offer the social and natural sciences. Conversely, the social and natural sciences will provide philosophy of science with the material for improved philosophical understanding of significant problems about these sciences.

Philosophical concepts in archaeology

Archaeology turned to philosophy of science at a time when philosophy itself was in a state of flux before the new directions were articulated or widely disseminated. It was quite natural, given both the reasons for discontent in archaeology in the early 1960s and the state of philosophy at that time, for the towering figures of Hempel and Popper to catch the archaeological imagination. For precise, rational accounts of science, the work of these philosophers has not been fully superceded, for no radically new rational account of science has emerged nor does it seem likely that such an account is "waiting in the wings." Other important contemporary philosophers, such as Wesley Salmon, who have offered alternative accounts of some of the central concepts dealt with by Hempel are still working sufficiently within the same mold that they do not provide a really radical departure (*see* Chapter 5).

In part, what archaeologists were seeking as they turned to the deductive-nomological model of explanation and the hypothetico-deductive method for confirmation were directions for improving inference in archaeology by ensuring that inferences were supported in some rational way beyond that of mere collections of positive instances. More importantly, the interaction with philosophy of science seemed to shift the basic disciplinary goals of archaeology from descriptive culture history to more theoretical and processual problems. That the dichotomies between old and new were vastly overdrawn is immaterial in this context.

After two decades of the New Archaeology, a number of archaeologists have taken stock of the results. Binford (1977) calls what might have been a revolution "a little rebellion," and Flannery (1982), in his latest parable, presents the "oldtimer" who still believes in the concept of culture in a much more favorable light than he does the archaeological philosophers of the 1960s or the "children of the '70s" who are in management. Other evidence of a reaction that is, in part, against the effects of philosophizing in archaeology can be seen in Schiffer (1981). These and other expressions of discontent with certain aspects of contemporary archaeology (and especially the New Archaeology) have several facets. One is that the concerns of archaeologists with philosophy have not appeared to provide new substantive understanding within the discipline. This worry seems to arise because of a misunderstanding of the nature of philosophy. In the beginning of the movement toward a New Archaeology, its proponents seemed to believe that philosophy could provide a clear set of rules that would allow them to make new discoveries with clear scientific status. But this is not the role of philosophy; philosophy does not provide ready made templates for directing us how to do science—it provides arguments for particular conceptions of what science is. Hempel's accounts of the logic of justification are attempts to formulate models of explanation and confirmation that reflect both the way science has actually proceeded and a more idealized conception of how science ought to proceed if it is to be rational.

For reasons that are not entirely clear, archaeologists have tended to focus more on philosophical authority than on broad philosophical issues. Psychologists have turned their attention primarily to questions concerning reduction and theoretical terms, physicists to theoretical entities, and biologists to functional explanation. Archaeologists have focused chiefly on deductive models of explanation and seem to have taken the views of Hempel, especially, as authoritative without delving into the extensive debates that ultimately resulted in philosophers' opting for new directions of inquiry into the nature of science. Hempel's models have been criticized on two fronts. First, because his formal models have not succeeded in capturing the logic implicit in our notions of explanation and confirmation (see Chapters 5 and 6); and second, because the principles he identifies are either not, or not the only, relevant principles (see Chapters 3 and 4). But the important point for archaeology is that the work of Hempel and related philosophers was meant to provide a model for the justification of scientific conclusions rather than recipes for their discovery.

Another facet of the current discontent with the role of philosophy in archaelogy is the worry that a bifurcation will develop between theory and practice, perhaps even with theoretically oriented archaeologists thinking of themselves as an elite, and this would surely be counterproductive. There is a concern that some archaeologists will engage merely in criticism rather than in the actual doing of archaeology with the result that two levels of archaeology will develop, with consequent misunderstandings. Clearly, however, such an outcome not only needs to be but can be guarded against.

A related view that is sometimes expressed is that archaeologists should get on with the business of doing archaeology without becoming bogged down in philosophical concerns, for these contribute little to the actual progress of the discipline. This is a perception with which we have far less sympathy, for it is hard to see how any field can develop—and archaeology is not unique in this regard—without careful critical evaluation of its directions and innovations in method and strategy. Whether these come from archaeologically sensitive philosophers, from philosophically minded archaeologists, or from the two working in concert seems less important than that this kind of work continue to be done and to inform the development of new method and theory in the discipline.

When archaeologists are considering which philosophical accounts of, say, explanation are most appropriate for their discipline, it would be useful to bear in mind that scientists in several other disciplines have undertaken the same task. The development of the philosophical foundations of psychology is an interesting case for archaeologists to consider. In the early 1900s, John B. Watson reoriented psychology from the introspective study of mind to the empirical study of behavior (Watson 1924). With that transition, the testing of hypotheses and questions of objectivity became important issues. Social psychology, long dominated by the behaviorism that Watson introduced, has been undergoing a dramatic shift away from that position in recent years. In rejecting behaviorism, psychologists, at least in part, have been rejecting logical positivism. In Chapter 5, in connection with Harré and Secord's (1972) view of explanation, we will discuss some of the reasons why social psychologists have rejected certain elements of positivism. Archaeologists, assuming that they share some of the social psychologist's concerns, may find some of those reasons cogent.

Archaeologists have relied too heavily on a limited range of philosophical questions. Explanation has been perceived as a basic goal without enough attention being paid to the twin question of confirmation, for example. We regard the scientific enterprise as multi-level and multi-dimensional. The ongoing business of archaeology is a progression toward improved knowledge and understanding. Sometimes this progression will advance through pattern recognition, at other times it will advance through the use of low-level inductive procedures, the hypothetico-deductive method, heuristic use of covering laws, classification, or just an ability to count: no one technique can be expected to work in all situations.

The deductive-nomological model of explanation not only has its own inherent and apparently insoluble problems, but even if it were internally above reproach, it seems inappropriate as a working model of explanation in archaeology precisely because of its exclusively deductive nature and its insistence that the law statements that form some of the premises of the explanatory inference be true (*see* Chapter 5). If any model of explanation were to be appropriate for archaeology it would far more likely be an inductive or statistical model, for there the law statements are statistical in form, as archaeological laws would likely be, if we could find them. Thus it is unfortunate that induction has come to be thought so disreputable among archaeologists, for this has sometimes prevented them from seeing the possibilities not only for inductive or statistical patterns of explanation in archaeology but also for the importance of inductive reasoning in extending knowledge (*see* Chapter 6).

Although archaeologists have found logical positivism attractive, and undeniably much can be learned from the works of the positivists, it is fairly clear that archeologists need to explore philosophy far beyond these rather narrow confines. Alison Wylie takes the position that positivism is inappropriate precisely because it is such a thoroughly empiricist philosophy. The New Archaeologists, she maintains:

> took "positivist" philosophy of science to define a non-empiricist alternative to traditionalism without realizing that "positivism" is itself an empiricist theory of science. They thus reproduce in their own program the epistomological conditions that they found to be responsible for the failure of traditionalism (1981:i).

Wylie believes that, where the New Archaeology was consistently positivist, it failed to accomplish its objectives; conversely, its successes can be attributed to the development of clearly non-positivist research strategies. Wylie also believes that the Realist philosophy of Harré and others can provide a promising framework for archaeological concerns in a way that traditional empiricism cannot; but scientific realism is by no means the only alternative theoretical framework open to archaeologists (*see* Chapter 5).

In his review of Watson, Le Blanc, and Redman's book, *Explanation in Archaeology*, David Clarke insists that:

> what is currently needed is the internal analysis (with external aid) and explicit development of archaeological reasoning (archaeological logic), the specification of the general nature and special qualities of archaeological information (archaeological epistemology) and the careful clarification of archaeological concepts and their limitations (archaeological metaphysics) (1970:238).

Clarke notes, "there is little reason to suppose that the positivist philosophy of physics is especially appropriate for archaeology—not least if, for example, it appears only weakly applicable even for biology" (Ibid.). Clarke argues for a discipline-specific, discipline-oriented philosophy of archaeology. We concur wholeheartedly with Clarke's position. At the present time, the attention paid to the logical positivists has obscured other potential avenues of exploration. Archaeologists won't know which concepts will work best in archaeology until they extend their vision; but the fact that the avenues that were explored in the early days of the New Archaeology did not yield all that was hoped should not be taken as a reason for rejecting their aspirations to try to show that archaeology was amenable to scientific study aimed at explaining, at uncovering patterns, order, regularities, and even predictability. These are among the aims of all scientific study, and the self-conscious attention of the New Archaeologists to hypothesis testing, to greater precision and quantification, were surely steps in the right direction even if the particular theoretical frameworks they relied upon were ultimately inadequate to the task set for them.

In the following chapters, some of the fundamental concepts in philosophy of science will be examined and their applicability to archaeology explored. This exercise will not, and given the nature of philosophical investigation could not, add to the body of substantive archaeological knowledge. Instead, it is aimed at explaining and illustrating how archaeologists can use the philosophy of science to better understand the theoretical foundations of their discipline.

Notes

1. There are many excellent sources for further information on the Vienna Circle and logical positivism. The account in the text follows in part that in the article "Logical Positivism" by John Passmore (1967). For additional information see also A. J. Ayer (ed.) (1959); P. Achinstein and S. F. Barker (eds.) (1969); and R. von Mises (1951).

2. Some more recent approaches to the problem of demarcation can be found in Marsha P. Hanen, Margaret J. Osler, and Robert G. Weyant (eds.) (1980). But the difficulty of drawing clear lines of demarcation is by no means unique to science. It is interesting to note that a similar fate appears to befall the attempt by legal positivists to state a clear criterion for what is valid law and what isn't. See H.L.A. Hart (1961: Chapters 5 and 6); Rolf Sartorius (1966); Ronald Dworkin (1977); and Rolf Sartorius (1975: Chapter 10).

2 / Some Fundamental Confusions

I n attempting to find a methodology for placing archaeological
inference on a sound basis—for making archaeology more
scientific—archaeologists selected certain aspects of philosophy of
science that appeared to lend rationality and rigor to the research
process. Thus the New Archaeologists turned to the literature of
philosophy of science and especially the literature on scientific
explanation for an account of the essential nature of science to
which they might try to make their discipline conform. From this
literature they placed an emphasis on such things as the deductive
nature of scientific explanation and the importance of testing.
Within the framework of the new borrowings from philosophy of
science, there arose several essentially related confusions about
philosophical positions, and these quickly became entrenched in
archaeology. There are probably several factors that are responsible for the development of these misconceptions. First, the pi-
oneers in the New Archaeology who consulted philosophy of
science focused on a limited area in the discipline—in particular,
the work of Hempel on explanation, with little consideration of
complementary discussions on related concepts such as confirmation and without exploring the degree to which Hempel's account
was controversial within philosophy. Second, only a few archaeologists actually investigated philosophy of science for themselves;
most simply followed the selections from that field made by the
first New Archaeologists, thus entrenching the rather narrow
focus and the misconceptions to which it gave rise. And third, the
timing of the effective impact of philosophical concepts on archae-

ology in the 1960s coincided with the beginning of major changes in the outlook of philosophy of science, which was leading to the replacement of the Hempelian position by quite a different perspective. This situation, unfortunately, left archaeology in the position of internalizing somewhat outdated philosophical positions.

Some of the misconceptions that arose concerned rather basic and not particularly controversial philosophical concepts such as induction, deduction, discovery, and justification. The difference between a deductive justification of a body of material and a deductive research strategy (and therefore the differences between the deductive-nomological model of explanation and the hypothetico-deductive method) was not fully appreciated. And the full import of the debates about objectivity and subjectivity in science were not recognized in archaeology. All these misconceptions occurred in the context of a deep desire to strengthen the nature of archaeological inference—a goal that cannot be faulted.

Archaeologists are quite properly concerned about the issues at stake in the foundations of science. To the extent that their understanding of these issues is clouded by misconceptions, the productiveness of reflective, philosophical discussion and the development of adequate archaeological research programs are compromised. One of our purposes in this chapter is to review the misconceptions about philosophical positions that have arisen in archaeology with a view to clarifying and correcting them so that archaeologists can make better use of existing research on foundational issues. Another is to place before archaeological audiences a wider range of philosophical positions and an understanding of the nature of major philosophical debates in order to remove the impression that philosophy is able to provide authoritative answers to methodological questions in archaeology or recipes for making archaeology more scientific. We begin the process by addressing some of the misconceptions which seem to be most fundamental and widespread.

Perhaps the most basic issue in the emerging methodological awareness in the archaeology of the 1960s and 1970s concerned the role of testing in scientific procedures. More than anything else, the question of whether archaeology could or should be a science was based on the question of whether or not there could be a systematic process of testing archaeological hypotheses. If archaeology was to be a science, it needed to incorporate a systematic program of testing of the kind exemplified in the natural sciences. The central concern with testing was a direct outgrowth of the more basic preoccupation with strengthening archaeological inferences—with finding a means of evaluating inferences that was more secure than intuition or what is sometimes referred to as narrow inductivism.

Testing

According to Binford: "The accuracy of our knowledge of the past can be measured; it is this assertion which most sharply differentiates the new perspective from more traditional approaches. The yardstick of measurement is the degree to which propositions about the past can be confirmed or refuted through hypothesis testing" (1968:17). The role of testing in archaeology has, in recent years, been profoundly contentious—a crucial point of division between those who champion an explicit view of archaeology as a science and those who are perceived to hold contrary views. As S. LeBlanc (1973:199–200) has said: "A principal tenet of the 'new' or scientific archaeology is that hypotheses must be explicitly tested."

The testing of hypotheses can be seen as a part of the larger philosophical area concerned with confirmation; for, of course, the point of testing hypotheses is to see whether or not they are confirmed by the available data. It seems clear that, given the fact that archaeologists want to provide the strongest accounts possible of the past, intersubjective checking of hypotheses—i.e., testing—is essential to this purpose and to the general purpose of the advancement of knowledge in the area. There are a number of alternative avenues to confirmation that need to be explored (see Chapter 6), but for the moment we will concentrate on the explicit testing of hypotheses.

The problem of testing has concerned philosophers of science of a number of different philosophical persuasions. Popper is particularly associated with the hypothetico-deductive model, emphasizing the falsification of hypotheses rather than their confirmation. Hempel, on the other hand, emphasized procedures of verification or confirmation. We use a well-known example from Hempel to illustrate the way hypotheses in science are tested.

Hempel (1966) draws this example from the work of the nineteenth-century Hungarian physician Semmelweis. Semmelweis attempted to explain why the death rates from childbed fever in two divisions of the same Vienna hospital were vastly different. As Hempel points out, sometimes the testing of possible explanatory hypotheses is quite straightforward. For example, in connection with Semmelweis' study "the conjectures that differences in crowding, or in diet, or in general care [accounted] for the difference in mortality between the two divisions" (Ibid.:6) were considered. But as Semmelweis discovered, these conjectures conflicted with the readily observable facts: there were "no such differences between the two divisions" (Ibid.). Thus, those hypotheses were dismissed as false.

Usually, testing is not so simple. For example, after having rejected the more obvious hypotheses concerning diet and crowding, Semmelweis postulated that the difference in mortality between the two divisions could be attributed "to the dread evoked by the appearance of the priest with his attendant" (Ibid.). His suspicion was that seriously ill patients interpreted the priest's visits to them as a sign that they were dying, and that the fear and horror brought on by that prospect actually worsened their conditions. The problem with testing such a hypothesis, Hempel says, is that the "intensity of that dread, and especially its effect upon childbed fever, are not as directly ascertainable as are differences in crowding or in diet" (Ibid.:6). We can count hospital beds and analyze food for its nutritive value, but dread is a rather more difficult phenomenon to observe. So, Hempel tells us, Semmelweis was forced to use an indirect method of testing. He asked himself whether there are any particular effects we should expect to be able to observe if the hypothesis were true: "And [Hempel says] he reasons: *if* the hypothesis were true, *then* an appropriate change in the priest's procedure should be followed by a decline in fatalities. He checks this implication by a simple experiment and finds it false, and he therefore rejects the hypothesis" (Ibid.).

In the more indirect of the two tests, Semmelweis deduced consequences from the hypothesis under consideration. Thus, if the hypothesis were true, the consequences would have to be true:

if the priest's behaviour was responsible for the difference in mortality rates between the two divisions, then "an appropriate change" in his behaviour should lead to a change in the mortality rate of the First Division. But the changes that would have to result if the hypothesis were true were not observed and, hence, Semmelweis had to reject the hypothesis.

There is an obvious similarity between the testing procedure that Hempel describes in the above and the hypothetico-deductive method which is often discussed in connection with Popper (*see* Chapter 3). In the latter, scientists deduce consequences from hypotheses they have formulated (which Popper calls "conjectures"), which they then proceed to try to refute. Hempel's procedure can be viewed as constituting part of the hypothetico-deductive method, but only part. For that method, especially in Popper's version, includes a prescription concerning the formulation of hypotheses, i.e., that they are not to be formulated in an inductivist way.

What is the relevance of this reference to Popper's well-known eschewing of the inductive method? Just this: that induction has played a large role in archaeology which can hardly be ignored despite the pronouncements of Popper. Generally, archaeologists have tended to adopt a view that induction is related to problem formulation and deduction to testing. Binford, for example, says: "It is maintained here that we must continually work back and forth between the contexts of explaining the archaeological record and explaining the past; between the contexts of proposition formulation (induction) and proposition testing (deduction)" (1972: 118–119).

Binford is surely right to stress the importance of working back and forth between the inductive and deductive contexts, but it is also important to be clear that there is no reason why results derived through inductive procedures cannot be subjected to testing. Several kinds of misconception occur in strongly equating deduction or deductive methods with testing or in denying that testing can occur within inductive frameworks. In the first place, induction is inevitably involved in the hypothetico-deductive method in at least two places. Popper to the contrary notwithstanding, the hypothesis formulation itself may be and often is inductive, and there are auxiliary assumptions involved which are also inductive. There may also be inductive elements in the actual testing. What is deductive is the deducing of the test implications and the form of the argument. Deductive reasoning can operate in the discovery process, as does induction. Lastly, testing is not incompatible with inductive methodologies or with induction as an explanatory model.

Views that link testing to deduction or to deductive methods to the exclusion of induction or inductive methods ignore alternative forms of establishing support for inferences. More fundamentally, there seems to be a lack of appreciation of the distinction between discovery and justification (*see* following section)—otherwise, the idea that the use of induction in archaeology precludes testing is inexplicable. As Hempel reminds us in the following often quoted statement: "What determines the soundness of a hypothesis is not the way it is arrived at (it may even have been suggested by a dream or a hallucination), but the way it stands up when tested, i.e., when confronted with the relevant observational data" (1965:6).

It is axiomatic in philosophy that a hypothesis discovered in even the most bizarre fashion may be shown to be justified. Richard Wasserstrom gives the example of a scientist who informs the scientific community that the way he has discovered a chemical formula that he now offers as a vaccine for immunization against cancer is by writing down 1,000 possible formulas on separate pieces of paper, putting them into a large hat, and pulling out one piece at random. Wasserstrom argues that the question of how the formula was selected is entirely different from the question of whether the vaccine will be effective. Furthermore, if the vaccine were tested and found to be effective, the formula's having been selected in an unorthodox way would not be grounds for rejecting it (Wasserstrom 1961:25–26). We may well doubt that a scientific hypothesis arrived at in this way will be justified because we think that there are other, far more efficacious ways of discovering good hypotheses, i.e., those that are eventually justified. But if, in this case, the vaccine is shown to be effective, then the hypothesis has been justified despite the unorthodox method of its discovery.

Hempel schematizes the reasoning that leads to the rejection of Semmelweis' "priest-behavior" hypothesis in the following, where he calls the hypothesis H, and the results to be expected if H is true the "test implication of the hypothesis" I:

If H is true, then so is I.
But (as the evidence shows) I is not true.

H is not true (Hempel 1966:7).

And as Hempel says of the above:

> Any argument of this form, called *modus tollens* in logic is deductively valid; that is, if its premises (the sentences above the horizontal line) are true, then its conclusion (the sentence below the horizontal line) is unfailingly true as well. Hence, if the premises of [the argument] are properly established, the hypothesis H that is being tested must indeed be rejected (Ibid.:7).

The scientist may not always conclude that the results of his experiments were "properly established." He might have reason to think, for example, that the change in the priest's behavior which was made for experimental purposes induced fear in the maternity patients in different, and unforeseen, ways. But the essential point is this: if the premises of a *modus tollens* argument are accepted as true, then the conclusion, as well, must be accepted as true. In this example, and in most such cases, the conclusion reached is that the hypothesis under test is not true; and this is the conclusion that Popper is aiming at with his falsification approach to science.

What conclusions can the scientist draw if the test implications of his or her hypothesis are observed upon experimentation? First, he or she cannot conclude from that experiment alone that the hypothesis is true, for to do so would be to commit the *fallacy of affirming the consequent*. Hempel schematizes this fallacious form of reasoning as follows:

If H is true, then so is I.
(As the evidence shows) I is true.

H is true (Ibid.:7).

Arguments of this form are deductively invalid because even if the premises are true, the conclusion may be false. As Hempel points out, this fact was

> illustrated by Semmelweis' own experience. The initial version of his account of childbed fever as a form of blood poisoning presented

infection with cadaveric matter essentially as the one and only source of the disease; and he was right in reasoning that if this hypothesis should be true, then destruction of cadaveric particles by antiseptic washing should reduce the mortality. Furthermore, his experiment did show the test implication to be true. . . . Yet, his hypothesis was false, for as he later discovered, putrid material from living organisms, too, could produce childbed fever (Ibid.:7–8).

It is not to be concluded from the above remarks that test results which bear out the implications of a hypothesis are useless to the pursuit of knowledge. Far from it. Such results constitute confirming instances of the hypothesis in question, and these are obviously relevant:

for each of our tests might conceivably have had an unfavorable outcome and might have led to the rejection of the hypothesis. A set of favorable results obtained by testing different test implications . . . of a hypothesis, shows that as far as these particular implications are concerned, the hypothesis has been borne out; and while this result does not afford a complete proof of the hypothesis, it provides at least some support, some partial corroboration or confirmation for it (Ibid.:8).

The reasoning used to draw conclusions from confirming instances of a hypothesis plays an important role in science and makes evident the close connection between explanation and confirmation. We need well-confirmed hypotheses to serve as lawlike statements in our explanations and, roughly stated, confirmable generalizations are those that could serve an explanatory role.

No number of positive instances of a hypothesis shows a hypothesis to be true. As we note in Chapter 6, we can never know with certainty that an empirical proposition is true. What we can know is that such a proposition has been well confirmed. Although it is very difficult to say exactly how much evidence, or what kind of evidence, constitutes the basis for rational acceptance of a hypothesis, we do regularly accept well-confirmed propositions into our body of knowledge, albeit usually tentatively.

To anticipate an important aspect of testing as it is related to another point we make in Chapter 5, it should be noted here that a promising approach to confirmation presently under discussion in the philosophical literature—the eliminative or selective approach—may be applicable to archaeology. It looks as though we ought to be looking to confirmation theory not so much for acceptance rules as for rules according to which we select hypotheses from among the alternatives considered: the selected, or best, hypothesis is the one which, in a given situation, survives elimination. This requires, then, that we test not just one hypothesis at a time, but plausible others as well, and "plausible" must be stressed. As W. Salmon (1966b:129) points out, while there are always "infinitely many possible hypotheses to handle any finite body of data, it does not follow that there is any superabundance of *plausible* ones." The question of which hypotheses are plausible in any particular evidence situation is a difficult one, for the issue does not reduce to a simply logical matter. As a first approximation, we will suggest in our fuller treatment of this topic that those hypotheses which should be seriously considered as providing alternatives are those which can reasonably be seen as *explanations* of the evidence (see Chapters 5 and 6).

We have argued that the testing of hypotheses is essential in archaeology, as it is in any discipline, and there would appear, in theory, to be no obstacle to the testing of archaeological generalizations. There are a number of difficulties, however. First, there is a serious paucity of data available against which archaeologists can test their hypotheses. This is a result of a number of factors, not the least important of which is the tampering with archaeological data that has occurred over a very long time. McGimsey (1972), in particular, has drawn attention to the destruction of these non-renewable resources. In *Public Archaeology*, he cites figures which suggest that development occurring in Arkansas has destroyed up to ninety percent of the archaeological record in or on the affected land. Further, he argues that his results should not be viewed as unique to Arkansas—the same is happening elsewhere. Others have noted that areas of intense prehistoric occupation unfortunately often coincide with sites of modern urban sprawl: the San Francisco and St. Louis areas are examples of this phenomenon. A further worry is that not only does such tampering with the archaeological record limit the quantity of data available to the archaeologist, but it also raises important questions about the representativeness of the data the archaeologists do have the opportunity to consider.

Some archaeologists (cf. Le Blanc 1973; Spaulding 1973:343) have argued that hypotheses need not be tested on a body of data different from that which was used to generate the hypotheses. Sometimes there is no other adequate body of data on which to test a given hypothesis, either because previously collected data which might be potentially relevant to testing the same hypothesis were not collected in ways that allow such testing or because there is a paucity of relevant cases. If the testing supports the hypothesis in question, what we have is one supporting instance, though we may also in some cases be in a position to make new observations on the existing body of data. For the hypothesis to achieve further support and more general usage, the ideal would be to have this hypothesis tested in a variety of contexts. As we argue in Chapter 4, the emphasis on novelty in archaeology has worked against routinizing certain aspects of the scientific method. Repeated testing of the same hypothesis on relevant bodies of data is less conducive to archaeologists' personal advancement in the profession than is originality. While testing of the hypothesis on the one body of data is reasonable, archaeology would move more rapidly toward the goal of improved knowledge if multiple and independent tests were performed.

A second difficulty with testing archaeological hypotheses revolves around the explanation of rare, or even unique, events. In a trivial sense, every event is unique—in some way different from all others. But in a significant sense, many events are relevantly similar to others. One crop of a particular species of peas is very much like another and so Mendel could easily replicate experiments to test his hypotheses concerning genetics, and the same is the case in most of the traditional sciences. Since human behavior and culture and their relationships to the environment are complex, with countless variables operating at any one time and over time, the question becomes one of identifying the relevant similarities. To identify other cases similar to the Chaco phenomenon (*see* Cordell 1984: 245–274), for example, do we pick out ecological similarities or similarities in population density, population distribution, exchange patterns, social organization—or combinations of these, assuming that there was agreement on the nature of each? Or do we regard the Chaco phenomenon as unique? Simply recognizing rare and unique phenomena is a problem: uniqueness or rarity may be a result of sampling procedures, for example, rather than being intrinsic to the case. By and large, archaeologists have tended to prefer dealing with more generally occurring phenomena, but this fact probably introduces its own biases into the data.

Other problems in formulating hypotheses in archaeology include deducing which consequences of a hypothesis require testing, and deciding what counts as support or failure. If a hypothesis has been formulated from the same body of data used for testing, it is quite likely that there will be some support forthcoming. Le Blanc (1973:199–200) argues that a hypothesis is logically independent of the data which may have been used to generate it. Although we agree with this point, we are less certain than he is that a hypothesis derived from a body of data can be easily falsified. As a safeguard against raising the possibility of falling into fallacious forms of reasoning such as affirming the consequent, the reasonable procedure seems to be the use of multiple, competing hypotheses (cf. Zubrow 1973 for examples of kinds of models that might be used; *see also* Hole 1973). More generalized models of hypotheses have a better chance of providing support than more detailed ones, and the complexity of the hypothesis should be considered in evaluating the results of testing.

This is not, of course, to claim that seeking multiple competing hypotheses is necessarily the best strategy for generating answers to archaeological questions. Sometimes, for example, it will be more productive to fasten on a single hypothesis and develop it as far as is possible, noting where it fails, if it does. Still, inference to the best explanation (*see* Chapter 6) represents one possible strategy, but more importantly, is a plausible desideratum for justification of hypotheses. That is to say, that a hypothesis survives when all plausible competing ones are eliminated is a reason for accepting it, even if not a conclusive one.

Since basic archaeological data are so malleable, the starting point of any inference or hypothesis is crucial. The underlying assumptions need to be assessed as carefully as the hypothesis itself and its test implications, a point stressed by Nicholas David (1971) in his Fulani Compound study and seen clearly also in countless other cases in ethnoarchaeology.

Hypothesis formation and testing employ both inductive and deductive methods, and there can be no doubt that archaeologists need to attend to both of these methods in coming to understand the reasoning used in the discipline. It is important as well, at least in the first instance, to be conscious of the difference between hypothesis formation or discovery and hypothesis testing, validation, or justification—a matter we take up in the next section.

The discovery/justification distinction

Archaeologists have paid attention to both discovery and justification procedures in their efforts to become clear about the foundations of the discipline. The fundamental point of the distinction was simply to make clear that the discovery of a hypothesis is logically independent of its justification and thus that we might be justified in accepting a hypothesis even though it has been arrived at in some entirely unorthodox manner. The emphasis on research strategies and problem orientation in archaeology essentially addressed discovery procedures, whereas testing, appeal to the D-N model of explanation, and concern with explanation as the basic goal of archaeology focused on justification. The distinction between the two was absolutely fundamental for some of the foremost spokesmen of the new methodological awareness. Nevertheless, some confusions arose as to precisely what the distinction was and what its importance was.

It is not difficult to understand why archaeologists have sometimes combined the processes of discovery and justification into a single package. It was felt by key movers in the New Archaeology that archaeology needed a major overhaul. Although most of the rhetoric was devoted to justification issues, a host of other changes occurred at the same time, many of them at the level of problem orientation and research design. In historical perspective, it seems to be the case that research strategies for discovery were based to some extent on the hottest item in the archaeological repertoire— namely the logic of justification, and this probably occurred because at that stage of the development of philosophical self-consciousness among archaeologists, it was not thought important enough to make the discovery/justification distinction explicitly.

The process of discovery may or may not be open to codification in terms of some sorts of objective logical consideration. But whether it is or not is largely independent of the question as to whether we can offer a reasonable reconstruction of scientific explanation or confirmation. The Hempelian proposals in which archaeologists have shown so much interest need to be seen as the latter sort of enterprise; they are not meant as recipes for scientific discovery. In his work on confirmation and explanation, Hempel is attempting to give an account of what it means to say that certain evidence confirms a hypothesis, and what it means to say that a hypothesis or theory or law explains certain facts. How we come to the hypothesis or theory or law is not in question. His emphasis is thus on testing hypotheses rather than on discovering them, on trying to formulate criteria for the acceptance or adequacy or reasonableness of hypotheses or theories. Any discussion of the discovery/justification distinction would be futile if it were not that the distinction, though widely recognized, is almost universally ignored, and certainly not just by archaeologists.

It is clear that philosophers of science are not offering a codification of rules that scientists follow or should follow when they go about their business of making scientific discoveries. Popper and Hempel are explicit on this point. According to Popper, it is not the task of the logic of knowledge to reconstruct the processes of inspiration and discovery:

> Such processes are the concern of empirical psychology but hardly of logic. It is another matter if we want to reconstruct rationally the *subsequent tests* whereby the inspiration may be discovered to be a discovery, or become known to be knowledge (1961:31).

Hempel (1965:5–6) makes this point as well, when he argues that we have no general rules of inductive inference that will inevitably direct us to the discovery of correct scientific hypotheses or theories.

Fritz and Plog (1970) are among the archaeologists who have been criticized for confusing the context of discovery with the context of justification. Correctly identifying Hempel's remarks in *Philosophy of Natural Science* (1966:10–18) as criticisms of a particular method of discovery, they proceed to contrast an "empiricist research design," which they claim is characteristic of traditional archaeological research patterns, with their "explanatory research design." Michael Levin argues that "Fritz and Plog and the empiricists are simply talking past each other for the two programs are designed to do two different jobs"; the reason the authors "believe they are offering an alternative program for discovery" is "that they confuse the context of justification with the context of discovery" (1973:393). Corroborating a law candidate is done through the hypothetico-deductive model. After noting that it is irrelevant where the laws used in explanation come from, Levin continues:

> Equally irrelevant is the problem of corroborating a law-candidate once it has been proposed, and this is the activity of which the hypothetico-deductive model gives an account. The hypothetico-deductive model of corroboration has as little to do with the *formation* of hypotheses as does Hempel's model of explanation (Ibid.: 393).

As Rogge says in his paper "Processual Archaeology and the Philosophy of Science: Three Problems," many archaeologists "have maintained mistakenly that the only difference between the D-N model of explanation and the H-D method of testing hypotheses is that the latter does not require that the major premise be a covering law, but merely any hypothesis." As Rogge continues, "This is certainly not the case" (1975:7). Rogge cites the Hole and Heizer (1973) textbook and an article by Sabloff, Beale, and Kurlan (1973) as other examples of the failure to distinguish adequately between justification and discovery.

Although we maintain, as have many philosophers, that the distinction between discovery and justification needs to be clearly stated, it is true in archaeology, as in other fields, that there are certain data requirements to be met for effective testing of a hypothesis or for showing that a particular case can be subsumed under a general category of phenomena. Research strategies, at least in part, need to be oriented toward the explanations or inferences to be drawn. Indeed, archaeological inferences are so tightly bound to the particular questions being asked that discovery and justification at some pragmatic levels, although not at the level of formal validation, are in fact closely linked. Justification procedures will be more effective if there is a strong continuity between the discovery and justification processes—between the provision of data for testing and the process of testing itself.

Archaeologists are not alone in suggesting that the distinction between discovery and justification needs to be blurred somewhat. As early as 1958, N. R. Hanson, in his *Patterns of Discovery*, was explicitly challenging the distinction. In historical and sociological views of science such as those of Thomas Kuhn (*see* Chapter 3), philosophers of science since the early 1960s have been turning their attention both toward history of science (often in the form of case studies) and toward the actual practice of science (cf. Giere 1973; McMullen 1975; Burian 1977; Maull 1976; Wartofsky 1976). Part of the point of this movement has been the recognition that scientific propositions or theories are not just atemporal entities whose justification has no connection with how they came to be discovered or proposed. Instead, the record of how science actually has been practiced in the past and how it is presently practiced are both seen to be relevant to the question of whether a particular scientific theory is acceptable; justification is not a purely logical matter, although of course logic is involved. Conversely, the process of discovery is not outside the bounds of rationality, even if capturing the process in purely logical terms is beyond traditional philosophy of science. As Patty Jo Watson notes in discussing the use of analogies in ethnoarchaeology: "Logically speaking, it does not matter where these interpretive hypotheses come from; what matters is how they stand up when tested against the archaeological record. I say *logically* it does not matter, but—as is so often the case—*practically* it does matter" (1979:277).

The situation, then, is that historical and descriptive considerations are relevant to, but do not entirely determine the answers to justificatory questions. As Nelson Goodman tells us in another context—that of the problem of justifying induction:

> Principles of deductive inference are justified by their conformity with accepted deductive practice. Their validity depends upon accordance with the particular deductive inferences we actually make . . . All this applies equally well to induction. . . . the traditional smug insistence upon a hard-and-fast line between justifying induction and describing ordinary inductive practice distorts the problem (1965:63–64).

Thus it seems reasonable to maintain that there is no hard-and-fast line between the context of discovery and the context of justification, that philosophy of science needs to take account of the actual history of science and of the values implicit in particular understandings of the scientific enterprise, and also that there are pragmatic reasons in archaeology for designing the discovery process so that the justification procedures can be effectively pursued. Nevertheless, the interests of clarity are served if we distinguish carefully, at least in the first instance, between questions of discovery and questions of justification, however much these two come together as the inquiry proceeds.

Inductive and deductive inference

One of the consequences of selecting deductive methods and forms of explanation as appropriate to archaeology, as did the New Archaeologists, has been the relegating of induction methods and inductive explanations to the realm of the disreputable. Induction was correlated with traditional methods in archaeology which were thought to be discredited in favor of more scientific procedures, and these were to be deductive. Along the way, some serious misconceptions arose about the nature of induction. A number of writers have commented on these misconceptions (*see* M. Salmon 1976), but it seems worthwhile nevertheless to discuss them in a systematic way.

The idea that deduction proceeds from the general to the particular and induction from the particular to the general is widespread. The popular notion that induction is simply generalization from collected data can be traced from Francis Bacon all the way back to Aristotle. This view has been introduced into archaeology, where it is to be found, among other places, in the writings of Watson, Le Blanc, and Redman in their book *Explanation in Archaeology* (1971) and Fritz and Plog in their article "The Nature of Archaeological Explanation" (1970).

While there are many differences between induction and deduction, that the latter leads us from the general to the particular and the former leads us from the particular to the general is not one of those differences. To see this we can look at a number of examples given by Brian Skyrms (1975:13–15):

> There are deductively valid arguments that go from general to general:
>
> > All gorillas are apes.
> > All apes are mammals.
> > _____
> > All gorillas are mammals.
>
> from particular to particular:

Ezekial is a wolf.
Ezekial has a tail.
___

Ezekial's tail is the tail of a wolf.

and from particular to general:

Frederick is an alderman.
Frederick is a thief.
___

Anyone who knows all aldermen knows some thief.

Similarly, inductively strong arguments do not fall into the narrow category of arguments having particular premises and a general conclusion. *Arguments by analogy* proceed from particular to particular, and although such arguments are often misused, some are quite strong. The following is an example of a strong argument by analogy:

1. Car A is a Hotmobile 66 and car B is a Hotmobile 66.
2. Car A has the super-zazz engine and car B has the super-zazz engine.
3. Car A's engine is in perfect condition and car B's engine is in perfect condition.
4. Both cars have the same type of transmission and the same final drive ratio.
5. Car A's top speed is over 150 miles per hour.
___

Car B's top speed is over 150 miles per hour.

Premises (1) through (4) are said *to set up the analogy,* that is, they describe relevant similarities between car A and car B which make it likely that their top speed is the same, and therefore license the move from premise (5) to the conclusion.

It is not difficult to find inductively strong arguments that have general premises and a general conclusion; for example:

All students in this class are highly intelligent.
All students in this class are strongly motivated to do well.
No student in this class has a heavy work load.
No student in this class has psychological difficulties that would
    interfere with his course work.
___

All students in this class will do well.

Inductively strong arguments with general premises and general conclusions play an important role in advanced science. For example, Newton's laws of motion were confirmed because they accounted for both Galileo's laws of falling bodies and Kepler's laws of planetary motion. We can give a rough approximation of the argument supporting Newton's laws of motion as follows:

> All bodies freely falling near the surface of the earth obey Galileo's laws.
> All planets obey Kepler's laws.
> ___
> All material objects obey Newton's laws.

We can also give an example of an inductively strong argument with a general premise and a particular conclusion:

> All emeralds previously found have been green.
> ___
> The next emerald to be found will be green.

People often use arguments of this type when they marshall generalizations about past experience in order to make a prediction about a particular impending event.

From the above examples it should be clear that the difference between induction and deduction has nothing to do with generality or particularity. Instead, the difference lies in the relation between premises and conclusions in inductive and deductive inference.

In a correct deduction the truth of the premises guarantees the truth of the conclusion, while in an acceptable inductive inference all the premises may be true and the conclusion yet be false. It is always possible, of course, for the premises of a deductive argument to be false (but in that case it is not said to be a sound deduction even though it may be valid). In other words, in a correct or valid deduction, if the premises are true then the conclusion must be true: if all gorillas are apes and all apes are mammals, then all gorillas must be mammals. In contrast, even if all emeralds previously found have been green it is not necessarily the case that the next one found will be green. It could be otherwise, unless greenness is used as a defining characteristic of emeralds. Similarly, we might say that all Folsom points have basal grinding, but the finding of one without basal grinding is possible (unless basal grinding is defined as a diagnostic characteristic of Folsom points).

In his article "On Explanation in Archaeology: A Rebuttal to Fritz and Plog," Michael Levin (1973) discusses a related confusion. There he criticizes the views attributed to Raymond Thompson (cf. Thompson 1958b), on the one hand, and Fritz and Plog (1970), on the other, on the role of logic in archaeology. While Thompson has argued that deduction is inappropriate to archaeology and Fritz and Plog that it is absolutely essential to the discipline, Levin claims that both views "are wrong for precisely the same reason," and that:

> Both mistakenly assume that deductive inference is a technique for arriving at *certainty*. Fritz and Plog find in it an explanation of the (supposed) certainty of classificatory ascriptions, while Thompson thinks that it is just this feature of logic which wars with the empirical and interpretational dimension of archaeology (1973:388).

Levin stresses the point that "logic is not a technique for arriving at certain knowledge" (1973:388), and this really cannot be overstressed. The conclusion of a deductive argument is neither certain, nor true, simply because it has been reached deductively. All that a deductively valid argument ensures is that if the premises of the argument are true then the conclusion must be true as well. Levin (Ibid.) uses an example which would look like this with its essential premises made explicit:

All items I have just found in the barrow are ceremonial daggers.
This object is one I have just found in the barrow.

This object is a ceremonial dagger.

The above is a deductively valid argument. But all this means is that, if the proposition "All items I have just found in the barrow are ceremonial daggers" is true (and the second premise is true as well), then the proposition "This object is a ceremonial dagger" must be true. Whether in fact all the items I have found in the barrow are ceremonial daggers is a question that cannot be answered through application of the rules of deductive logic alone. Most archaeologists would surely feel more comfortable in identifying objects from barrows on some other, inductively derived, grounds.

Another aspect of the difference between induction and deduction is that the conclusion of a valid deductive inference is said to be implicit in the premises of the inference. Similarly, it is often said that the conclusion of a valid deductive inference is "contained in" the premises of the inference, or that nothing can appear in the conclusion that does not already appear, at least implicitly, in the premises. In contrast, inductive inference is said to be "ampliative," or, as Skyrms writes:

> The conclusion of an *inductively strong argument*, . . . ventures beyond the factual claims made by the premises. The conclusion asserts more than the premises, since we can describe situations in which the premises would be true and the conclusion false.
>
> If an argument is inductively strong, its conclusion makes factual claims that go beyond the factual information given in the premises.
>
> Thus an inductively strong argument risks more than a deductively valid one; it risks the possibility of leading from true premises to a false conclusion. But this risk is the price that must be paid for the advantage which inductively strong arguments have over deductively valid ones: the possibility of discovery and prediction of new facts on the basis of old ones (1975:8–9).

Whatever else may be true of inductive inference, it should be apparent from Skyrms' final remark above that it is absolutely essential to scientific progress. Without it we could not expand the body of knowledge of any discipline, for we could never justifiably project beyond already identified facts.

Deductive inferences fall into two categories: valid and invalid (or those in which *if* the premises are true the conclusion must be true, and those in which the truth of the premises does not ensure the truth of the conclusion). On the other hand, inductive inferences are evaluated according to their strength. Thus, whether an inference is deductively valid (or invalid) is an all-or-nothing matter, whereas whether an inference is inductively strong is a matter of degree. As Skyrms says, "An argument is *inductively strong* if and only if it is *improbable* that its conclusion is false while its premises are true, . . . The *degree* of inductive strength depends on how improbable it is that the conclusion is false while the premises are true" (1975:7).

One might even want to say, along with W. Salmon (1971:79), that a valid deductive inference is a limiting case, i.e., the strongest kind of inductive inference, since the premises of a valid deductive inference confer a probability of 1—the highest probability possible—on the conclusion. Other writers strongly resist this way of treating deductive inference, insisting that it is a different *kind* of inference entirely from inductive inference. One could also say that inductive inferences, the premises of which confer a probability of less than 1 on their conclusion (as is usually the case with inferences we call "inductive"), are deductively invalid. Although there are these overlaps, the fact remains that inductive and deductive logic are different in this essential respect: in a valid deductive inference the truth of the premises guarantees the truth of the conclusion, while in an acceptable inductive inference all the premises may be true and the conclusion still be false.

Another way to think of inductive logic is as a system of rules consisting of the rules of deductive logic with something added, that is, as deductive logic plus one or more rules of inductive inference; but even if we adopt this model, it does not at all follow that principles which hold in deductive logic hold in a straightforward way in inductive logic. For example, in deductive logic, if all A's are B's and all B's are C's then all A's are C's. (*See* Skyrms' gorilla example above.) In inductive logic, however, analogous relationships do not hold. Thus, even if most A's are B's and most B's are C's it does not follow that most A's are C's (but it does follow that over 1/4 of A's are C's). For an obviously farfetched archaeological example, suppose that A's are Pueblo IV people of New Mexico, B's are males, and C's are Marpole Phase people of British Columbia. It clearly does not follow from the facts that some Pueblo IV people of New Mexico are males and some males are Marpole Phase people of British Columbia, that some Pueblo IV people of New Mexico are Marpole Phase people of British Columbia.

As W. Salmon and other philosophers have noted, inductive inferences suffer from ambiguity in a way in which deductive inferences do not. Incompatible inductive inferences can be drawn from two sets of true premises. Further, both incompatible inferences in such a case may be quite strong. Hempel offers the following example (which comes from Toulmin 1958:109) of this phenomenon:

> Petersen is a Swede.
> The proportion of Roman Catholic Swedes is less than 2 percent.
> So, almost certainly, Petersen is not a Roman Catholic (1965:55).

Even if the premises of the above are true, the premises of the following may be true as well:

Petersen made a pilgrimage to Lourdes.
Less than 2 percent of those making a pilgrimage to Lourdes are not
    Roman Catholic.
So, almost certainly, Petersen is a Roman Catholic (Ibid.:55).

And Petersen cannot be both almost certainly not a Roman Catholic and almost certainly a Roman Catholic.

Philosophers concerned with this issue are generally agreed that in order to solve this problem of ambiguity a condition of total relevant evidence must be imposed on inductive inference. Roughly, this condition requires that all the evidence relevant to drawing the inductive inference must be considered. If we want to draw a justifiable conclusion about Petersen's religious affiliations, it won't do just to consider the fact that he is a Swede, for some Swedes are and some Swedes are not Roman Catholic, or so we may assume for the purposes of this example. All other relevant, available facts must be considered. Even though only a small percentage of Swedes are Roman Catholics, if a Swede has made a pilgrimage to Lourdes it could be said that it becomes likely that he is one of that small percentage. We note that Petersen is in a two percent minority both as a potential Roman Catholic Swede and as a non-Roman Catholic visitor to Lourdes. It might be that the relative population sizes of Sweden and Lourdes would become relevant if the example were pursued, but this information is not given.

We do not propose to consider the total evidence requirement in any detail. To do so would most assuredly take us too far afield, for there are interesting problems with the requirement and both Salmon's (1971) and Hempel's (1965) formulations of it are highly technical. Rather, we introduce it to illustrate a difference between induction and deduction: while the total evidence requirement must be imposed on inductive inferences, it is automatically satisfied in valid deductive inference. For in a valid deductive inference, true premises ensure a true conclusion. If it were true that no Swedes are Roman Catholic then to the argument

No Swedes are Roman Catholic.
Petersen is a Swede.
Therefore, Petersen is not Roman Catholic.

any number of premises may be added—for example, that Petersen made a pilgrimage to Lourdes, or even that he was blessed by the Pope. If the premises of the above argument are true, then no number of additional premises can vitiate that argument's validity. The same is not true of inductive inferences, as we have seen.

The essential point about the confusion of induction and deduction is that these are different systems of logic, or at least of inference, providing us with criteria according to which we can evaluate arguments. That is, we can evaluate the relationships between claims (conclusions) and the reasons given for the claims (premises). If the premises guarantee the conclusions—if the conclusions are "contained in" the premises—then the relationship is deductive in form. If the conclusions go beyond the premises, or are not guaranteed by the premises (i.e., the conclusions can be false even when the premises are true), then the relationship is inductive in form. Although induction and deduction are two forms of logic, they are not two different methods of scientific enquiry. Logic provides us with rules for evaluating the conclusions. It does not, and cannot, provide us with the procedure for obtaining those conclusions or results.

Given that sound deductions require true premises, the question arises whether that form of logic, and deductive models of explanation, are appropriate for archaeology. Was Raymond Thompson right when he argued that induction was what was appropriate for archaeology? If induction is appropriate, why has it fallen into such ill repute among archaeologists? In fact, both forms of logic are required. Thompson was partly right in urging archaeologists to turn to induction. It is unfortunate that a more balanced view of the two forms of logic has not been maintained in archaeology, because inductive methodology other than the hypothetico-deductive method has not been exposed to the scrutiny it deserves within archaeology. The reason for disregarding the entirely necessary role of induction in archaeology is tied to the desire to make archaeology more scientific, more theoretically sophisticated—a science of (relatively) "certain" knowledge. It is hardly surprising that, in reaching for such goals, the strongest form of logic was thought to be appropriate.

Nondeductive considerations and confusions between induction and deduction versus inductivism and deductivism

Most of the primary writers on philosophical matters in archaeology—Watson, Le Blanc, and Redman (1971) among them—have recognized that the field requires both inductive and deductive methods. Binford has written that "[Archaeologists] must continually work back and forth between . . . the contexts of proposition formulation (induction) and proposition testing (deduction)" (1972:118). Paul Martin has said that "Our knowledge of the past can only be increased by these procedures of interplay and feedback of deduction-induction, formulating hypotheses concerning human behavior and then testing them by relevant archaeological data" (1971:6). Dunnell (1982:6) asserts that "Science is a dualistic enterprise, involving both induction and deduction in the strict senses." The actual depiction of inductive and deductive methods by these several authors is not always as clear as one might wish, but they did not carry the burden of dotting every *i*. Other users of these concepts have not always been as perceptive about this issue, and a number of unclarities and confusions about induction and deduction have been introduced into the archaeological literature.

While it is clear that philosophers have been concerned primarily with the logic of justification, some philosophers as well as many nonphilosophers have addressed more general problems of scientific method (*see* Chapter 3). The views taken on this topic in archaeology have advocated one of two investigative procedures—inductivism or deductivism—and this is unfortunate for two reasons. First, it is not at all clear that there are actually two contrasting procedures that are distinct from one another in any meaningful way; indeed, we would argue that the controversy over the appropriateness of one or the other of these methods is essentially empty. Second, the names which have become attached to these two approaches have no doubt been the source of much of the confusion we have noted in the previous two sections of this chapter. This is the case because the names of two systems of logic or inference used in the process of justification of scientific results—"induction" and "deduction"—have become associated with two putatively different approaches to the discovery of scientific results in only a slightly modified form.

Deductivism, or the hypothetico-deductive method, is described by Achinstein in this way: a hypothesis, or law, is formulated to deal with some scientific problem, and

> Consequences are derived deductively from the law, possibly in conjunction with other assumptions, and are subjected to empirical tests . . . When a law is proposed its deductive consequences are tested by seeing whether they are compatible with the observation statements. If they are not the law is refuted (1971:110).

On the other hand: "With respect to the context of discovery, the inductivist position is that scientific inquiry is a matter of inductive generalizations from the results of observations and experiments" (Losee 1972:148). So, contrary to deductivism, inductivism holds that the scientist does not begin his investigations with a hypothesis in mind, but only arrives at one through inductive generalization from the results of those investigations.

Inductivism is usually associated with the names of Francis Bacon and, especially, John Stuart Mill. Indeed, John Losee, in *A Historical Introduction to the Philosophy of Science*, describes Mill as "an effective propagandist on behalf of certain inductive methods." "So much so," Losee continues, "that these methods came to be known as 'Mill's Methods' of experimental inquiry" (Ibid.:148). Popper and his followers have been the most vociferous advocates of deductivism in the twentieth century; however, William Whewell, a contemporary of Mill's, was one of the first philosophers to clearly advance a noninductivist account of scientific method. In so doing, Whewell directly challenged Mill's views, and, in fact, the two men had occasion to debate many of the differences between them. Noting that the views of Mill and Whewell are frequently contrasted, Losee says, "Often Mill is presented as identifying scientific discovery with the application of inductive schema [sic], whereas Whewell is presented as viewing scientific discovery as a free invention of hypotheses" (Ibid.:152).

As we have already suggested and will argue below, these are really not two complete and distinct scientific methods. Interestingly, as Losee points out, it seems that even Mill recognized this:

> No doubt Mill did make incautious claims for his inductive methods. . . . But despite the comments that Mill directed against Whewell on this issue, Mill clearly recognized the value of hypoth-

esis-formation in science. It is unfortunate that subsequent writers have overemphasized the incautious claims that Mill made in his debate with Whewell (Ibid.:152).

Despite the paucity of historical support for the thesis that inductivism and deductivism are mutually exclusive approaches to science, debates as to which is the correct approach to scientific investigation are common. For example, some New Archaeologists seem to think that deductivism is more respectable from a scientific point of view than the putative alternative to it. Thus Fritz and Plog criticize what they call the empiricist or narrow inductivist approach. Inductivism, they claim, "assumes that facts will speak for themselves and that explanations or laws are summaries of facts. If one collects sufficient data, analyzes, classifies, and otherwise juggles it sufficiently, explanations or laws are supposed to begin to propose themselves" (1970:410).

Fritz and Plog rightly claim that no archaeologist does or could proceed in this way. Probably no scientist could. The caricature of scientific method which suggests that we go around collecting instances of hypotheses that haven't yet occurred to us and generalizing when we've collected enough is absurd. An obvious difficulty with this view is that, without a hypothesis in mind, we could never know when enough data was enough! We know of no scientist or philosopher of science in recent times who has held that this was the correct scientific method. As Popper says in arguing against inductivism:

> Without waiting, passively, for repetitions to impress or impose regularities upon us, we actively try to impose regularities upon the world . . . Without waiting for premises we jump to conclusions . . . The belief that we can start with pure observations alone, without anything in the nature of a theory, is absurd (1962:46).

Unfortunately, in the archaeological literature, questions of testing and of explanation have been conflated, and the emphasis on deductive procedures in each seized upon, with a consequent confusion of the issues in these two areas and a resulting overemphasis on deductive method.

Not that the deductive method is unimportant, for neither in theory nor in practice do we find a coherent alternative to what has come to be called the deductivist approach to scientific discovery.

Nor does there appear to be a genuine alternative to deductivism for the purposes of testing hypotheses. It would seem that what we mean by "testing" is what the hypothetico-deductive method prescribes: the deduction of certain observational predictions from a hypothesis which is refuted if the predictions fail to be fulfilled in the appropriate circumstances. There are other grounds for rejecting a hypothesis—for example, because it resembles in some important ways other hypotheses which have been falsified. But clearly, if a scientist is going to engage in hypothesis testing, he or she must resort to deductivism.

It is clear, however, that deductivism alone does not tell the whole story. For example, one needs to have some reasons for proposing particular hypotheses for evaluation in the first place. Scientific activity does not take place in a vacuum. Rather, we have a background of myriad hypotheses and much data, and we are faced with the problem of determining which hypotheses or conclusions are correct, are likely to be correct, or are even worth further investigation. This determination cannot be a purely deductive matter.

As to reasonable grounds for proposing a particular hypothesis, arguments based on induction by enumeration (e.g., "All *examined* emeralds have been green, so probably all emeralds whatever are green," or, to take an archaeological example, "Known tribal level societies lack urbanism, so probably all tribal level societies lack urbanism"), constitute one category. Others include models and analogies based on similarities between the features of the hypothesis and some already well-established theory, and sometimes even arguments from authority—the fact that the theory was suggested by a brilliant scientist who has never been wrong in the past. Similar arguments can be adduced for failing to give serious consideration to some hypothesis that may have been proposed. A theory which contradicts a large body of accepted scientific theory is prima facie suspect, as is one put forward by someone we know to be completely uninformed in the area in question. Considerations of this sort are, of course, not at all decisive, but they do operate to a greater or lesser degree in science.

Hypotheses which, at some point in history, appear unlikely—for example, that the earth is not flat, but round—often turn out to be correct. And the converse also holds: many of those which look exceedingly promising are eventually falsified. Thus, the probability of its truth does not alone determine whether a hypothesis should be proposed; and, indeed, there is no single criterion for determining this. Simplicity and explanatory power, although both difficult to measure, also come into play; all other things being equal, the hypothesis that explains the most with the least is the most desirable. The problem here is that of deciding when all other things are equal, and simplicity alone is not an adequate criterion. Other considerations also bear on the reasonableness of putting forward hypotheses, but the general point is that while we may jump to conclusions, to use Popper's phrase, it is unlikely that we will do so without any reason at all or without any thought of the plausibility of the hypotheses advanced.

Scientists approach their research with at least vague expectations of, if not clear ideas about, what they may find. Most do not operate only on the basis of discrete hypotheses, one for each separate experiment or research task. Each, as well, holds a general view about how the entities which his or her discipline studies are related to one another and to entities outside his or her field. Such a view gives scientists a framework to which they can relate their own findings, and this undoubtedly directs their research as well as determining to some extent which theories will be found worthy of further investigation.

Clearly, then, scientific investigation involves nondeductive elements, and this is made even more obvious when it is realized that there are inductive aspects to the hypothetico-deductive method itself. The hypothetico-deductive method involves the deduction of a prediction from a set of premises, for example:

All men are mortal.
Socrates is a man.
_____

Socrates is mortal.

Since by definition the above is a deductively valid inference, if the premises of it are true then the conclusion must be true, and if the conclusion is false then at least one of the premises must be false. Thus, assuming Socrates is a man, if the prediction "Socrates is mortal" turned out to be false, then the generalization "All men are mortal" must be false. Little information about the truth of that generalization can be drawn from the truth of that prediction, however. That Socrates is mortal does not establish that all men are mortal, but neither, of course, does it disprove that claim.

Thus, inductive elements come into play when an event predicted by deduction from a hypothesis being tested actually does occur. We are not always significantly enlightened if a prediction turns out to be true; while a predicted event which fails to occur at least prima facie eliminates from further consideration the hypothesis from which it was deduced, true predictions are not as uniformly helpful. As true conclusions of a deductively valid argument, true predictions could have been deduced from either true *or* false premises. There are, however, some circumstances in which we are justified in saying that a true prediction lends support to the hypothesis from which it was deduced. It is not at all easy to say when we are so justified, for, as we discuss at length in Chapter 5, "In what cases can we say that certain evidence *e* confirms some hypothesis *h*?" is a vexing question in confirmation theory. The essential point here, however, is that sometimes we have reasonable grounds for claiming that a true prediction lends support to the hypothesis from which it was deduced, and when we do the logic of that support is inductive.

An example may help to show why confirmation of the sort described above must be inductive. Consider the following:

> H: Range conditions for bison deteriorated between 1450 and 1550 in the Pecos Valley.
> This site is a bison kill of that period.

> Prediction: The bison killed at this site had been subjected to poor range conditions.

Assuming we found evidence that the bison killed were in poor condition, we have support for the initial premise. The initial premise could still be false, or incomplete, however, for there could be other reasons for poor condition, such as seasonality. Still, it is clear that whatever support is lent to the hypothesis is inductive rather than deductive.

A caveat is perhaps in order here. We have suggested, for reasons Popper and others have given, that no scientist ever could actually employ an inductivist approach to collecting data. This is perhaps overly strong. There probably are stages in the work of most scientists where data are treated in an open-ended, inductive way, and certainly the early development of sciences such as biology is often thought to be essentially inductive in form. Nevertheless, we would still want to claim that no science could proceed in a thoroughly inductivist manner throughout its history. Scientific investigation no doubt requires a constant readiness to shift attention back and forth between inferences and the data, further considerations in the one giving a fresh perspective in the other. Raymond Thompson has written of "indications in the data" that cause an archaeologist to pursue a certain direction in subsequent analysis and interpretation, and this occurs frequently in actual practice. Notwithstanding the fact that scientists' accounts of their own intellectual procedures are often untrustworthy, from many of these accounts it emerges clearly that both inductivist and deductivist approaches played a role in the work being discussed. For example, Darwin's descriptions of the scientific method he adopted are notoriously unreliable, but they do suggest the extent to which both inductivism and deductivism may have been informing his thought. In one and the same work, Darwin said of himself that he "worked on true Baconian principles, and without any theory collected facts on a wholesale scale," and that he "could not resist forming a hypothesis on every subject" (F. Darwin, ed. 1887).

What must be remembered is that induction and inductivism are two different things, as are deduction and deductivism. We have deliberately made these points several times, believing that it is virtually impossible to overstate them, for while they are absolutely fundamental, they nearly always go unappreciated. The consequences of overlooking their significance can be disastrous, as a passage from Rogge makes clear:

> As the label "induction" came to be used more and more it acquired negative connotations. Meanwhile the new paradigm came to be called "deductive" archaeology. . . . [m]ost archaeologists came to accept the fact that both induction and deduction played necessary roles in scientific archaeology . . . but despite the compromise, there has been little indication that most archaeologists understand the basic difference between the two logics . . . Many seem to erroneously equate induction with a lack of problem orientation and deduction with specific problem-oriented research. As I have . . . pointed out, this confusion stems in part from the *failure to dis-*

tinguish between the contexts of discovery and justification (Rogge 1975:4, emphasis added).

The essential point is that induction and deduction are systems of logic with rules for evaluating the worth of the conclusions we draw from data (premises). Induction and deduction, in other words, are the tools we use for the job of justifying our claims. "Inductivism" and "deductivism," on the other hand, are the names that have been given to two methods of discovery of data and hypotheses. As we have previously argued, methods for discovery should not be confused with standards of adequacy or justification, however much the two may come together in actual practice.

3 / Two Views of Science

Archaeologists initially turned to a limited range of philosophers for methodological guidance, with the main focus of
the early New Archaeology on the work of Hempel. There was
much to be gained in clarifying the foundations of archaeology by
attending to Hempel's writings, but this early exclusivity also had a
distorting effect. Had archaeologists delved more deeply into a
broader range of philosophical positions, they would have found
divergent views of science and controversies of great significance
to archaeological concerns. To counteract the distortion brought
about by a too narrow focus on the work of Hempel, we present in
this chapter the views of two other philosophers of science, Thomas Kuhn and Karl Popper. This is useful because both present
significantly different alternatives to the received view (and the
view to which archaeologists tended to gravitate), and both have
been taken seriously by practising scientists in recent years.

Until quite recently most philosophy of science has been concerned primarily with issues surrounding the justification of results in science, but some philosophical attention has been paid to
the process of scientific investigation and to the question of
whether there can be a *logic* of scientific discovery. Kuhn and
Popper are probably the philosophers whose views on this subject are best known. At least at first glance, their accounts of how
science does and should proceed differ radically in some respects.
A comparison of the two accounts highlights many of the basic
issues at stake in the resolution of questions such as "What is
science?" and "What is scientific activity?," questions that are
surely of great interest to archaeologists. Both Kuhn's and Pop-

per's accounts have been tremendously influential across a broad range of disciplines, archaeology among them. A focus on Kuhn and Popper allows us to present some of the central views without much technical philosophical vocabulary and thus to introduce some of the important concepts in a way that is more directly accessible to nonphilosophers than it might otherwise be. It also permits us to raise the question early on of why there was so much focus in the New Archaeology on Hempel, on explanation, and on deduction—a question to which we offer some tentative answers in the next two chapters.

Neither Kuhn nor Popper is unfamiliar to archaeological audiences. Archaeologists have been deeply interested in the presence or absence of Kuhnian-style paradigms in archaeology, though, on the whole, they have tended to pay more attention to Kuhn's sociological views of science than to his philosophical position. It is not always clear that archaeologists fully appreciate that Kuhn's views of science depart in some radical ways from those of Hempel, for example.

Popper, although very popular among physical scientists who have an interest in foundational questions, has been considerably less influential than either Hempel or Kuhn among archaeologists. Nevertheless, his work is somewhat familiar to archaeologists, and indeed, Robert Butler (1965) advanced Popperian methodology as relevant to archaeology (as we noted in Chapter 1) prior to Binford's appeal to Hempel that archaeologists took so much to heart. It is not entirely clear, then, that archaeologists fully appreciate the relationship of either Kuhn's or Popper's views to those of other philosophers. In any event, this century's most important debate over the nature of science has been conducted between philosophers who are not unknown to archaeologists, so a basis exists for appreciating alternative accounts. We hope that this will make it easier for archaeological readers to progress through the more purely philosophical sections that follow.

There are a number of different ways in which these two views of science could have been presented. The one we have settled upon results in some redundancy, but this is perhaps not a bad thing, for it also seems to increase the clarity of the presentation. There is some overlap with the preceding chapter where misconceptions about the nature of inductive and deductive research strategies were discussed, but this, too, seems to promote clarity.

The following sketches are, of course, incomplete presentations of the contrasting views. The sections on similarities and differences between the two views not only sum up the two positions on major philosophical points but they also address issues raised independently in archaeology about the nature of science. Perhaps the most important of these issues is whether science is, as Kuhn holds, a progression away from primitive beginnings, or, as Popper holds, a progression toward truth. Thus, this chapter is aimed at an evaluation of this century's best publicized, and probably most important, debate over the nature of science.

Kuhn's view

The fullest exposition of Thomas Kuhn's concept of science is found in his book *The Structure of Scientific Revolutions* (1970a, 2nd edition). Originally published in 1962, *The Structure of Scientific Revolutions* has been widely read and discussed. As a result, the impact of Kuhn's views has been considerable in archaeology and elsewhere. It is interesting to note that Kuhn's book was published in the same year as Binford's first article which introduced philosophical concerns to archaeology in a definitive way, directly shaping the next sequence of methodological events in archaeology.

Fundamental to Kuhn's conception of science is the distinction he draws between what he calls "normal" and what he calls "revolutionary" science. Normal science, according to Kuhn, involves "puzzle-solving," the extension of established principles to cover new problems. In the Kuhnian view, scientists are not typically engaged in a search for new theories; ordinarily, they occupy themselves with expansion of the existing body of knowledge through application of accepted principles to novel circumstances. Where they run into snags with such applications, it is the competence of the scientist, and not the adequacy of the principle, that is called into doubt. As Kuhn says: "I use the term 'puzzle' in order to emphasize that the difficulties which ordinarily confront even the very best scientists are, like crossword puzzles or chess puzzles, challenges only to his ingenuity" (1970b:5 n. 1). In Kuhn's view, the current theory couldn't be in question because it is *presupposed* by scientific investigation: the identification of both puzzles and their solutions is only possible against the background of some theory or other, and at any particular time it will be the theory in vogue which plays the role. The testing of theories—contrary to what many would have expected—is thus no part of normal science for Kuhn.

Any discussion of Kuhn's concept of normal science would be incomplete without some reference being made to his sociology of science (cf. Chapter 4). Throughout his analysis of science, Kuhn's attention is focused on the "scientific community," the unit he characterizes as consisting of "the practitioners of a scientific specialty" (1970a:177). It is from these communities that the sorts of results we expect of science emerge; it is only—and this is absolutely central to his account of science—among the members of such groups that we find a sufficient commonality of views to permit the solution of puzzles. The work of a group of scientists can only be directed towards puzzle-solving "when," as Kuhn has written, "its members take the foundation of their field for granted" (Ibid.:178).

What precisely it is that a scientific community's members share that "accounts for the relative fulness of their professional communication and the relative unanimity of their professional judgments" (Ibid.:182) is a question that Kuhn has addressed on a number of occasions. The first edition of *The Structure of Scientific Revolutions* suggested the answer: "a paradigm or set of paradigms." Kuhn is no longer content with that response, however, for, of the concepts central to the thesis developed in *The Structure of Scientific Revolutions*, none is more obscure than that of a "paradigm." As critics have pointed out and Kuhn himself has written:

> [I]n much of the book [*The Structure of Scientific Revolutions*] the term "paradigm" is used in two different senses. On the one hand, it stands for the entire constellation of beliefs, values, techniques, and so on shared by the members of a given community. On the other, it denotes one sort of element in that constellation, the concrete puzzle-solutions which, employed as models or examples, can replace explicit rules as a basis for the solution of the remaining puzzles of normal science (Ibid.:175).

In the "Postscript" to the second edition of *The Structure of Scientific Revolutions*, Kuhn reserves "paradigm" for the latter of these two uses. For the former, he has coined the term "disciplinary matrix": " 'disciplinary' because it refers to the common possession of the practitioners of a particular discipline; 'matrix' because it is composed of ordered elements of various sorts, each requiring further specification" (Ibid.:182). The main components of a disciplinary matrix are (1) symbolic generalizations ["those expressions, deployed without question or dissent by group members, which can readily be cast in a logical form like $(x)\,(y)\,(z)\,\phi\,(x,y,z)$"[1] (Ibid.:182)], (2) shared commitments ["to such beliefs as: heat is the kinetic energy of the constituent parts of bodies; all perceptible phenomena are due to the interaction of qualitatively neutral atoms in the void, or, alternatively, to matter and force, or to fields" (Ibid.: 184)], (3) shared values ["for example, [that] science should (or need not) be socially useful" (Ibid.:185)], and, finally, (4) exemplars [or "concrete puzzle-solutions" (Ibid.:187)].

According to Kuhn, normal science—as its name would suggest—makes up most of what goes on in science; thus, on the Kuhnian model, archaeologists, if they are scientists, spend most of their time solving puzzles. But what would a puzzle-solving activity look like in the archaeological context?

Without suggesting here that we share Kuhn's concept of science, a possible example of puzzle solving in North American archaeology is the dating and placing of previously undescribed or undefined southwestern pottery remains through typological analysis. The typologies used for classification are firmly entrenched in the discipline and virtually taken for granted by practicing archaeologists. The working assumption is that the typologies are adequate for identifying the pottery from the region. Thus, the archaeologist's task in this activity might be seen as simply applying current ideas to novel situations, i.e., newly discovered remains. Another example of an archaeological area of inquiry that behaved very much like Kuhn's formulation of "normal science" is the post-Folsom discovery developments in the North American Early Man field, as we discuss elsewhere. In fact, numerous examples of "normal science"-style puzzle solving can be presented for archaeology, though there remain disputes about whether any genuinely "normal" science could occur in archaeology just because of the lack of an integrating, overarching theory in the discipline. "Normal" science seems to require a theoretical framework in terms of which to solve puzzles, and many would argue that this is precisely what is lacking in archaeology.

The contrast Kuhn offers to normal science is "revolutionary science"—the periods of activity in science when individual scientists' puzzle solvings *are* tests of current theory. It is only during periods of revolutionary science, according to Kuhn, that fundamental commitments are at stake, and, since normal science is by far the more common state of affairs, revolutionary science and, hence, theory testing only rarely occur.

One example of revolutionary science which Kuhn offers is the discovery of oxygen, usually attributed to Lavoisier in 1777. Actually, in Kuhn's view, it was not so much the discovery of oxygen as Lavoisier's oxygen theory of combustion that was the startling innovation (cf. Kuhn 1970a:53–56). Until Lavoisier's discoveries, the generally accepted account of combustion was the phlogiston theory, according to which phlogiston was regarded as:

> an actual physical substance—solid and fatty, though apparently impossible to secure in isolation. It was given off by bodies in the process of combustion, or by metals in the process of calcination, and it went out in flame to combine with air, or perhaps deposited at least a part of itself in an unusually pure form as soot (Butterfield 1957:207).

Despite its popularity, Lavoisier had believed for some time prior to 1777 that the phlogiston theory was inadequate; in Kuhnian terminology, that the normal science of the day was not satisfactorily solving the puzzles presented to it. As Kuhn has written:

> Long before he played any part in the discovery of the new gas [oxygen], Lavoisier was convinced both that something was wrong with the phlogiston theory and that burning bodies absorbed some part of the atmosphere. . . . What the work on oxygen did was to give much additional form and structure to Lavoisier's earlier sense that something was amiss (1970a:56).

Thus, in Lavoisier's experiments, a fundamental tenet of eighteenth-century chemistry—the phlogiston theory—was at stake. According to Kuhn, it was Lavoisier's realization that the received theory was in trouble that made it possible for him to entertain a new hypothesis, namely the oxygen theory of combustion. Correspondingly, the inability of Lavoisier's rival contemporaries, one of whom was Priestley, to appreciate the inadequacies in the theory contributed to their failure to recognize combustion for what it was. According to Kuhn, Lavoisier's advance awareness of difficulties [with the phlogiston theory] must be a significant part of what enabled [him] to see in experiments like Priestley's a gas that Priestley had been unable to see there himself. Conversely, the fact that a major paradigm revision was needed to see what Lavoisier saw must be the principal reason why Priestley was, to the end of his long life, unable to see it (Ibid.:56).

Again, without committing ourselves as to whether Kuhn's concept of science is a good one, it is of interest here to consider what a revolution in archaeology might look like. One archaeologist, Don Dumond, has suggested that the most likely candidate for a Kuhnian revolution would be the new definition of culture adopted during the middle of this century by both anthropologists and archaeologists. In his words: "from a view of culture as a collection of 'shreds and patches' in the expression of Robert Lowie, when Roland B. Dixon (1928) could title a book *The Building of Cultures*, there has been a shift to an explicit view of culture as systematic adaptive strategy, with books titled, for example, *Man in Adaptation* (Cohen 1968–71)" (1977:346). Over the last thirty years, Dumond says, this change has fundamentally altered the view archaeologists and anthropologists have of their disciplines' theoretical foundations and, relatedly, of how the study of culture should proceed. Although Dumond believes this is a likely candidate, for reasons we discuss in the next section he does not accept it as heralding a Kuhnian revolution. The most publicized "revolution" in archaeology, and one that has been specifically called a revolution (Martin 1971), is, of course, the New Archaeology. Though its revolutionary status has been much debated, there is no very conclusive reason to believe that it represents a revolution in the Kuhnian sense.

Much of Kuhn's analysis of science is devoted to developing an account of the relationships among different periods of normal science, and some of his most provocative results arise in this context. One of these is that since the fundamental presuppositions are, Kuhn claims, at stake during periods of revolutionary science, there is, at least in one sense, no basis for comparing two periods of normal science. Because there is no common standard according to which the truth of each period's dominant theory can be judged—as Kuhn would say, "truth" has no inter-theoretic application—it makes no sense to compare them. While Kuhn can agree with more traditionally minded philosophers of science, such as Popper, that one theory can sensibly be said to be better, or worse, than another "as a tool for the practice of normal science" (Kuhn 1970c:264), there is a further step, he writes:

> which many philosophers of science wish to take and which I refuse. They wish, that is, to compare theories as representations of nature, as statements about "what is really out there." Granting that neither theory of a historical pair is true, they nonetheless seek a sense in which the latter is a better approximation to the truth. I believe nothing of that sort can be found (Ibid.:265).

Kuhn's relativism is, of course, well known. While Kuhn would in no way want to deny that some aspects of his thesis are relativistic, however, he does insist that his views do not support a radical subjectivism in the way that some have thought they did. It is important in this context to remember that Kuhn is by no means the most relativistic contemporary philosopher of science. The most radical on the Kuhnian side is perhaps Paul Feyerabend. In his aptly titled work, *Against Method* 1975, he claims the history of science shows that any effort to impose philosophically coherent conceptual structures on the practice of science would, if adopted by scientists, actually thwart scientific progress (cf. Chapter 2). To return to Kuhn, he argues that it does not follow from the impossibility of comparing the truth of theories that the choice of one theory rather than another can have no rational basis:

> It is emphatically *not* my view that "adoption of a new scientific theory is an intuitive or mystical affair, a matter for psychological description rather than logical or methodological codification." On the contrary, the chapter of my *Scientific Revolutions* from which the preceding quotation was abstracted explicitly denies "that new paradigms triumph ultimately through some mystical aesthetic," and the pages which precede that denial contain a preliminary codification of good reasons for theory choice. These are, furthermore, reasons of exactly the kind standard in philosophy of science: accuracy, scope, simplicity, fruitfulness, and the like (1970c:261).

It is thus clear that Kuhn does not want to be seen as a thoroughgoing relativist. Whether his theory allows a retreat from complete relativism or whether any interestingly relativistic position is ultimately defensible in any case, is not obvious. These questions are nonetheless important, and we will consider them in some detail below.

A related aspect of Kuhn's account of the relationship between different periods of science concerns what Kuhn calls the "incommensurability" of theories. Kuhn argues that, for a detailed comparison of two theories, we would require "a language into which at least the empirical consequences of both [theories] can be translated without loss or change" (Ibid.:266). Many philosophers, Kuhn claims: "continue to assume that theories can be compared by recourse to a basic vocabulary consisting entirely of words which are attached to nature in ways that are unproblematic and, to the extent necessary, independent of theory" (Ibid.:266). Such a vocabulary is simply "not available":

> In the transition from one theory to the next, words change their meanings or conditions of applicability in subtle ways. Though most of the same signs [i.e., words] are used before and after a revolution—e.g., force, mass, element, compound, cell—the ways in which some of them attach to nature has somehow changed. Successive theories are thus, we say, incommensurable (Ibid.:266–267, footnotes deleted).

Some difficulties with the Kuhnian view

Especially when it first appeared, Kuhn's view of science was seen as a radical departure from the traditional account, and so it is not surprising that, although many have adopted his position, numerous objections have been made to it. One of these we have already touched upon: Kuhn's use of the word "paradigm" in *The Structure of Scientific Revolutions* was not unequivocal. Kuhn later claimed that through the introduction of the term "disciplinary matrix" these problems were overcome, and even though the difficulty was largely terminological—and thus probably not as serious as some of those revealed by other objections—it was nonetheless important (1970a:182). The identification of paradigms, assuming for the moment that there are such things, would have been difficult enough, especially in fields such as archaeology where theoretical foundations are the subject of vigorous and continuing debate, but the lack of clarity surrounding the word "paradigm" in Kuhn's original text made the task virtually impossible.

Identification of a disciplinary matrix by a member of a group steeped in the matrix is presumably not at all easy. If such a matrix defines a discipline or a part of a discipline in the way Kuhn suggests, it would seem almost impossible for such a group member to recognize those constellations of beliefs, values, and so on that control that person's professional behavior. It is almost as though one must be able to step outside of oneself and one's beliefs and view them from the outside. To identify X as an A is, in part, to see X as an A *as opposed to* a B, C, or what have you, and to do this one must be able to step back from X to observe its distinctive, i.e., A-like, characteristics. This is precisely what a member of a discipline cannot do, for if that were possible he or she would no longer be dominated by the matrix in the way he or she is by definition, according to Kuhn's account.

What this may mean is that Kuhn's view of normal science is not definitive of what happens in science. For example, one archaeologist who has attempted to describe the matrices that are operative in contemporary archaeology (although he uses Kuhn's earlier term, "paradigms") is Stanley South. In his *Method and Theory in Historical Archaeology*, South claims that "three paradigms under which archaeological excavation is conducted can be defined. These are archaeology in the humanities, particularistic archaeology, and scientific archaeology" (1977:5). South examines each of these alleged disciplinary matrices through "the attitudes characteristic of each, as revealed by those conducting archaeology on historic sites" (Ibid.:5). For the reasons we have already discussed, in Kuhn's view, South could not have perceived the matrix which circumscribes his own professional behavior, other than by accident. Nevertheless, South, as well as other archaeologists who have attempted to identify paradigms or matrices in archaeology, has undoubtedly perceived some important differences in intra-disciplinary viewpoints. Whether these are properly to be thought of as paradigms or matrices in Kuhn's sense is much more problematic.

The entire question of whether there is anything in archaeology that can properly be described as a paradigm, or whether archaeology has undergone or is undergoing a revolution, has been much debated in the recent archaeological literature. Much of this debate arises from an influential paper by David Meltzer (1979) which we discuss in some detail in the next chapter, and it is interesting that the archaeological case can, in some respects, be seen as exemplifying many of the standard philosophical objections to the Kuhnian model. For example, the view that there may be several paradigms operating in archaeology (Clarke 1972) is developed more broadly for social science in general by David Thomas (1979), who argues that mature social science need not be practiced under the control of a single monolithic paradigm but rather can develop in parallel under several different paradigms. Thomas further suggests that the use of a number of paradigms may be particularly advantageous for social science in that it allows us to take account of the unique complexity of the subject matter of these fields by allowing us to consider many dimensions that could not be encompassed within a single paradigm.

Kuhn's conception of normal science thus may not fit archaeology at all well; but this lack of fit may be even more general and pervasive. John Watkins is among those who have made interesting and penetrating general criticisms of Kuhn's account of science.[2] His comments are aimed at showing that scientists are not as blindly accepting of the basic tenets of their disciplines as they are made out to be in Kuhn's view. According to Kuhn, revolutions occur in a particular area of science when normal science consistently fails to solve puzzles of a kind relevant to that area (1970a:67–68). The crises that precipitate revolutions in science are thus brought about by a "[b]reakdown of the normal technical puzzle-solving activity" (Ibid.:69). Kuhn would grant that other, for example, social and economic, factors influence scientific revolutions, but he claims that in mature sciences "external factors like those . . . [i.e., social and economic influences] are principally significant in determining the timing of breakdown, the ease with which it can be recognized, and the area in which, because it is given particular attention, the breakdown first occurs." "Technical breakdown," Kuhn says, "would still remain the core of the crisis" (Ibid.).

Watkins' attack on Kuhn is a straightforward one: scientific crises, he claims, may have "theoretical rather than empirical causes" (1970:31):

It seems that a dominant theory may come to be replaced, not because of growing empirical pressure (of which there may be little), but because a new and incompatible theory (inspired perhaps by a different metaphysical outlook) has been freely elaborated . . . If that is so, there is more free thinking in science than Kuhn supposes (Ibid.:31).

Watkins would insist that scientists can entertain doubts which may eventually culminate in a new theory. Further, they do and must have such doubts, for contrary to Kuhn's picture of the emergence of a new theory, Watkins claims that the formulation of a radically new idea in science takes time (*see also* Thomas, 1979).

Kuhn has written that "The new paradigm, or a sufficient hint to permit later articulation, emerges all at once, sometimes in the middle of the night in the mind of a man deeply immersed in crises" (1970a:89–90). In the same vein, he has said that theories are "invented in one piece" (1970b:12). Watkins finds what he calls Kuhn's "Instant-Paradigm thesis" little short of preposterous on psychological as well as other grounds. Watkins' conclusion is that Kuhn's "single-paradigm dominance theory" is wrong. He writes:

> I have maintained that since it takes time—a matter of years rather than of hours—to develop a potential new paradigm to the point where it may challenge an entrenched paradigm, heretical thinking must have been going on for a long time before paradigm-change can occur. This means that it is not true that a reigning paradigm exercises such a monopolizing sway over scientists' minds that they are all unable to consider it critically, or to toy with (without necessarily embracing) alternatives to it (Ibid.:37).

Watkins' view also accords with Stanley South's perception of archaeology.

A related worry is raised by those who ask whether a distinction between normal and revolutionary science, of the kind Kuhn postulates, can be drawn. If, as Watkins claims, even during periods of relative calm some scientists are looking critically at current theory, then is there really a difference in kind, as opposed to one of degree, between times of crisis and of complacency in science? Many think not, and one claim made is that in at least this respect Kuhn's theory doesn't fit the historical facts.

An argument in support of the above objection runs as follows. If the theories which have dominated different eras in science are incommensurable, then rational discussion of scientific revolu-

tions among advocates of competing positions is impossible. Historically, however, such debates have gone on, and, therefore, an
examination of the history of science shows that theories are not
incommensurable and the distinction between periods of science is
not what Kuhn thinks it is. In fact, an argument of this form
appears in the archaeological literature as a refutation of the claim
that there has been, during the last thirty years or so, a Kuhnian
revolution in archaeology. In his article, "Science in Archaeology:
The Saints Go Marching In," Don Dumond suggests, as we have
already briefly discussed in the previous section, that the "closest
thing" to a revolution in twentieth-century archaeology has been
the turmoil brought about by the fairly recent adoption of a new
definition of culture by archaeologists. He argues, nevertheless,
that important though it was, that change did not constitute a
revolution in the Kuhnian sense of the term. He acknowledges that
theories in anthropology and archaeology, and indeed all social
sciences, have changed considerably over the past thirty years or
so. He suggests, however, that it is

> unlikely . . . that there is any absolute barrier to communication
> between the several generations of archaeologists now alive, al-
> though certainly some of the "Newer" ones comprise quoting cir-
> cles that are exclusive enough. It seems also clear that the most
> partisan of the New Archaeologists would hold that the New Ar-
> chaeology does not reject the old completely, but rather subsumes
> it; it must, for without it, it would be woefully short both of data and
> of the chronological background that is so necessary to any sophisti-
> cated diachronic study.
>   If this is revolution, then it is a revolution that virtually all Ameri-
> can archaeologists participate in on a conceptual level, although
> they may not all feel particular affection for all self-designated New
> Archaeologists. If it is a revolution, then all evolution in a social or
> intellectual sense must also be called "revolution." And if this is so,
> the term revolution has little meaning (1977:346).

Thus, in Dumond's view, to claim that the change in definition of
culture in twentieth-century archaeology brought about a Kuhnian
revolution in that discipline is to rob Kuhn's concept of revolution
of whatever meaning it has.
  Notwithstanding criticisms of his position, Kuhn still adheres to
an incommensurability thesis; however, he now denies that peri-

ods of science are completely divorced from one another. Indeed, on the subject of change during normal and revolutionary science, Kuhn has written that "the two sorts of change interpenetrate: revolutions are no more total in science than in other aspects of life" (1970c:250). While Kuhn claims that this view is merely an elaboration of his original account and not a retreat from it, critics have argued that it is the latter. Stephen Toulmin has gone so far as to suggest that it is a concession which completely undermines the distinction Kuhn initially drew, and still wants to maintain exists, between normal and revolutionary science.

In Toulmin's view, the account of science Kuhn gave in *The Structure of Scientific Revolutions* "went too far by implying the existence of discontinuities in scientific theory far more profound and far less explicable than any which ever in fact occur" (1970:41). According to the argument presented in that book, Toulmin claims, "the differences between the kinds of change taking place during 'normal' and 'revolutionary' phases of scientific development are, at one intellectual level, absolute" (Ibid.). The effect of Kuhn's apparent withdrawal from his original, rather extreme position to a less extreme one is, in Toulmin's view: "to demolish entirely his original distinction between 'normal' and 'revolutionary' phases" (Ibid.:40).

To justify his claim that Kuhn's later remarks destroy the initial distinction which is central to his account of science, Toulmin uses an interesting analogy from the history of palaeontology between 1825 and 1860, when the opposition between the uniformitarian geology and paleontology of Hutton and Lyell clashed with the theory of Cuvier and Agassiz, who held that the many observations of discontinuities in these disciplines were evidence of "catastrophes" which were "intellectually unbridgeable" (Ibid.:42). The eventual resolution was accomplished through a gradual recognition on the uniformitarian side that the changes they had believed to be very small were actually more drastic and catastrophic; and on the other side, that the catastrophes used to explain the evidence were actually very minor, so that they began to appear to be uniform and law-like. Toulmin argues that, just as the distinction between two radically different geological theories began to evaporate, so does the distinction between normal and revolutionary science.

Kuhn, not surprisingly, does not accept Toulmin's analogy, for he insists that the result of the debate was actually to accept "two sorts of geological change, no less distinct because both [are] due to natural causes; one acted gradually and uniformly, the other suddenly and catastrophically. Even today we do not treat tidal waves as special cases of erosion" (Kuhn 1970c:250). It is not at all clear that Kuhn has successfully defended his theory against Toulmin's claim that there remains only a difference in degree between normal and revolutionary science. One can readily accept that, just as there may be two kinds of geological change, there may be two kinds of scientific change—gradual and not gradual. Nevertheless, none of the interesting aspects of Kuhn's theory would seem to follow from this observation.

The final objection to the Kuhnian thesis that we will discuss in this section is one which can be made against any relativistic theory. The difficulty arises when the principles which underlie Kuhn's account of science are applied to that account itself: if what Kuhn says about theories is true (and there is no obvious reason why his views should be restricted to science), then his own theory of scientific revolutions cannot be said to be true or false, nor can it be said to be closer to an accurate account of science than any other. Thus, there is little basis for accepting it, or even for taking it very seriously. On the other hand, if Kuhn's views on change in science are more than that—that is, better than the competing accounts in some important sense of "better," then his account of which theories are incommensurable is false.

The paradox revealed by the objection we are considering is analogous to the dilemma connected with the objectivity/ subjectivity debate. As Israel Scheffler says:

> [T]o put forth *any* claim with seriousness is to presuppose commitment to the view that evaluation is possible, and that it favors acceptance; it is to indicate one's readiness to support the claim . . . as being correct or true or proper. For this reason, the particular claim that evaluation is a myth . . . is obviously self-destructive. If it is true, there can be no reason to accept it; in fact, if it is true, its own truth is unintelligible: What can truth mean when no evaluative standard is allowed to separate it from falsehood? (1967:21).

These remarks seem to be directly applicable to Kuhn's position: if his theory is right, then there is no reason to accept it as such.

The Kuhnian reply is that it was never claimed there are *no* rational bases for choosing one theory rather than another (we have mentioned this point earlier). Kuhn would have us accept, as compatible with his account of science, that the criteria for selection of a theory—accuracy and scope are two he suggests—remain constant. No theory can be said to better represent what is the case than any other, but one can be said, for example, to have greater scope than another. How can this be? Presumably the concept of "scope" arose in the context of a particular theory. If so, why should it be applied to another? Can it be assumed that all scientists, regardless of the theory within which they are working, will value "scope"? Also, even if alternative theories employ the term "scope," if theories are incommensurable how can we ever know whether the concept of "scope" has the same meaning across those theories? Finally, if no theory can represent what is the case better than any other, then what possible sense can be made of the claim that one is more "accurate," "simple," or "fruitful" or has greater scope than another? These assertions would seem to presuppose a reality apart from and against which theories can be evaluated and a methodology that provides some form of direct access to it. But such a theory-independent standard is precisely what Kuhn has argued against. To suggest that his form of relativism need not imply a complete absence of evaluative standards, though obviously a move in the right direction, is not yet to provide us with a sufficiently precise characterization of his view that we can be sure we know what he is claiming.

Popper's view

Karl Popper's account of science is one of the most influential ever put forward, and also one that many scientists have adopted as an accurate model of their discipline (*see* Medawar 1969). Thus it is not surprising that we have had cause to touch upon aspects of it in several chapters of this work. In Chapter 1 we discuss his claim that falsifiability is the criterion for distinguishing scientific from nonscientific hypotheses. In Chapter 2 we refer to Popper's views on the discovery/justification distinction; we discuss his remarks on the impossibility of doing science in the inductivist way; and we discuss the similarity between his view of testing and that of Hempel. Finally, in Chapter 5 we note the strong similarity that exists between Popper's account of causal explanation and Hempel's deductive-nomological (D-N) model. In this chapter we primarily treat issues connected with his "deductivist-falsificationist" picture of science—especially those features of his theory that serve as a foil for Kuhnian criticism.

There are many reasons that Popper's views have been so influential. For one thing, they have represented a persistent foil for the positivist form of empiricism and thus a challenge to the emphasis in philosophy of science that has been placed on verification and confirmation. More recently, his emphasis on falsification as the main ground for rejecting scientific hypotheses has been developed into a challenge to the Kuhnian conception of scientific change. We focus on the first of these in this section. This is particularly relevant for archaeology as a way of clarifying some of the confusions that have arisen about the roles of induction, deduction, and testing in the discipline.

Popper's account has been seen by many as, in general, an alternative to the traditional "inductivist" conception of science, a conception which Popper has said characterizes scientists as "waiting, passively, for repetitions to impress or impose regularities" upon them (1962:46). Inductivism, according to Popper, claims that scientific investigation can begin "with pure observations alone, without anything in the nature of a theory" (Ibid.:46). Against what he says is the "absurd" inductivist position, Popper argues for his model of conjectures and refutations which has come to be known as the hypothetico-deductive method (*see* Chapters 2 and 6). In this account, scientists deduce consequences from conjectures (hypotheses) they have formulated and then attempt to refute those predictions. It should be noted here that the hypothetico-deductive method does not require *pre*dictions in the strict sense; the consequences deduced from the conjecture could be *retro*dictions. The aim of the scientific endeavour, according to Popper, is to show the conjecture false by conducting a test which fails to achieve the predicted result. If the test does not fail and the conjecture is thus not falsified, in Popper's view there is no confirmation because it is still an open question whether the hypothesis under examination is true, as it may yet be falsified by further tests. Popper does allow, however, that while hypotheses are not confirmed by testing, in some circumstances they can be said to have been strengthened by the results of a test. When this is the case, Popperians say a hypothesis has been "corroborated" (*see* Popper

1961, Chapter X and Appendix LX). Other writers often speak of the hypothetico-deductive method as a method of confirmation (*see* M. Salmon 1982: 34–36). The hypothetico-deductive method as well as Popper's negative views toward inductivism have been wholeheartedly accepted in archaeology in the sense that the New Archaeologists forcefully and explicitly rejected the inductivism of traditional archaeology, though they did not do so under the explicit influence of Popper.

It is a fundamental tenet of Popper's account that scientists are not looking to confirm their conjectures, and we have earlier discussed the logical basis for this view (*see* Chapter 1). Scientists seek reliable generalizations about the world, and universal generalizations can never be conclusively verified: even if all observed emeralds have been green, the next emerald observed might be of another color. Such generalizations are always open to falsification: if the next emerald observed is fuchsia, then "All emeralds are green" has been refuted. Thus, for Popper falsifiability was the distinguishing characteristic of scientific hypotheses.

There were additional reasons for Popper's rejection of the inductivist approach, one of which was that he saw the inductivist's theory of confirmation as an absurdity. According to Popper, inductivism was characterized by the view that "any positive instance of a hypothesis also necessarily qualifies as a *supportive* instance of the hypothesis" (Grünbaum 1976d:218). This doctrine is sometimes referred to as "the instantiation condition," and as Adolf Grünbaum has said of it: "In an *unbridled* use of the instantiation condition, any positive instance of a hypothesis—e.g., one resulting from the hundredth repetition of the same experiment— will *increase* the credibility of the hypothesis *as much as any other*" (Ibid.:219).

There are profound difficulties in specifying precisely what we mean when we say that some evidence *e* confirms some hypothesis *h* (*see* Chapter 6). Are we saying that *h* is virtually certain in light of *e*, or instead that *h* is merely more likely in light of *e* than without it? In either case, a confirming instance must increase the probability of *h*, but this is precisely the requirement for confirmation that the instantiation condition—at least the form of it that Grünbaum describes—does not meet. The repetition of an experiment simply cannot generate a result which can be said to *increase* the likelihood of any hypothesis. (*See* our discussion of this issue in Chapter 2). It is thus clear that the instantiation condition will not do as the foundation of an acceptable approach to confirmation.

One of the most radical of Popper's views concerns the procedure he recommends for choosing hypotheses for testing. According to Popper, highly probable theories are uninteresting because they have little content and little explanatory power—they aren't risky enough to be scientifically informative. Among the many hypotheses compatible with the facts, Popper recommends choosing the most falsifiable and, therefore, least probable. If this hypothesis is then subjected to severe testing and survives unfalsified, it is corroborated. The greater the number, variety, and severity of the tests to which the hypothesis is subjected, the more highly corroborated is the hypothesis that survives them.

In this view, then, the degree of corroboration of a hypothesis need not be the same as its probability. Since interesting and powerful statements always have low probability, precisely by virtue of having a high risk of falsifiability, it will only be hypotheses with low probability that receive corroboration. Another interesting feature of Popper's view is that it constitutes a method for selecting, among unfalsified hypotheses, some—i.e., those with the highest degree of corroboration—that can be accepted, albeit tentatively (as always). Obviously, acceptance in this theory has very little to do with high probability.

Some difficulties with the Popperian view

We will discuss here only two weaknesses in the Popperian account, both of which stem from what looks to be rather pointless pitting of inductivism against deductivism. As we have already argued (see Chapter 2), the putative contrast between the two is illusory because there is no genuine alternative to something like the hypothetico-deductive method for the purposes of testing. Since testing is essential, science cannot rely exclusively on inductive inference. This, however, is too simple in several respects. First, the testing of hypotheses that results in their falsification is actually far more complex than the simple model discussed in the previous section would suggest. Almost no hypotheses alone deductively imply observation statements; they do so, rather, in conjunction with numerous auxiliary hypotheses about background conditions, reliability of equipment and observers, and so on. What a negative observation falsifies, then, is *either* the hypothesis in question *or* one or more of the auxiliary hypotheses; and the argument that it is the hypothesis under test that is falsified is actually an inductive one. Second, the hypothetico-deductive method is often proposed as a method of confirmation wherein a hypothesis subjected to test and not falsified is thereby confirmed. Again, this is an oversimplification, as we discuss in Chapter 2, but suitably amplified it does represent a plausible method of confirmation that is inductive in form. Thus, in at least these two respects the hypothetico-deductive method is not exclusively deductive. It is also necessary to science that there be some basis for initially proposing new hypotheses, and this cannot be a purely deductive matter. As a result, induction is integral to various phases of scientific investigation.

One objection, then, is that Popper has mistakenly attacked "inductivism." Of course, as Popper and others have argued, the view that arbitrarily taken observations will somehow "speak for themselves" is ridiculous; but the vital point is that no scientist or philosopher in recent times has claimed that this was a plausible account of scientific procedure.

Another way in which Popper has been said to have wrongly criticized inductivism concerns his characterization of the inductivist approach to confirmation. As we noted in the previous section, Popper claimed that inductivists typically have adhered to the instantiation condition for confirmation, but Grünbaum (1976a; 1976b; 1979), for one, argues that on this point Popper is mistaken. While some inductivists have held that view, many, and certainly the best known and most influential, have rejected it. Indeed, writes Grünbaum:

> It was none other than the confirmed inductivist J. S. Mill who emphatically rejected the instantiation condition after Bacon had rejected induction by simple enumeration of positive instances as puerile. Thus, when discussing the case of a *spurious* plurality of causes in his *Logic*, Mill *denies* the probative value of *mere* repetitions of positive instances just as much as Popper does, and indeed for much the same reasons (1976c:107, reference omitted).

The second, related objection is that, Popper's protestations to the contrary notwithstanding, there must be, and are, inductivist elements in his own account of science. There must be, because, while there are always many unfalsified alternatives to a given falsified hypothesis, there are occasions on which we must choose one as more likely than others, worth predicting from, or more useful in a particular context. Predictions have to be made both in science and in the more mundane areas of our lives; but predictions necessarily go beyond the evidence on which they are based, and thus necessarily involve some sort of inductive inference.

That there are inductive elements in Popper's analysis is shown by his talk of "degree of corroboration" as "degree to which a statement h is supported by a statement e," and this is certainly at least part of what is usually meant by "degree of confirmation" as well. A standard approach to confirmation takes as the main inductive question: "What do we mean when we say that some evidence e confirms some hypothesis h"? As will be clear from our discussion of confirmation in Chapter 6, an unexceptionable answer to this question is not available and will be much harder to come by than one might have expected. Although it seems clear that an important part of what we mean when we say "Evidence e confirms hypothesis h" is that the hypothesis statement h is supported by the evidence statement e, we do not have a clearly acceptable account of what constitutes such support, or of what kind or degree of support should be required for us to accept the hypothesis.

Grünbaum, in addressing Popper's remarks on the choice of hypotheses for testing, has pointed to a quite different way in which Popper's account of science may be inductivistic. In Popper's view, scientists should choose the most risky—that is, the least probable of the hypotheses compatible with the facts. As Grünbaum writes:

> Popper contends that the risky predictions of a *false* theory are more likely to fail than its *non*-risky ones. This contention is the core of his *avowedly deductivistic* rationale for conjecturing that severe tests have a better chance of weeding out false theories than non-severe ones. And since he conceives of the scientific enterprise as a search for truth, the latter conjecture, in turn serves as one of his major reasons for advocating severe or "crucial" tests of bold theories (1976c:107).

What Grünbaum challenges is not Popper's conclusion concerning the most productive strategy for testing, but instead Popper's claim that that conclusion can be justified deductively. Grünbaum's argument is more technical than it is useful, but it is his view both that Popper has not provided an acceptable rationale appropriate to his deductivist program for advocating severe tests as more promising than nonsevere ones, and that it remains an open question whether such a rationale can be given at all. It seems likely, he thinks, that we will not be able to avoid a resort to inductivistic devices if we are to claim epistemological sanction for the belief that severe tests are appropriate.

Kuhn and Popper: A comparison

While the differences between Kuhn's and Popper's views of science have more significance for our purposes, any comparison of the two accounts which did not touch upon the elements shared by them would be seriously incomplete. From Kuhn's perspective (and numerous philosophers would agree with him), the similarities between the two are sufficiently extensive that they place Popper and himself "in the same minority among contemporary philosophers of science" (Kuhn 1970b: 2). For example, one important feature of both accounts is that they focus on the problem of discovery, not—as is usually the case in the philosophy of science—on justification. Although Popper claims to be interested in justification and not discovery, there are actually both strands in his thought, and we think Kuhn is correct in saying that "We are both concerned with the dynamic process by which scientific knowledge is acquired rather than with the logical structure of the products of scientific research" (Ibid.:1). Indeed, it is interesting to note that many scientists have interpreted Popper as addressing discovery and not primarily justification procedures in spite of Popper's claims to the contrary. Kuhn sees their joint interest in discovery as leading both to "emphasize, as legitimate data, the facts and also the spirit of actual scientific life, and both of us turn often to history to find them" (Ibid.).

Many philosophers of science, including Hempel and Braithwaite, for example, rely upon the actual practice of science as a source of "data" for their investigations. What sets Kuhn and his followers apart from the rest of twentieth-century philosophers of science is the way in which they make use of the facts of scientific life. For Kuhn, the manner in which science actually proceeds provides the standard; unlike Hempel, he is not looking for any more absolute norm for scientific investigation. The Hempelian approach, of course, was based on a fairly rigid distinction between descriptive and normative aspects of science; once this line was blurred somewhat, it was a short step to the view that the norms of science arise solely from a description of how science is actually done rather than from some idealization of how it ought to be done. This blurring of once-sharp distinctions is at least partly a sign of the times, but in this case it has had the interesting effect of bringing about a closer integration of history and philosophy of science.

A further, related point should be noted here. Even though Kuhn might argue the contrary, Popper cannot be seen as having taken an essentially non-normative position. In fact, while some of Popper's followers are more historically oriented than he is, Popper himself is probably just as normatively inclined as someone like Hempel. The main difference between Popper and Hempel lies in the former's falsificationism, in opposition to the Vienna Circle's verificationism (*see* Chapter 1, where this contrast is discussed). Interestingly, Popper himself viewed his work as less normative than it is. He appreciated that his demarcation criterion for science—falsification—was a standard, but, he would have said, not a standard *imposed* on science, and therefore not a norm. Popper believed that his analysis, central to which is the claim that falsifiability is the bench mark of science, had actually captured the essence of science as it is practiced.

One of the more important of the views Popper and Kuhn share is a rejection of the claim that science progresses by accretion (Kuhn 1970b:1–2). Against a picture of science which suggests that results of investigation are added, bit by bit, to an ever-expanding core of knowledge, Kuhn and Popper argue for an account which emphasizes instead "the revolutionary process by which an older theory is rejected and replaced by an incompatible new one" (Ibid.:2). The circumstances under which each would say that a theory could be replaced differ: for Kuhn, theories are threatened only during periods of revolutionary science, while for Popper each scientific experiment is a test of the current theory.

Nevertheless, Kuhn's and Popper's views of scientific progress are similar. At least, Kuhn claims that they are, and in respect to the rate, or pattern, of progress in science this seems true. Still, what each means by "progress" is, surely, quite different. For

Popper, science makes gains in the sense that contemporary science has a more accurate picture of the world than did earlier varieties—for example, the Aristotelian. Kuhn could never accede to this notion, for he explicitly denies that theories can be compared on the basis of how well they mirror reality. For Kuhn, "truth" has no transtheoretic application and, thus, Popper's concept of progress is nonsensical.

The notion of progress to which Kuhn does subscribe is one he calls "evolutionary":

> [M]y view of scientific development is fundamentally evolutionary. Imagine, therefore, an evolutionary tree representing the development of the scientific specialties from their common origin in, say, primitive natural philosophy. Imagine, in addition, a line drawn up that tree from the base of the trunk to the tip of some limb without doubling back on itself. Any two theories found along this line are related to each other by descent. Now consider two such theories, each chosen from a point not too near its origin. I believe it would be easy to design a set of criteria—including maximum accuracy of predictions, degree of specialization, number (but not scope) of concrete problem solutions—which would enable any observer involved with neither theory to tell which was the older, which the descendant. For me, therefore, scientific development is, like biological evolution, unidirectional and irreversible (1970c:264).

At least two things should be noted about Kuhn's concept of progress. The first is that it is concerned (as it must be, given Kuhn's account of "truth" in science) with the form and not the content of successive theories. And the second is that any notion of progress presupposes a standard against which change can occur. Kuhn denies that there is such a standard for the *truth* of theories, but clearly accepts that there is one for the practical utility of them. He says: "One scientific theory is not as good as another for doing what scientists normally do. In that sense I am not a relativist" (1970c:264). But why allow the possibility of progress in this respect and not the other? On what grounds can we say that our concept of what science is about (i.e., what scientists normally do) remains constant, thereby providing a basis for evaluating progress, while our concept of what science is investigating changes, thereby precluding such an assessment? Furthermore, surely any particular notion of who is to count as a scientist requires some argument. For example, would the same people receive this label in both the sixteenth and twentieth centuries? To raise such questions is not to argue that Kuhn is wrong on this issue; rather, it is to demand a justification for the distinction he draws. If no such justification can be given, then his partial relativism is called into question and it appears that either the incommensurability thesis or the evolutionary concept of science must be mistaken.

Other similarities between the Kuhnian and Popperian accounts of science include their rejection of "a number of classical positivism's most characteristic theses" (Kuhn 1970b:2). They both, Kuhn writes, stress

> the intimate and inevitable entanglement of scientific observation with scientific theory; we are correspondingly skeptical of efforts to produce any neutral observation language; and we both insist that scientists may properly aim to invent theories that *explain* observed phenomena and that they do so in terms of *real* objects, whatever the latter phrase may mean (Ibid.:2).

Actually, this similarity between Kuhn and Popper does not set them apart from the current mainstream of contemporary philosophy of science, much as it would have distinguished them from the mainstream thirty years ago. Classical positivism has been superceded, and virtually all philosophers would agree with their claim about the inevitable entanglement of observation and theory. Also, their concern with explanation in science is shared by many—most notably, of course, Hempel, and their emphasis on realism puts them on one side of what is perhaps the most hotly debated issue in contemporary philosophy of science (*see* Chapters 5 and 6).

While Kuhn and Popper may be united on these issues, they seem irreconcilably opposed on others. One of the most obvious points at which their accounts of science conflict involves the significance of testing in normal (ordinary) science. For Popper, the aim of each application of the hypothetico-deductive method is falsification of the theory from which the tested consequences were deduced. The dominant theory is thus continually threatened. But in Kuhn's view, it is the individual scientist, not the prevailing theory, that is tested during periods of normal science; only the scientist's conjecture, the mere existence of which necessarily presupposes the theory, is at stake—not the theory itself. According to Kuhn, the testing of basic commitments happens rarely, and when it does, then it constitutes revolutionary or extraordinary science.

That Kuhn and Popper disagree on the significance of testing in normal science is clear, but the exact nature of that dispute may not be so obvious. Kuhn and Popper are, to a great extent, agreed on what actually happens both at the operational and logical levels of testing in ordinary science. In other words, there is little controversy between them over what testing in science *is*. The disagreement, instead, concerns how test results are viewed by the scientific community. The dispute, then, is one of interpretation, as is suggested by John Watkins when he writes of Kuhn's account:

> His idea is this. So-called "testing" in Normal Science is *not* testing of theories. Rather, it is part of puzzle-solving activity. Normal Science is governed by some paradigm [which] . . . is taken as guaranteeing the existence of a solution to every puzzle generated by apparent discrepancies between it and observations. Hence, although the "tests" carried out within Normal Science may *look* like tests of the prevailing theory if viewed through Popperian spectacles, they are really tests of something else, namely, the experimenter's puzzle-solving skill (1970: 27).

A related point of contention concerns Popper's demarcation criterion for science. As we have discussed earlier (*see* Chapter 1), Popper has suggested that what characterizes scientific propositions is their falsifiability. A theory is scientific, in this view, if it is vulnerable to refutation by empirical testing. Kuhn dissents from this position on two grounds. First, the prevailing theory is not at stake in normal science; instead, it is presupposed by it. Kuhn argues that if a demarcation criterion of the sort Popper proposes exists, then it must operate within normal science. Thus, if anything, it is the individual scientist's conjecture, not the current theory, that is falsified or refuted by testing. Second, Kuhn claims that, since no conclusive disproof of a theory can ever be given, Popper's reliance on falsifiability as a demarcation criterion for science is pointless. Interestingly, he further claims that Popper is equally aware of the impossibility of such disproof. He writes:

Sir Karl is not, of course, a naive falsificationist . . . very early in his *Logic of Scientific Discovery*, for example, he writes: "In point of fact, no conclusive disproof of a theory can ever be produced; for it is always possible to say that the experimental results are not reliable or that the discrepancies which are asserted to exist between the experimental results and the theory are only apparent and that they will disappear with the advance of our understanding." Statements like these display one more parallel between Sir Karl's view of science and my own, but what we make of them could scarcely be more different. For my view, they are fundamental, both as evidence and as source. For Sir Karl's, in contrast, they are an essential qualification, which threatens the integrity of his basic position (1970b:14, footnote deleted).

Again, we see how a disagreement between Kuhn and Popper turns on the *interpretation* of phenomena that both accept as relevant to the debate.

Popper and Kuhn also hold conflicting views on the adjudication of competing theories. Kuhn argues, as we have seen, that there are no methodological rules on the basis of which the choice between alternative theories can be made, because all such putative rules are themselves theory based. According to Kuhn, Popper cannot be offering methodological prescriptions for theory choice—such as, that the best theory is the one that explains the most facts—for even the notion of what is to count as a fact, or how facts are to be individuated, is inescapably theory ridden. Instead, in Kuhn's view, Popper must be seen as advocating an ideology, or a set of "procedural maxims" (Ibid.:15). Ultimately, according to Kuhn, theory choice must be made on the basis of sociological and psychological considerations, but the choice must be made by the scientific community, not the individual, so the relativism of the decision procedure is tempered.

It should be noted again here that Kuhn is very eager not to be seen as suggesting that the choice between competing theories is necessarily irrational. While he argues against Feyerabend that logic plays a role in such deliberations, claiming that the essential problem lies in the incommensurability of fundamental premises—incommensurable just because they are fundamental, because there is nothing more basic against which they can be evaluated—adjudication between alternative theories occurs, in Kuhn's view, according to values rather than rules. As we have already discussed, he does not want to deny that there are good reasons for choosing one theory over another, but he does insist that "such reasons institute values to be used in making choices rather than rules of choice" (1970c:262).

Kuhn's most fundamental criticism of Popper's conception of science is that it mistakenly imposes aspects of routine science on extraordinary science. He has written that:

> The criteria with which scientists determine the validity of an articulation or an application of existing theory are not by themselves sufficient to determine the choice between competing theories. Sir Karl has erred by transferring selected characteristics of everyday research to the occasional revolutionary episodes in which scientific advance is most obvious and by thereafter ignoring the everyday enterprise entirely. In particular, he has sought to solve the problem of theory choice during revolutions by logical criteria that are applicable in full only when a theory can already be presupposed (1970b: 19).

Again, Kuhn attributes to Popper awareness of many of the same features of science as have motivated his own account of science. He says that: "Though [Popper] insists he is writing about the logic of knowledge, an essential role in his methodology is played by passages which I can only read as attempts to inculcate moral imperatives in the membership of the scientific group" (Ibid.:22). The example from Popper that Kuhn cites is the following:

> "Assume," Sir Karl writes, "that we have deliberately made it our task to live in this unknown world of ours; to adjust ourselves to it as well as we can; . . . and to explain it, *if* possible (we need not assume that it is) and as far as possible, with help of laws and explanatory theories. *If we have made this our task, then there is no more rational procedure than the method of . . . conjecture and refutation . . .*" (Ibid.).

Commenting on the above, Kuhn continues:

> We shall not, I suggest, understand the success of science without understanding the full force of rhetorically induced and professionally shared imperatives like these. Institutionalized and articulated further (and also somewhat differently) such maxims and values may explain the outcome of choices that could not have been dic-

tated by logic and experiment alone. The fact that passages like these occupy a prominent place in Sir Karl's writing is therefore further evidence of the resemblance of our views. That he does not, I think even see them for the social-psychological imperatives that they are is further evidence of the gestalt switch that still divides us deeply (Ibid.).

The last issue dividing Popper and Kuhn that we will address concerns the role of "critical discourse" in science. According to Popper, because in every test the current theory is at stake, critical discourse goes on continually in science. Kuhn disagrees, saying that "it is precisely the abandonment of critical discourse that marks [a discipline's] transition to science" (Ibid.:6). He argues further that such discourse only occurs when the bases of a field are threatened, i.e., during periods of revolutionary science. As John Watkins has said of this controversy between Popper and Kuhn:

> Thus we have the following clash: the condition which Kuhn regards as the normal and proper condition of science is a condition which, if it actually obtained, Popper would regard as *un*scientific, a state of affairs in which critical science had contracted into defensive metaphysics (1970:28).

For Popper, the mark of the scientific is the ruthlessness with which proposed theories are challenged through empirical testing, while Kuhn sees such a view as hopelessly misguided. Since, according to Kuhn, progress is only possible against a stable framework, he believes Popper is simply wrong in exhorting scientists to be forever questioning theories and proposing alternatives.

An assessment of the debate

Kuhn is representative of a movement, sometimes called "the new philosophy of science," which emphasizes the history of science, and without doubt this school of thought has made its mark. It is no longer believed that a strictly logical account of science is what we are after, and virtually all philosophers now agree that an appreciation of history is necessary to significant work in the philosophy of science. Notwithstanding this unanimity, however, questions remain as to precisely how to make use of the historical facts. A deep division still exists between the historicists, including, along with Kuhn, philosophers such as Toulmin (see, for example, 1953 and 1961), and those who take a more normative approach, such as Popper, Hempel, and Grünbaum (see Grünbaum 1976a, 1976b, and 1979). Shapere (1969) offers a useful commentary on these two approaches to philosophy of science.

Two questions often put to the historicists by the more normatively inclined philosophers of science are: (1) whether Kuhn's account of how science has proceeded is right, that is, accords with history; and (2) even if his account is accurate, whether that is how science should proceed. There are, as one would expect, serious disagreements over both, and an interesting—although counterproductive—aspect of the debate is the frequency with which its participants appear to confuse the factual and normative issues at stake. On the one hand, Kuhn is wont to suggest that his opponents misinterpret the practice of science, however unintentionally, because they view it from a normatively tainted perspective. That is to say, they can't see the scientific past for what it really was because they are obsessed with what it should have been. On the other hand, Kuhn has great difficulty in confronting directly the question concerning the historical accuracy of his account. When it is claimed by his critics that science has not proceeded in the way he suggests, Kuhn typically responds with arguments aimed at establishing that it must, that is, should, have proceeded in that way.

In our view, both Kuhn and Popper, and therefore the approaches they represent, are mistaken in some ways. As we have already indicated, their emphasis on history is helpful—especially Kuhn's, since his is the more forceful, and Popper's normative approach is a useful tempering of Kuhn's relativism. Nevertheless, the objections to their accounts which we have discussed in the preceeding sections seem well founded for the most part. Thus, the task of evaluating the relative merits of their views remains.

We start this assessment with Kuhn's notion that scientific activity is divided into normal and revolutionary phases which can reasonably easily be distinguished from one another. This point, besides being central to Kuhn's account of science, is also fundamental to his debate with Popper, for Kuhn does not flatly deny that what Popper points to as distinctive of science can be found in science. Instead, he argues that Popper misunderstands the significance of those features—that Popper imposes aspects of routine or revolutionary science on all science because he does not see the essential distinction between ordinary and extraordinary scientific activity. For example, Kuhn could hardly challenge the claim that prevailing theories are at least sometimes at stake in testing. What he does say, against Popper, is that they are threatened only during the rare periods of revolutionary science. Likewise, Kuhn argues that if there is a falsifiability criterion which somehow, as Popper claims, sorts science from nonscience, then it must operate within normal science, not science generally—for there can be no such adjudication during scientific revolutions.

Kuhn also claims that, in attempting to "solve the problem of theory choice during revolutions by logical criteria that are applicable in full only when a theory can already be presupposed" (1970b: 19), Popper has mistakenly imported "selected characteristics" of normal science into revolutionary science—mistakenly, according to Kuhn, because it is only during periods of normal science that logical criteria can be presupposed. As a final example, Kuhn criticizes Popper's encouragement of "critical discourse" in science, arguing that controversy impedes scientific advancement and only occurs during phases of revolutionary science when progress, as a result, is impossible.

It is essential, then, to determine how sound Kuhn's distinction is between normal and revolutionary science. When Kuhn first put the notion forward he was understood as claiming that the line between the two was a clear one—indeed, an absolute one, but critics were quick to pounce on that idea, since it seemed impossible to find in the history of science even one hiatus of that kind. Kuhn responded by saying, as we noted earlier in this chapter, that he never intended to suggest that normal and revolutionary science were completely divorced from one another, but instead that "the two sorts of change interpenetrate: revolutions are no more total in science than in other aspects of life" (1970c:250).

The question is, can this less sharp distinction bear the theoretical weight Kuhn tries to put on it? It would seem not. Once revolutions are conceded not to be absolute, or "total," the essence of Kuhn's account of science is destroyed. If change in science is gradual, then routine and frontier-breaking types of science are at least often, and probably always, going on simultaneously. At some times relative calm will prevail and at others there will be rapid and extensive change. The sort of upheaval that would justify Kuhn's incommensurability thesis, for example, just cannot occur in such a model of science. Since it never happens that everything in science changes at the same time, there is always a methodological basis of some sort for evaluation of competing theories, and this is so even during periods of radical change.

If change in science is variable but often gradual, then disciplinary matrices are not the rigid thought and behavior patterns that Kuhn has described. There simply are not insuperable obstacles to bridging the differences among perspectives on doing science; instead, one approach to science shades into another, and scientists are not ignorant of the alternatives to the approach they have chosen. This is not to say that one cannot find in science the "concrete puzzle-solutions," for which Kuhn now reserves the word "paradigm," but the fact that the validity of routine procedures, such as litmus-paper tests in chemistry, is taken for granted has little significance for the debate over the nature of science.

Furthermore, if the two sorts of change in science "interpenetrate" as Kuhn concedes they do, then not only is "revolutionary" science not purely revolutionary, but "normal" science is not purely normal. Even during periods of relatively little creative activity, the dominant theory is open to question, with some alternatives to it contemplated and even explored. It is one thing to say, with Kuhn, that some science involves more or less straightforward application of accepted theory; but it is quite another to say, as he does, that for the most part the business of science is *just* that. Surely, a discipline in which questioning has ceased, or has been suspended for any appreciable length of time, has lost whatever claim it may have had to scientific status. Thus, we concur with Popper in finding that the "science" which Kuhn in his more extreme moments claims is standard is simply unscientific.

Earlier we considered what a Kuhnian, normal scientific activity might look like in archaeology, and gave as a possible example the dating and placing of southwestern pottery remains through typological analysis. There we suggested that such archaeological work might be viewed as puzzle solving, i.e., as simply the application of accepted typological classifications or other concepts to archaeological remains. But is it correct to view this activity as *only* that, and does typological analysis, for example, therefore constitute a counterexample to our argument that there is no such thing as Kuhnian normal science? For three reasons, we think not. First, archaeologists don't blindly adhere to the established theory when classifications look inadequate to the tasks at hand; instead, under such circumstances they search for more appropriate labels. As we discuss in greater detail in Chapter 4, typological classification is used extensively in archaeology, but its frequent application is just as much a matter of the success archaeologists have had with it as it is a matter of rigid dogma. Second, it cannot be said that when snags are encountered in typological analysis it is always, or even generally, the archaeologist, and not the theory, which is in trouble. Even though practicing archaeologists take the prevailing typological theory for granted until problems arise, the discipline is open to its refutation, although perhaps not as much so as a Popperian might wish. Finally, typological analysis is hardly a science in its own right, but only one aspect of a much broader discipline which is in the throes of an explicitly self-critical phase. No matter how strictly defined a pursuit it may be, therefore, typological analysis does not constitute the sort of example of "normalcy" that is needed to save the Kuhnian account of science.

We thus concur with Toulmin when he says:

> the "absoluteness" of the transition involved in a scientific revolution provided the original criterion for recognizing that one had occurred at all. And, once we acknowledge that *no* conceptual change in science is ever absolute, we are left only with a sequence of greater and lesser conceptual modifications differing from one another in degree. The distinctive element in Kuhn's theory is thus destroyed, and we are left looking beyond it for a new sort of theory of scientific change (1970:41).

It follows then, in our view, that Kuhn's criticisms of Popper are off the mark: Popper cannot have failed to appreciate the unique characteristics of normal and revolutionary science if such periods in fact do not occur. On the other hand, at numerous places in the text we have indicated problems with Popper's view of science, some (but by no means all) of them articulated by Kuhn. As for Kuhn's views, though his account does not succeed in detail in most of the ways he hoped it would, his contributions—especially his very valuable emphasis on history—have been extremely influential in redirecting research in history and philosophy of science toward theories that take account of actual examples of scientific change.

Significance of the debate for archaeology

Many of the topics debated by Kuhn and Popper have been involved in the archaeological concern over methodology and the nature of science, although they have not usually been expressed with specific reference to the Kuhn-Popper debate. Whether inferences are progressively closer approximations of past reality (or truth), or have progressively evolved away from their primitive beginnings, is a matter archaeologists have independently considered. The degree of relativity involved in science, the role and limits of objectivity, and the role of testing have also been scrutinized by archaeologists within their own discipline.

Archaeologists have also been concerned with the discovery process, as well as with the logic of justification. The emphasis placed on problem orientation and research strategies, as well as that placed on explanation, were all directed toward ways of improving the caliber of archaeological inferences and making the discipline more scientific.

Two other aspects of the debate between Kuhn and Popper seem to us to be particularly important to archaeologists. First, our consideration of that controversy shows, once again, that philosophical positions cannot be adopted wholesale. No discipline, archaeology included, can expect to find—even in sophisticated theories such as those of Kuhn and Popper—a complete and readymade answer to its conceptual problems. All theories, at least all the interesting ones, are in some way mistaken or incomplete, or fail to capture the intricacies of scientific pursuits. Further, none of the leading accounts of science were developed with archaeology specifically in mind; therefore, it is not surprising that none explicitly address all the points of special concern to that field.

Second, the debate between Kuhn and Popper has, of course, centered on the question "what is science?," an issue in which archaeologists have shown a great deal of interest. In our view, the fact that there is no single hallmark of the scientific emerges clearly from our discussion of that debate, as it does in Chapter 5, where we address the same topic from a different perspective. Thus, as an example, it follows that Popper's claim that falsification of the prevailing theory is the aim of *every* test in science is just wrong. Science is not nearly so much in the grips of dogma as Kuhn claims, but neither is it as unwaveringly rigorous as Popper would have us believe; critical discourse is sacrificed for the sake of progress. In evaluating the scientific status of their discipline, archaeologists must consider not just one factor—such as the purpose of testing in archaeology but a host of them, including those that we have already discussed in this chapter and will discuss also in Chapters 5 and 6.

Actually, archaeologists concerned with the debate over the conceptual foundations of their discipline should be aware of another philosophical school of thought, according to which there neither is, nor should be, methodology in science. In this approach, the imposition of any procedural constraints at all in the practice of science is seen as totally disadvantageous—in fact, as hindering the advancement of science. The leading exponent of this view is Paul Feyerabend (1975), who has argued that what he calls "theoretical anarchism"—which, for example, countenances the use of hypotheses that contradict well-confirmed theories and counterinduction (on which it is assumed that the future will not be like the past)—is the most fruitful approach to science and the one which has historically yielded results. According to Feyerabend:

> one of the most striking features of recent discussions in the history and philosophy of science is the realization that events and developments, such as the invention of atomism in antiquity, the Copernican Revolution, the rise of modern atomism (kinetic theory; dispersion theory; stereochemistry; quantum theory), the gradual emergence of the wave theory of light, occurred only because some thinkers either *decided* not to be bound by certain "obvious" methodological rules, or because they *unwittingly broke* them (1975:23).

Feyerabend has even gone so far as to say that the only methodological prescription which does not stifle the progress of science is "anything goes"—a complete abandonment of traditional ideas of method, objectivity, and truth.

There are many problems with "theoretical anarchism," but one of the most serious arises from the claim (which is also an element of the approach under consideration) that one can distinguish between good and bad science. The difficulty is this: if methodologically "anything goes" as Feyerabend argues it does, then it is simply impossible to see how a piece of scientific work can be evaluated on any formal basis. One could say, of course, in the anarchistic view and in many others, that good science is that which yields results when applied to practical problems, and bad science is that which does not—but that would be to make the judgment on the basis of consequences, not methodological considerations. What is interesting, and problematic, is that Feyerabend insists that the latter is possible notwithstanding his other arguments in support of the abolishment of methodological strictures. He has written that:

> [I]t should be pointed out that my frequent use of such words as "progress," "advance," "improvement," etc., does not mean that I claim to possess special knowledge about what is good and what is bad in the sciences and that I want to impose this knowledge upon my readers. *Everyone can read the terms in his own way* and in accordance with the tradition to which he belongs. . . . *And my thesis is that anarchism helps to achieve progress in any one of the senses one cares to choose* (Ibid.:27).

But is this position coherent? If methodology is counterproductive, then how can it be progress to replace a theory, the basic assumption of which cannot be tested, with one the basic assumptions of which can be? After all, the requirement of testability is one of the usual methodological requirements of empiricist science. A scientist from that tradition may think a testable hypothesis is better than an untestable one, but according to the central tenet of "theoretical anarchism," such a view is mistaken. We are thus forced to conclude that on this issue, at least, theoretical anarchism is self-contradictory and, therefore, necessarily wrong.

On the fundamental question as to the role of methodology in science, we take a middle-of-the-road position. On the one hand, the view that science can be neatly and completely codified seems hopelessly inadequate. As we have already suggested regarding, for example, the positivists' search for a criterion of cognitive meaningfulness (*see* Chapter 1) and as we will suggest for Hempel's deductive nomological model of explanation (*see* Chapter 5), it seems clear that the extent to which the intellectual processes of science can be formalized is far more limited than, for example, the logical positivists thought. On the other hand, it seems equally clear that theoretical anarchism will not do. We have discussed one difficulty we see with it, and there are others. As we argue in connection with radical subjectivism elsewhere (*see* Chapter 2 and Chapter 5), in the final analysis there is something self-defeating about a claim that there are no standards for the evaluation of claims. Even if true, there is no reason to accept such a claim.

It seems especially strange to argue that methodology is inappropriate to science, since there would appear to be no way of identifying what is scientific apart from methodological criteria. If even art history can be seen as a science under the appropriate conditions—and we would argue that it could be, then it isn't the subject matter of an inquiry that determines whether or not it is a science. If this is so, then it must be the way in which the inquiry is conducted—that is to say, its *methodology*—that determines the issue. The anarchist, of course, can reply that each person is free to decide for him- or herself what science is, and that, therefore, the question of identifying science is a spurious one. But to so respond would be to undercut the entire exercise, for if we are not agreed— at least roughly—on what is to count as science, then there is no basis for the debate over the role of methodology in activities of which we have entirely different conceptions.

Notes

1. The expression "(x) (y) (z) φ (x,y,z)" is used to symbolize the sentence "All x's, y's and z's satisfy the function φ" where 'φ' can stand for any relation involving three variables. One such function is the three-place 'taller than' relation—if x is taller than y and y is taller than z, then x is taller than z. When this φ is substituted into the original expression, the latter is then translated as "For all x, y and z, if x is taller than y and y is taller than z, then x is taller than z."

2. The volume in which Watkins' criticisms appeared resulted from a conference in which Popper and a number of his students and followers articulated what has come to be thought of as the Popperian response to the first edition of Kuhn's book. Watkins thus provides a link to our discussion of Popper's views.

4 / The Social Context of Archaeology

Transition from the positivist framework

The shift away from logical positivism toward a broader conception of scientific inquiry bears many similarities to Kuhn's account of a paradigm shift and is sometimes thought to be a rather tidy example of that phenomenon within philosophy of science itself. The kinds of studies that either became relevant or were undertaken in the new context include actual case studies of theory construction by Galileo and Newton (*see* Kuhn 1957, 1970a; Feyerabend 1975; Butts and Davis 1970; Westfall 1971); studies of structural and cultural features of science such as the functionalist account of Merton (1970); the paradigm-spread view of Holton (1962); various accounts of differentiation of disciplines (*see* Hagstrom 1965; Barnes 1972a); discussions of how science becomes institutionalized in particular settings; and studies of social and ideological factors influencing the development of science (cf. Kuhn 1970a).

A sociological approach in archaeology, which provides a perspective that is crucial for understanding the discipline, will be explored in this chapter. The effort has been made to provide quantitative data and specific examples when possible, though all too often, data are not available. There are no comprehensive sociologies of archaeology on which to base our remarks, of course; however, there have been attempts to fit events in archaeology into a Kuhnian framework, and thoughtful evaluations of those attempts (*see* Meltzer 1979). There have also been humorous but

99

insightful commentaries on archaeology (cf. Sellars 1973), parables on the archaeological scene (cf. Flannery 1981), biographies of archaeologists, histories of the discipline (cf. Daniel, ed., 1981; Willey and Sabloff 1974), and other literature that provides glimpses into the social framework of archaeology (*see* MacNeish 1978). The professional associations have been concerned with problems of ethics, employment, training, and other matters that have less to do with science as traditionally conceived than with politics and economics. In spite of archaeological awareness of these contextual matters, the relationship of such factors to the actual practice of archaeology appears to be only dimly perceived by most archaeologists. Recently, the impact of such external factors on the scientific aspects of the discipline has begun to be studied (cf. Miller 1980; Woodall and Perricone 1981). A symposium devoted to these factors entitled "The Socio-Politics of Archaeology" at the 1982 Society for American Archaeology meetings in Minneapolis was noteworthy for the empirical basis of the studies presented (cf. Gero, Lacy, and Blakey, eds., 1983).

In spite of the lack of systematic research, it seems useful to identify and present a preliminary exploration of some of the factors that might be considered in a more empirically based study. Our purpose, then, is to point the way toward what a sociology of archaeology might encompass, as well as to look at possible relationships between sociological and scientific matters. Clearly, neither archaeology nor any other science is a closed system unrelated to the societies in which it is embedded. Social and other contextual considerations affect both discovery and justification procedures, exercising a particularly significant influence on the selection of approaches, problems, and methodologies. If the basic goal of archaeology is to achieve improved levels of inference grounded in scientific methodology, it is important to recognize the role of nonlogical aspects of science as they influence the achievement of that goal. As we will argue more fully in the last section of this chapter, what is at stake, ultimately, is objectivity, for the question to be addressed is still whether there is some objective measure of what is true (or at least what is rationally acceptable) in archaeology or whether truth is simply a matter of what is believed by the main practitioners of the discipline (cf. F. Plog 1982:26).

It seems advisable to introduce considerations of the social context of archaeology by presenting one standard sociological view of science in order to anchor the following discussion in a widely recognized account. Kuhn's view is selected for two reasons: first, his work is at the core of philosophical debates, and many of the developments of the last twenty years in philosophy of science depart in some fashion from Kuhn's several publications; second, of all sociologists of science, Kuhn's work has received the most attention from archaeologists. As our own presentation of selected aspects of the sociology of archaeology proceeds in subsequent sections, we will draw again on Kuhn's views as well as on those of other sociologists of science.

Kuhn's sociology of science

Kuhn's account of science is deeply antithetical to more traditional philosophical views (*see* Chapter 3). Unlike many other philosophers of science, Kuhn does not take the position that features of the practice of science such as how scientists are trained or how they communicate with one another are irrelevant to issues of justification in science. Questions as to which of competing theories should be accepted at any given time, for example, can be affected by these nonlogical aspects of science. According to Kuhn, it is with just these sorts of data that the construction of an adequate account of theory choice in science must begin: "Some of the principles deployed in my explanation of science are irreducibly sociological, at least at this time. . . . Whatever scientific progress may be, we must account for it by examining the nature of the scientific group, discovering what it values, what it tolerates, and what it disdains" (1970a; 237–238). Kuhn clearly focuses his attention on the behavior of scientific groups, or communities, as being of primary and often unrecognized importance. One of the most important characteristics of these groups is the extent to which their members hold a virtually identical set of commitments (i.e., beliefs, attitudes, values) that Kuhn now calls a "disciplinary matrix." It is only against such a background of conformity that progress in science is possible. No scientific group could "practice its trade without some set of received beliefs" (Kuhn 1970b: 4). It is the disciplinary matrix, as common denominator, that gives rise to and keeps intact the extraordinary cohesiveness necessary for scientific advancement which occurs through the solution of puzzles. It is understandable that, in the Kuhnian approach, challenges to the stability of any scientific group are undesirable, for, if at all successful, they impede normal scientific activity. As Kuhn has written, "in the developed sciences occasions for criticism need not, and by most practitioners ought not, deliberately be sought. When they are found, a decent restraint is the appropriate first response" (1970b: 247).

Not only is a high degree of cohesiveness necessary for scientific progress, but Kuhn claims that its presence allows scientists to communicate among themselves far more efficiently than would otherwise be the case: "When the individual scientist can take a [disciplinary matrix] for granted, he need no longer, in his major works, attempt to build his field anew, starting from first principles and justifying the use of each concept introduced. That can be left to the writer of textbooks" (1970a: 19–20).

A correlative aspect of the Kuhnian account of science is the emphasis it places on specialization: the more focused and, therefore, narrow a scientific group's interests, the more feasible it is for its members to share completely the common basis which is essential for the success of their puzzle solving. Kuhn realizes that with greater specialization, scientific results become directly accessible to fewer and fewer people. He acknowledges that, with the "rigid definition of the scientific group" which is an inevitable consequence of specialization, the scientists' researches are no longer as a rule:

> embodied in books addressed, like Franklin's *Experiments . . . on Electricity* or Darwin's *Origin of Species*, to anyone who might be interested in the subject matter of the field. Instead, they will usually appear as brief articles addressed only to professional colleagues, the men whose knowledge of a shared [disciplinary matrix] can be assumed and who prove to be the only ones able to read the papers addressed to them (1970a: 20).

This and the other disadvantages of specialization are somewhat irrelevant in the Kuhnian approach, which is ostensibly descriptive and not normative; that is, Kuhn is not attempting to provide us with a characterization of how science ideally might be carried out but rather of how it actually does proceed in historical examples with which we are familiar. According to Kuhn, it is an observable fact that the natural development of science is toward greater degrees of specialization: "Although it has become customary, and is surely proper to deplore the widening gulf that separates the professional scientist from his colleagues in other fields, too little attention is paid to the essential relationship between that gulf and the mechanisms intrinsic to scientific advance" (1970a: 21).

Kuhn's account of science also addresses the role that professional training plays in the structure of science as a kind of indoctrination or socialization process. Not surprisingly, he sees that training of new scientists as being aimed at developing exactly the sort of conservatism he believes is necessary to the doing of normal science: "At least in the mature sciences, answers . . . to questions [such as: what are the fundamental entities of which the universe is composed?] . . . are firmly embedded in the educational initiation that prepares and licenses the student for professional practice" (1970a:5).

Kuhn further argues that highly conservative forces are at work on the practice of normal science. These serve to preclude the possibility of challenges to the received views localized in the disciplinary matrices—challenges that might otherwise threaten the stability of the scientific community. Conservatism is especially important in the choice of research problems, for it is when scientists begin to frequently run into difficulties solving puzzles that dissatisfaction with the prevailing theory develops and revolution becomes possible. It is through the operation of the disciplinary matrices themselves that this result is usually avoided. Normal science insulates itself by actively discouraging its members from tackling potentially dangerous questions: in Kuhn's words, "normal-scientific research is directed to the attention of those phenomena and theories that the [disciplinary matrix] already supplies" (1970a: 24).

Two caveats regarding Kuhn's account of science are in order here. First, our use of Kuhn's sociology of science in this chapter should not be seen as a retreat from the view we expressed in the previous chapter that much of Kuhn's philosophy of science is mistaken. One could agree with Kuhn that scientists practicing in well-established fields typically form highly cohesive groups and still reject his incommensurability thesis or his claim that prevailing theories are tested only during what he says are the very rare periods of revolutionary science. Much can be learned from Kuhn's often astute observations on the actual process of scientific investigation; nevertheless, we would argue that his philosophy of science—the central tenet of which is the claim that science divides into identifiable, if not absolutely distinct, "normal" and "revolutionary" periods—does not adequately portray the way sciences behave.

Second, we do not mean to suggest by considering Kuhn's sociology of science in the context of archaeology that archaeologists should try to shape their discipline to fit the Kuhnian model. Some of the features of archaeology discussed here may reveal problems specific to contemporary social science, and some may even be peculiar to archaeology. Although Kuhn insists on the importance of increased specialization as a science evolves, it may be that in certain fields, of which archaeology could well be one, this direction for development is simply undesirable; as a matter of fact, archaeology seems to offer a number of examples in which increasing specialization has been combined with interdisciplinarity. Kuhn's sociological views thus provide a useful framework within which we can discuss a number of features of the development of archaeology that have not usually been treated in the literature on philosophy of science; but, in many of their details, these are not applicable to archaeology. We need both to look carefully at the way in which archaeologists have tried to use Kuhn's work and to be aware that there are a number of aspects of the sociology of archaeology important to an understanding of the discipline which Kuhn has not discussed.

Archaeological responses to Kuhn

Kuhn's discussion of the revolutionary aspects of science has risen to prominence in the archaeological literature in an atmosphere of discontent with traditional archaeology. The emphasis has been on the positive aspects of change, especially with respect to methodology. A general consensus about whether Kuhn's view of science is applicable to archaeology appears to be lacking; where it is perceived to be applicable, there is no agreement on where archaeology stands with regard to revolutions, paradigms, and normal science.

Quite clearly, archaeological interest in Kuhn's views is closely connected with his ideas about revolutions. Archaeologists have often perceived themselves to be participating in a revolution, and the New Archaeology, particularly, was often seen in this light. Much less has been said about normal science in archaeology. Most scientific revolutions, Kuhn notes, "have customarily been viewed not as revolutions but as additions to scientific knowledge" (1970a: 136), and "there are excellent reasons why revolutions have proved to be so nearly invisible" (1970a: 136). These include the enormous authority granted to science textbooks, which "address themselves to an already articulated body of problems, data, and theory, most often to the particular set of paradigms to which the scientific community is committed at the time they are written" (1970a: 136), with the result that changes do not make themselves especially evident. In addition, it was not altogether clear how large a change had to be to merit the name "revolution," though the impression conveyed by the first edition of Kuhn's book seems to have been that a new paradigm represented the sort of substantial change brought about by a Kepler, a Galileo, or an Einstein. In the postscript, Kuhn corrects this misconception, making it clear that he is especially interested in little revolutions: "A revolution is for me a special sort of change involving a certain sort of reconstruction of group commitments. But it need not be a large change, nor need it seem revolutionary to those outside a single community, consisting perhaps of fewer than twenty-five people" (1970a: 180–181).

Of course, if this is all that is meant by "revolution," then it is probably correct to say that such changes have occurred in all disciplines, including archaeology, with some regularity. Once Kuhn includes these relatively small changes as revolutionary and also admits that periods of normal and revolutionary science are not completely divorced from each other but that "the two sorts of change interpenetrate," it would appear that the distinction between the two has, for all practical purposes, collapsed and that the talk of Kuhnian style "revolutions" takes on quite a different meaning.

In discussing paradigms and community structure, Kuhn (1970a: 176) pointed to links between his views, the literature on social networks in science, and invisible colleges. Scientific revolutions sometimes begin with individuals who may be lone workers or with small groups that may be nonconforming vis-à-vis the existing paradigm: "Probably the single most prevalent claim advanced by the proponents of a new paradigm is that they can solve the problems that have led the old one to a crisis" (1970a: 153). But while such arguments are appealing to scientists, they are frequently not compelling. Other factors, including those Kuhn calls aesthetic and subjective, may be at least as important. Sometimes, two competing paradigms both solve the problems equally well, so that the criterion for choosing between them must be something older than the ability to resolve the crisis-producing difficulties. As Kuhn tells us, "paradigm debates are not really about relative problem-solving ability, though for good reasons they are usually couched in those terms. Instead, the issue is which paradigm should in the future guide research on problems many of which neither competitor can yet claim to solve completely" (1970a: 157).

The use of paradigms for problem solving occurs within a framework that remains open to more than one possibility. As Barry Barnes tells us: "The paradigm is a resource for the scientist, not a determinant of what he does. . . . They cannot instruct the scientist how to carry out his research; they merely remain available as resources for use in that research" (1982: 49). This notion of the paradigm's operating not to determine the research but to direct it is an important feature of scientific reasoning that we discuss in detail in Chapters 6 and 7. Not surprisingly, this open-endedness, along with Kuhn's view of the way normal science operates, with "conversion" to a new world view having a prominent role in the transition from one paradigm to another, has led a number of critics to charge that Kuhn is overly relativistic and subjective. Such views are seen to threaten the very basis of scientific objectivity. Kuhn replies that his position is not mere relativism; and that to the extent relativism is involved, he cannot see "that the relativist loses anything needed to account for the nature and development of the sciences" (1970a: 207). Indeed, the kind of relativism at issue here scarcely arises as a problem until we can identify some clear examples of paradigms and paradigm shift for whatever discipline is under study—something that is by no means unproblematic for archaeology.

It is not surprising that there is no general consensus about what, in archaeology, is to be identified as a paradigm or about the stage of development of any paradigm that is claimed to be identifiable, especially in light of the fact that one of the most common criticisms of the Kuhnian notion has been that it was unclear just

what was meant by the key term "paradigm." Margaret Masterman (1970: 58–59), for example, claimed to find some twenty-one
different uses of the term "paradigm," including uses of it as a
universally recognized scientific achievement, a myth, a constellation of questions, a textbook, a tradition, an analogy, an accepted
device in common law, a set of political institutions, and a new way
of seeing. It should be noted that, notwithstanding her objection
that Kuhn had used the term "paradigm" in a variety of ways,
Masterman regarded the basic concept as one of supreme importance to an understanding of how science is done and thus worthy
of more precise elucidation; and, of course, it is clear that, of the
many uses of the term that she identified, some can be ignored and
others can be grouped into just a few fundamentally different
meanings. Kuhn acknowledged, in his 1969 postscript to *The Structure of Scientific Revolutions*, that he had been using the term in two
different senses. The first sense he termed the sociological; the
second, deeper sense was the notion of paradigms as exemplary
past achievements.

In terms of certain criteria and processes which Kuhn associates
with revolutionary and paradigm changes, there is some evidence
that archaeology has exhibited these characteristics in abundance.
The problems addressed by the old paradigm have been seen as
uninteresting, new directions of research have been proposed,
conversions have occurred, and the new points of view have
steadily extended their influence, especially among younger professionals. There is a diverse body of literature belonging to the
New Archaeology; several textbooks expound its views (cf. D. H.
Thomas 1974); archaeologists devoted to the older paradigms can
be seen as throwing up a series of last-ditch defenses. Chester
Chard's *Man in Prehistory* (1969), for example, was so regarded by
Edwin Wilmsen when he wrote: "It is ironic that so many books on
the theme man in prehistory should have appeared just at this time
when archaeology holds the promise of offering more than what
Radcliffe Brown rightly called "conjectural history" (Wilmsen
1970:689).

To suppose that there is unanimity about the present state of
archaeology in relation to paradigms would be a considerable
oversimplification. Leone (1972:14–27), for example, argues that
anthropology and archaeology are "between" paradigms. Two of

anthropology's incomplete paradigms at the present time are cultural materialism (evolutionism) and structuralism. Of the change in archaeology, he asks:

> Is it a scientific revolution, or a prelude to one? Or is it best character-ized as the completion of a paradigm already present for decades in the field? Archaeology at any given time has acknowledged the theoretical requirements of the current anthropological paradigm, but in practice has actually addressed its substantive contributions to the preceding paradigm. This is a paradigm lag (1972: 16).

Although he believes that paradigm lag still operates, he identifies three candidates for change that would be more than change through gradual evolution (i.e., in the direction of a paradigm shift): the explicit use of evolutionary theory, an increasingly so-phisticated cultural ecology, and the advent of systems theory (1972: 18).

Leone further argues that two of archaeology's central goals are "outworn": "One, the reconstruction of events in the past, is nearly complete; it offers little in the way of challenge today . . . The other aim, a picture of past lifeways, is a scientific impos-sibility." A third aim, the development of "cultural analogues to theories like natural selection, remains both incomplete and poten-tially fruitful." This last aim may be successfully pursued, in Leone's opinion, through the activities of "three marginally affili-ated types of archaeology" (historical archaeology, ethnographic archaeology, and the science of material objects, or material cul-tures) (1972: 26). According to Leone, a successful paradigm should be able to resolve a fairly comprehensive range of prob-lems, and by this criterion, the paradigms of archaeology remain incomplete—potential rather than actual paradigms.

In the 1970s, other archaeologists conceived of the field as torn between two paradigms—the old and the new, while still others spoke of multiple paradigms. David Clarke (1972: 6–7) argued, for example, that we have a traditional paradigm and four competing ones—the morphological, the anthropological, the ecological, and the geographical. At times one suspects the term "paradigm" is used where another term such as "approach" would do as well, or better, as it would be free of the connotations connected with Kuhn's usage.

Along with the post-Kuhnian rush to praise change and to see archaeology in terms of paradigms, there were those who took a dim view of the actual accomplishments of the New Archaeology. Flannery (1973), for example, suggested that the search for laws in archaeology has so far yielded only trivial laws; Binford (1977: 1–10) argued that we experienced not a revolution but a "little rebellion"; he spoke of the "nonparadigmatic freedom" of the 1960s events that would be wasted unless archaeology moved toward theory building. He still awaits a really New Archaeology, and suggests that all that has occurred to date is antitraditional archaeology. Neither Flannery nor Binford, we suspect, would be inclined to argue that there have been paradigm shifts of any great magnitude in archaeology, although both have pioneered alternative approaches to traditional archaeology.

For Schiffer, on the other hand, the new paradigm had arrived in 1976. He opened his book *Behavioral Archaeology* with this sentence: "From the perspective of the mid-1970's, the 'new' or 'processual' paradigm finally has reached maturity" (1976: 1). Badges of this maturity are texts (including introductory ones), collections of classic principles and applications, and lengthy case studies. The new paradigm, he tells us, has entered a period of normal science in which the new "basic concepts and principles are beginning to submerge into the murky inaccessible depths of the discipline, where they no longer can be easily questioned or challenged—until, of course, the next major paradigm clash" (1976: 1–2). Schiffer seems to want to combat complacency when he asks whether the new paradigm: "under which we now labor . . . is fully adequate for genuine scientific studies of the past" (1976: 2). Some doubts about the completeness of the paradigm shift are suggested by the following statement: "Although archaeology is no longer divided into warring camps, the impression remains that we are in a state of fatigue-induced cease-fire rather than one of peace" (1976: 3–4).

Both Schiffer and David Clarke see reintegration of new and old as a normal part of a paradigm's developmental cycle. While this view might be defended by selectively quoting Kuhn, it seems contrary to his overall message of the way paradigms work. He seems to suggest instead that new paradigms expand by conversion and recruitment while the old guard loses members and loses ground through attrition. What does not emerge clearly from

Kuhn is a picture of reintegration, an intertwining of perceived or potential paradigms and non-crisis-related change. It is precisely these aspects of change in archaeology that strike us most forcefully.

If these aspects fit easily into Kuhn's scheme, it is at the pre-paradigmatic stage that they do so. Should we, then, think of archaeology as fitting the Kuhnian model but as being at a pre-paradigmatic stage? The difficulty is that archaeology seems not to fit this conception any more neatly than it does the conception of being in the midst of a revolution or of being in a new paradigm. Meltzer (1979) argues that the reason for this is that, although there is general agreement that archaeology, in recent years, has gone through significant changes, these changes have been largely methodological. Methodological changes are relatively routine, according to Meltzer, and it is only the deeper metaphysical changes that are revolutionary. Disagreement among the commentators, though it could be attributed to a misuse of the Kuhnian model or to the fact that we are too close in time to the events to be able to interpret them objectively, is more likely, according to Meltzer, to be a result of the fact that there "simply has been no archaeological revolution in a Kuhnian sense" (1979:652). He argues: "revolutionary change is one of structure, not method. It is neither simply the manner in which the discipline approaches its data nor the goals of the discipline itself that must change. It is, instead, change in the underlying metaphysic that is required to produce a revolution" (1979: 652). Whether or not Meltzer is right in claiming that archaeology has not undergone a Kuhnian revolution, he is essentially correct in noting that the changes have concerned methodology. As least these are the changes that have captured the archaeological imagination and that have been prominent in the debates.

The lack of agreement over what constitutes recognizable paradigms in archaeology is not surprising, given the flexibility of the concept in the parent source and the fluid state of archaeology. Although some of Kuhn's points about revolutions (particularly about the "little revolutions") show a remarkable fit with events in archaeology, the overall congruity between what has happened in archaeology and the Kuhnian view is not compelling. We suspect that the Kuhnian view of science cannot usefully be applied to archaeology in a holistic manner. In particular, the tendencies in archaeology to interweave the old and the new, to combine increasing specialization with interdisciplinary studies, and to make conscious efforts to counteract splintering and fragmentation through restructuring of the professional societies seem quite un-Kuhnian. It may even be that in archaeology (and in other social sciences as well) it is natural for several approaches, which may or may not legitimately be regarded as paradigms, to coexist at once with no clear means or need for resolving the conflict or choosing among them (cf. David Thomas 1979: 161–180). Although Kuhn's account can be applied only selectively to archaeology, it provided directions for exploring the social and cultural context of the discipline.

A model for archaeology

An alternative to recent attempts to characterize archaeology in terms of the Kuhnian concepts of paradigm and revolution might be to try to view the discipline in a way that expresses the dynamic relationships among various aspects of the discipline. For this purpose we imagine a sphere, which is meant to represent the totality of archaeological beliefs, views, methods, techniques, and information. Other geometric or dendritic forms could be used for creating an analogue of this sort, but, for reasons that will emerge shortly, the sphere seems most appropriate to our purposes. At the center is what we might call the Core System (CS), constituted of those views, beliefs, and methods shared by virtually all practitioners of the discipline. As one moves outward from the core, views, beliefs, and methods become progressively less universally held until one encounters, resting lightly on the surface, views that are shared by very few professionals in the field. Among the latter are recent innovations which might be expected to move inward toward the core as they diffuse through the discipline, but also to be found on the surface are long held notions that have failed to gain a significant following.

The sphere analogy is one way of expressing the dynamic and ever-changing nature of archaeology as we conceive of beliefs or methods or technologies moving toward or away from the center. Thus, we could trace the acceptance of Carbon 14 as a method of dating that rested on the surface of our belief system in 1949, moving toward the core rapidly after its usefulness and reliability were demonstrated and its respectability thus assured. When the dating technique was first developed (*see* Taylor 1985), carbon readings were correlated with reasonably well-known dates for Middle Eastern artifacts, and at first the ages produced by the method appeared reliable enough. Problems soon developed, and, as David Wilson (1975: 95) has written:

> One of the biggest sources of trouble came in comparing radio-carbon dates with the dates provided by the Egyptologists from the huge mass of well-preserved Egyptian material from the years of the pharoahs. Here the new scientific methods were plainly open to some very destructive criticism. Broadly speaking, the radio-carbon method gave dates that were pretty regularly two hundred or three hundred years too low.

As a result, various modifications were made in the technique, including the very important recalibration of the dating scale based on tree-ring dates obtained from the bristlecone pine (cf. Renfrew 1976). Technological advances in the method itself have also occurred (cf. Bennett 1979).

Many archaeologists tend to be somewhat cavalier in their acceptance or rejection of Carbon 14 dates, and one sometimes has the impression that acceptance or rejection is unduly influenced by how well the dates fit preconceived ideas. Chronometric measurements are one line of evidence which must be evaluated in the light of other evidence, and selection of acceptable dates is inevitable; nevertheless, the reliability of Carbon 14 dates as expressed in probabilistic terms seems to be frequently ignored (cf. Polach and Golson 1966; Ralph 1971).

In spite of these modifications, in spite of the fact that there is reason to believe C_{14} dates are often misinterpreted, and in spite of the fact that there are still those who feel C_{14} lacks sufficient strength to overturn other dating methods (as recently as the 1970s, several leading Japanese archaeologists held this view; cf. Ikawa-Smith 1975, especially pages 15–17), the method is nonetheless an integral part of the CS. As a part of CS, the technique does not have to be described anew every time it is used and the results are included in a report, and this is consistent with Kuhn's views on abbreviated communication among those who share basic ideas. Indeed, our characterization of the way in which the core system is shared has much in common with Kuhn's notion of the shared paradigm that is part of normal science.

Some of the archaeological core beliefs in the CS are what David Clarke calls "Controlling Models" of substantive views about the nature of the past or of culture concepts:

> The nature of these cognitive or controlling mind models is both complex and composite. Through exposure to life in general, to educational processes and to the changing contemporary systems of belief we acquire a general philosophy and an archaeological philosophy in particular—a partly conscious and partly subconscious system of beliefs, concepts, values and principles, both realistic and metaphysical . . . Archaeologists, like most other practitioners have the greatest difficulty in believing that their own perceptions are controlled to this degree—they may always apply to other archaeologists, of course (Clarke 1972:5).

The discipline of archaeology is dynamic over time and, as Clarke (Ibid.) and Ascher (1970) have demonstrated, things change their relative position in the sphere. The CS at any particular time is occupied by widely accepted views of culture history, by commonly used and accepted methods and techniques, by shared basic values, and so on.

The place occupied by various elements in the sphere affects the way they are treated by archaeologists. More rigorous standards of evidence are required and higher standards of testing are demanded of things outside the CS, whereas items firmly entrenched in the CS are subjected to less questioning. In fact, it may be quite difficult to challenge a core belief to which there is a deep commitment.

Attitudes about the antiquity of man in the New World provide one kind of example, though an imperfect one, of the way firmly entrenched views can set very high demands for evidence. This is an imperfect example for several reasons. First, it is, historically, a complex issue with several competing views in evidence during the late nineteenth and early twentieth century. Second, the conservative positions often attributed to the key actors (W.H.H. Holmes and Ales Hrdlicka) have tended to be oversimplified. Hrdlicka's name is often associated with the dictum that man has only been present in the New World for 3,000 years, but it is not clear that he ever set such a figure. Rather, for Hrdlicka, the phrase "Early Man" referred to the Pleistocene and older geological periods. At that time, the entire Pleistocene Period was believed to be about 300,000 years long, and its end was seen as occurring considerably before 10,000 years ago. On that scale, the post-Pleistocene period was a significant enough proportion of the postulated period of human evolution that "evolutionary progression" was deemed to be a significant variable (Hrdlicka 1907, 1912, 1918).

Without entering fully into a reanalysis of the Holmes/Hrdlicka position as it has been portrayed, it is true to say that Hrdlicka demanded very high levels of evidence for purported Early Man finds. As a minimum, he demanded marked serial somatological distinctions (especially for the skull), impeccable stratigraphic evidence documented photographically, and geological evolution; and fossilization was regarded as desirable. These levels of evidence, in Hrdlicka's opinion, were never successfully met, and he questioned or rejected a lengthy list of finds. Willey and Sabloff (1974:58), paraphrasing Frank H.H. Roberts (1940), note that Hrdlicka's rejections of cases were so strongly worded that "many workers [were] afraid to face Hrdlicka's attacks."

Although contrary beliefs existed throughout the historical period under consideration, the short chronology for the antiquity of man in the New World dominated. High standards of evidence were required to overturn that view. It was, of course, the Folsom discovery that convincingly provided the evidence that established a new threshhold for the antiquity of man in the New World. A new core belief had replaced the old. The evidence in this particular case was a convincing association of man-made artifacts with extinct fauna. Although we would not call this new threshold a new paradigm, this example shows stronger parallels to Kuhnian views of what happens in a paradigm shift than can be found in contemporary archaeology, because the older, formerly dominant view was effectively replaced, losing not only its position in the CS but being moved right out of the sphere. Once this new threshold of belief about the antiquity of man was accepted, one could argue that that part of archaeology devoted to Early Man studies began to operate very much within a "normal science" or "puzzle-solving" framework. That is, the archaeologists concerned acted to "fill-in-the-paradigm" by filling in the blanks in the cultural historical record. Early Man sites within the acceptable time period were no longer required to carry the burden of changing a widely accepted core belief, and as a result, acceptable sites were rapidly investigated in growing numbers. Should the currently competing views for an even greater antiquity of man be successfully demonstrated, it seems reasonable to predict that a similar situation will prevail. Before leaving this example, it is worth noting that the instrument of opinion change (the Folsom discovery) was actually quite chaotic by today's standards of fieldwork. The relevant evidence needed to shift beliefs across the discipline, however, was captured.

Not all changes in belief and attitude result from the use of relevant evidence to support a competing view which ultimately results in replacement of one belief with the successfully presented alternative. Some changes are simply added to the existing repertoire without displacing items already present. Technological changes such as the acceptance of Carbon 14 are essentially additive. Still other kinds of changes occur gradually as knowledge expands. Opinion-changing tactics, which also can result in major attitude shifts, are sometimes employed: an example is the use of clearly drawn dichotomies or bipolar oppositions. In the early days of the New Archaeology, several dichotomies were widely and effectively used to direct change in certain directions: old vs. new, induction vs. deduction, subjectivity vs. objectivity, culture history vs. processual archaeology were all set up as oppositions in which the second member of the pair was clearly to be preferred. Conversion of a prominent figure to a new point of view is yet another means of effecting shifts in the belief system. Paul S. Martin's (1971) account of his own changing outlook is a vivid case in point—one that was significantly influential in changing beliefs for some archaeologists.

Since factors of the sort discussed above are clearly important in shaping the total belief system and, within that, the CS, it seems reasonable to suggest that they deserve serious investigation. Not enough is known about the sociological and psychological processes that govern change in a discipline, and this is at least as true of archaeology as of other disciplines. Some things are obvious, such as that new ideas that do not seriously compete with well-entrenched beliefs are likely to be accepted more rapidly across the discipline than are those that do compete, and this, of course, is entirely consistent with Kuhn's conservative view of science. With respect to those essentially more threatening innovations which *do* compete with accepted beliefs, it appears that the better respected the archaeologists are who advocate their adoption, the more readily they will become part of the discipline's CS (cf. F. Plog 1982).

This latter point again raises the issue of objectivity in social science. Within archaeology, the subjective/objective debate has centered most frequently on evaluation of archaeological results through the evaluation of the previous track record of the archaeologist whose work it is. Raymond Thompson (1956), for example, wrote favorably of evaluating archaeological results in this way, and his comments are frequently criticized by those who advocate greater objectivity. Binford is among those who have vigorously denied Thompson's statement that "the final judgement of archaeological reconstruction must be based on the appraisal of the professional competence of the archaeologist" (Binford 1967:10); and it seems clear that this cannot serve by itself as the criterion for appraising archaeological work. Surely, however much an archaeologist's competence and reputation may provide a clue to the assessment of his or her conclusions, even the best investigator is capable of making foolish judgments or unwarranted inferential leaps in support of a pet theory. Ultimately it is the work rather than the scientist that must be evaluated. It is perhaps worth noting that in other disciplines, problems of objectivity more frequently center on issues such as repeatability of results and experimental controls rather than on the competence of investigators.

That archaeologists, their opinions, and their reputations at times play an active role in the archaeological endeavor is suggested by the not uncommon practice of having other archaeologists observe certain parts of an archaeological investigation. Particularly in threshold cases, when some previously unknown or unaccepted "fact" or inference is suggested by a discovery, other archaeologists are invited to witness and thereby validate the relevant evidence—a practice of both sociological and methodological interest. The individual archaeologist who extended the invitation is saying in effect: "I know this find is potentially controversial, and I need your independent testimony that I found what I said I did." The visitors (if they concur) offer their reputations and expertise to the profession as a sign of their own and the investigator's good faith; the profession often accepts such validation procedures as supporting evidence which might be viewed as part of the methodology of confirmation or which is at least difficult to untangle from the confirmation process. The "validation by witness" process is undoubtedly related to the fact that archaeology is not a laboratory science producing results under laboratory conditions which can be replicated experimentally. Such visitations do not always bestow validation, but some do, as can be seen from the case of the discovery of the San Isabel Ixtapan mammoth in Mexico during the early 1950s. The primary Mexican investigators (including Luis Aveleyra Arroyo de Anda and Manuel Maldonado-Koerdell) extended invitations to Marie Wormington, Alex Krieger, and E.H. Sellards, all of whom were notable Early Man authorities at that time. Not all of the authorities were able to visit the site, but those who did agreed that the find was bona fide. In this case, the threshold was not temporal but spatial, and was related to the important side issue of reassessing the earlier and highly controversial nearby find of Tepexpan Man. The procedure of partial validation through visitation of finds by highly regarded authorities can be defended as a useful time-saving device (archaeologists simply don't have the time to follow every argument presented by every archaeologist), and as a probable kind of indicator. This form of evaluation is never sufficient by itself, however. Ultimately, any ideas advanced must be established through other means.

Consideration of the reputation of archaeologists as relevant to evaluation of their work shows some interesting parallels to the differential treatment of different levels of beliefs, and indeed they may only be different aspects of the same phenomenon. Just as core beliefs are less likely to be challenged than are more peripheral beliefs—and when they are successfully challenged, the competing view must provide exceptionally compelling and relevant justification, so the ideas advanced by sociometric "stars" in the archaeological world are likely to be unusually influential. For example, it seems fair to regard such diverse personalities as Lewis Binford and Kent Flannery as archaeological "stars." One manifestation of their status is that their new work is rapidly and widely disseminated, and many archaeologists are prone to accept ideas presented by such personalities with few quibbles. Still, it should be noted that their pronouncements also elicit extensive criticism and debate, indicating that reputation provides at most a prima facie reason for accepting the views of an archaeologist. If every idea in the CS had to be reargued each time it was used, archaeology, like any other science, would simply disappear under a sea of uncertainty. Similarly, it is not only reasonable but even necessary to assume, at least provisionally, that people who have done good work in the past will probably continue to do good work; but this assumption is always provisional, and open to correction on the basis of careful evaluation of the work itself.

The question arises as to whether, given that every science needs a core of reasonably stable concepts, methods, and beliefs in order to get on with the business of solving its problems, it is appropriate to view the archaeological CS as what Kuhn calls the "entire constellation of beliefs, values, techniques and so on" (1970a: 175) necessary for the practice of normal science. Certainly there are strong similarities between Kuhn's constellation and our CS. The significant difference between the two schemes lies in their setting. Our spherical model allows relative degrees of professional acceptance to be portrayed by position; dynamic movement of things within the sphere can also be portrayed. Things that Kuhn would have to differentiate as normal vs. revolutionary can be presented less oppositionally. For example, we can place a number of subcenters within the sphere. At these centers, we find subgroups of archaeologists who share more values and attitudes with each other than with the discipline as a whole. Centers can be conceived at different levels of magnitude, ranging, for example, from those deeply involved with the longitudinal research of the University of Arizona field school at the Grasshopper site to those who might be broadly designated as Cultural Resource Management people.

In a different vein, the CS can be seen as normally exercising a conservative influence on the discipline, which in turn promotes stability and impedes change, at least in some areas. If we want to draw a parallel to Kuhn, it would be at the locus of the conservatism found in *normal* science; but this is not a terribly good fit, because it appears that individual archaeologists usually hold a range of beliefs or use a range of methods and techniques that are not found in the CS. Certainly, innovators are producing things that may or may not enter the CS.

With regard to the conservatism of the CS, one difficulty is that it may restrict alternative accounts of archaeological data. As Thomas Kuhn and Wesley Salmon, among other philosophers, have made clear, alternatives must be restricted or chaos will ensue; however, the CS, or the core beliefs and practices within a particular context, may restrict alternatives in an undesirable way. Consider the history of ceramic analysis within the American Southwest. Viewed cumulatively, a disproportionate percentage of the literature from that area is devoted to ceramic analysis. Further, the most common means of dealing with this pottery is, or has been, to assign it to types (or some related classification unit). The typological approach has paid handsome dividends; it is unquestionably a highly successful procedure, especially for problems requiring temporal and spatial definition. As long as southwestern archaeology was primarily concerned with culture history, this form of classification was eminently suitable.

The very success of southwestern pottery typology has impeded the development of alternative means of dealing with and extracting information from this class of artifacts. Classifications or concepts that have proved to be useful elsewhere, such as in Meso-America, Peru, or the eastern United States, do not seriously compete with the entrenched system unless they are effectively phrased as an extension of that system. By now, so much information is tied up in typological designations that if archaeologists are to continue to provide comparable background information, it would seem that pottery types should and will be perpetuated.

Assigning sherds to a type does not provide the most useful data from this artifact class for many archaeological problems. We might have predicted that as different approaches and problems became significant, means of ordering the data at hand would be altered to better correspond to the new needs. Interestingly enough, this has been slow to develop, although the last few years have seen significant moves in new directions. These show increased attention to ceramic manufacturing and distribution systems, following Anna Shepard's leads of several decades ago (cf. Shepard 1936). Examples of such recent studies include ceramic exchange patterns in Chaco Canyon (Toll, Windes, and McKenna 1980) and Helene Warren's (1967, 1969, 1970) several petrographically based analyses. Within a relatively short period of time, petrographic analysis of ceramics has moved from an esoteric to a more routine procedure in southwestern archaeology. Stylistic analysis is another direction enjoying increased popularity. Stephen Plog's (1980) recent book, Dorothy Washburn's (1977) study of Upper Gila Area Ceramic Design, and Joe Stewart's (1979) stylistic study of Lincoln Black-on-Red are examples of this approach.

Although these nontypological approaches to ceramic analysis were pioneered long ago, their increased popularity at this time reflects the growing tendency to exploit more fully the informational potential of ceramics for a wider range of problems. This, in turn, is related to the nature of the problems of interest to archaeologists. As problem orientation has shifted from culture history to, for example, exchange networks, the means of handling basic evidence such as ceramics shifts to those analytical procedures that are most informative for the new problem orientation. In spite of these adaptive analytical strategies, however, the fact remains that ceramic typology in the Southwest has exerted a powerful conservative influence on the forms of ceramic analysis practiced in this area, and some potentially useful concepts have either never been tried or failed to compete with entrenched ways of dealing with pottery.

As a final point about our spherical model, it appears that recent disciplinary changes in archaeology can, to some extent, be understood in terms of challenges to the traditional CS as it stood in the 1950s. Although the overt context of the debates was methodological, the attacks on selected aspects of the CS were quite fundamental, and this is one reason why some archaeologists urging change were attracted by Kuhn's depiction of revolutions in science. A less extreme view—even one that de-emphasizes the contrast between "normal" and "revolutionary" periods in science—might be more plausible, and is supported by various recent references in the archaeological literature to a reintegration of old and new archaeological paradigms.

The spherical model, then, can be viewed as an attempt to develop a fresh perspective on archaeology—one unencumbered by the preconceived notions inherent in models formulated for other sciences. It is obviously only preliminary, but may provide a "way of seeing" which allows us to examine certain dynamic aspects of belief and change.

Change through borrowing and some consequences

The rate of change in archaeological beliefs is a function of, among other things, the rate at which new ideas are presented to the discipline. Since the 1950s, many methods, techniques, models, and ideas have been borrowed from other academic fields. Archaeology has not been the only discipline to use this means of extending its horizons during this period. Prior to the Second World War academic disciplines were inwardly oriented, seeking to define their own boundaries and to develop their own procedures. Since that time, disciplinary boundaries have lowered, while cross-disciplinarity and inter-disciplinarity have steadily grown. Von Bertelanffy's (1972) history of the emergence of Systems Theory successfully captures some of the trends of those times; and the success of this movement can be seen to some extent in the proliferation of combined disciplines such as astrophysics or biochemistry, as well as in interdisciplinary programs, projects, and research.

Presumably, at least in principle, one could chart the degree to which particular disciplines borrow and/or serve as donors. We might expect that more mature disciplines, especially if we were taking a Kuhnian perspective, would serve as donors with greater frequency than less mature disciplines. Similarly, the more mature disciplines might borrow less often than less mature disciplines. Whether or not this donor-recipient relationship is seen in terms of relative maturity of the respective disciplines, it seems to be true that archaeology borrows more from other disciplines than it exports. Archaeology has borrowed heavily from ecology, systems theory, philosophy of science, and statistics, to name a few donors; technologically, our debt to chemistry, physics, engineering, and computer science is substantial. Archaeology is thus much more of a recipient than a donor, and this openness to input from other disciplines, as well as the fluidity of the framework within which archaeology is practiced, suggest that, in Kuhnian terms, archaeology is not a mature science. On the other hand, what this may suggest is that the Kuhnian model is not altogether appropriate for archaeology.

Also contributing to the rapid rate of change in archaeology is an increased interest on the part of archaeologists in innovation. They, like most of the rest of contemporary society, place a relatively high value on novelty and originality. This, too, says something about the scientific status of archaeology in the Kuhnian model, for science, as it is usually practiced, is seen by Kuhn as an exceedingly conservative activity—an enterprise devoted to forcing "nature into the preformed and relatively inflexible box that the paradigm supplies" (1970a: 24). Further, he claims that:

> No part of the aim of normal science is to call forth new sorts of phenomena; indeed those that will not fit the box [provided by the paradigm] are often not seen at all. Nor do scientists normally aim to invent new theories, and they are often intolerant of those invented by others. Instead, normal-scientific research is directed to the articulation of those phenomena and theories that the paradigm already supplies (1970a: 24).

There can be no doubt that, for Kuhn, science is a pursuit in which innovation is to be avoided, and he has emphasized this point by saying that "Perhaps the most striking feature of . . . normal research problems . . . is how little they aim to produce major novelties, conceptual or phenomenal" (1970a: 35).

That the high rate of borrowing has stimulated archaeological research in profitable ways is beyond question, and one need only think of the utility to archaeology of optimal foraging strategy and locational geography to acknowledge this point. There remain two main concerns. First, as long as archaeology places such a high value on innovation, the chances that many of the innovations will be appropriately tested and evaluated on multiple cases before passing into usage would appear to be disappointingly low; many may remain isolated without either further testing or usage. Second, we need to remember that borrowed innovations always bring potential dangers with them. For example, where ideas developed in one discipline are applied in another, those importing the innovative work all too often ignore valid criticisms of those ideas which have been made in the parent field, and in this way misconceptions are further propagated. The way in which concepts from philosophy of science have sometimes been incorporated into archaeology has drawn critical responses from philosophers (cf. Morgan 1973 and 1974–75), and ecological imports do not always arrive with the full range of debates from ecology but are used in an isolated way (*see* Keene 1983: 137–155 for a critical discussion of the role of borrowing in the case of optimal foraging theory; *see also* Hardesty 1980 and Dunnell 1982:2–3). Care must always be taken that borrowed ideas are appropriately adapted to the new discipline; otherwise, the innovation is unlikely to constitute an advancement in the borrowing discipline.

Cultural influences on the discipline of archaeology

Modern archaeology derives from the European intellectual tradition (cf. Rowe 1965). Despite the broadening of the archaeological base on a global scale that has occurred in recent years (especially since World War II), the European and North American centers of archaeological inquiry are still remarkably influential on archaeology worldwide. Although it seems clear that archaeology has to some extent been pursued from a particularly European or European-derived point of view, it is exceptionally difficult to identify what this might mean in terms of theoretical positions adopted in archaeology. Even if Euro-centrism in archaeology is admitted, this does not necessarily vitiate the results of archaeological investigation, because the choice of subject matter for study is a separate issue from that of how the study is carried out. Science must be carried out from some point of view, and the problem of ethnocentrism would not be less a problem had archaeology developed from a Hindu base, for example.

Participants in the Euro-centric tradition will find it difficult to identify its influences. In Kuhn's view of science this is to be expected, since all cultural biases, including those bestowed during professional training, are extremely difficult for the holders of those biases to perceive. Nevertheless, some suggestions have been made about concepts and attitudes held in archaeology that may have their roots in Euro-centrism. Evolutionary theory has been seen as arising from nineteenth-century European ideas about progress (*see* Chapter 7). Drawing on the work of Thomas Gladwin (1958), we will suggest in Chapter 6 that Euro-centric views toward small boats and large bodies of water might have affected archaeological attitudes towards transoceanic contact. Warren DeBoer (1982) has called archaeology a "myth-making" profession, noting that the "myths" we create parallel the major concerns of European thought, such as the origin of man. Blakey (1983) has argued that North American archaeologists tend to select their research locations outside of North America by their proximity to our Judeo-Christian, Euro-centric value system. Ford (1973) has noted that the development of prehistory coincides with the development of nationalism. Leone points out that nations would only spend large sums on archaeology for compelling reasons—suggesting that this important purpose is to obtain "an empirical substantiation of national mythology" (Leone 1973:129). Glock says:

> There is a growing awareness in American archaeology that bias affects all stages of inquiry. Bias is not only in the data but also in the people who collect and interpret yet, at the level of cultural and historical interpretation, few have the intellectual self-awareness that allows for procedures to balance bias (Glock 1985:469).

The possible impact of deep-seated legacies of our European intellectual heritage upon our scientific activities immediately raises the spectre of objectivity in science, which we discuss elsewhere. Our purpose here is to look at a range of sociological, psychological, and other contextual factors that affect archaeological research.

Within the areas of the psychology and sociology of science, some tantalizing studies suggest that different disciplines attract a different range of personality types and, at least in fairly stable times, they may attract individuals from a particular range of socioeconomic backgrounds. If this is true, as seems reasonable, the possibility immediately arises that there is a feedback reaction to the disciplines arising from the nonrandom range of individuals recruited that could be reflected in problem orientation or preferences for mathematical and technical aids. We have no such empirical information from North American archaeology, but it is worth looking at two of the sociological/psychological studies to see what an empirically based sociology of archaeology might encompass.

Kenneth Hardy's (1974) study deals with the backgrounds within the United States population at large from which academics tend to be drawn. He found that "scholarly doctorates come disproportionately from religious groups having certain beliefs and values" (Ibid.: 497), listing liberal and secularized Protestants and Jews as "Highly Productive," moderately liberal and dissident antitraditional Protestants as "Productive," traditional Protestants as having "Fair Productivity," Fundamentalist Protestants as showing "Low Productivity," and Roman Catholics as exhibiting "Very Low Productivity." His findings suggest that, in generic terms, academics are far from being a microcosm of the general population, and he identifies some first-level selection principles.

Anne Roe (1953) undertook "A Psychological Study of Eminent Psychologists and Anthropologists, and a Comparison with Biological and Physical Scientists," in which she investigated "the existence of relationships between life histories, intellectual functions or personality characteristics, and the selection and pursuit of a particular science as a profession" (Ibid.: 1). This was but one of many studies of this sort that Roe carried out in the 1940s and 1950s. Her samples were small (eight anthropologists and fourteen psychologists) because of the intensive and long-term nature of her research with her subjects. Among her many findings, the following are characteristic and worthy of brief mention. Anthropologists showed a higher incidence of feelings of personal and family superiority than any other group. Anthropologists characteristically disliked mathematics. Physical scientists and social scientists showed markedly different early interests. Social scientists were always more interested in and involved in social relationships than physical scientists, who tended to be socially isolated. Experimental physicists, particularly, seem to have formed early, direct relationships with objects rather than people. One might speculate that the high incidence of feelings of personal and family superiority noted among the anthropologists (one of whom was an archaeologist) was involved in developing the perception of cultural differences that seems to be fundamental to acquiring an anthropological view.

What can be said about the socio-economic background of North American archaeologists? Lacking empirical data, our comments must be based on informal observation of that part of the North American archaeological scene that has passed before us. From this limited perspective, it appears that archaeologists in the United States and Canada have until quite recently been drawn heavily from white Protestant backgrounds. Although anthropology in North America attracted liberal and secularized Jews throughout this century, and some of these like Franz Boas did some archaeology, it is our impression that liberal Jews only entered Americanist archaeology in the years after World War II. In general, we suspect that anthropology has consistently attracted more of the "highly productive" identified by Hardy (1974) than has archaeology.

In Roe's (1953) study, the eight anthropologists came from higher socio-economic backgrounds than did the other scientists, a factor that may have contributed to their feelings of personal and family superiority. A certain amount of archaeological folklore supports a view that archaeology is or has been an elitist occupation. A great many forces external to archaeology have acted to spread the socio-economic background from which archaeologists are recruited, however. Among these can be mentioned the depression of the 1930s with its enormous archaeological projects, World War II, the opportunities for higher education provided in the United States by the G.I. Bill, the later Civil Rights movement, and the growth of archaeology as a profession offering competitive advantages in employment. Although archaeology in the United States continues to be dominated by individuals from white Protestant backgrounds, the range of socio-economic origins has broadened so that there are now archaeologists from Wyoming ranches, West Virginia coal mines, Iowa farms, Detroit assembly lines, Newfoundland fishing villages, and the East Coast upper crust.

In contrast to the gradually diversifying recruitment base for archaeology, the recruitment base for anthropology has changed more rapidly. Since World War II, minority group members have been increasingly attracted to anthropology, although a headline over a report on the 1982 Ph.D. survey results in the May 1983 *Anthropology Newsletter* proclaimed "Anthropology Ph.D's still white, male, and employed in Academe." The *Report of the 1984 Survey of Anthropology Ph.D's* prepared by the American Anthropological Association shows that, of the respondents to the 1983–84 questionnaire who identified their ethnic group affiliation, ninety-one percent classed themselves as White, one percent as Black, two percent as Hispanic, five percent as Asian or Pacific Islander, and one percent as American Indian or Alaskan native.

It is hard to evaluate the effects of a broadening of recruiting on anthropological theory, but some consequences are quite apparent. Research areas of particular interest to minority groups have been defined (Black, Chicano, and Women's Studies, for example), and while none of these is unique to anthropology, they have affected the structure of course offerings and programs in universities as well as the problem orientation of anthropological research. When minority group members opted to study their own group, the relationship of the investigator to the people being studied was changed. A white Protestant Columbia graduate student from the Midwest studying Harlem Blacks would bring a different perspective to that study than would a Harlem Black who won a Harvard scholarship and returned to Harlem for dissertation work. This factor has sensitized anthropologists anew to problems of investigator bias. Other problems are discussed by Swallow (1974) and Hsu (1979).

Sociologically, hiring practices in anthropology have altered significantly. Minority and special-interest-group caucuses are regularly held at the American Anthropological Association meetings. The very existence of the AAA Ethics Committee is owing in no small measure to the decreasing homogeneity of anthropology and the resultant diversification of values and attitudes—a point stated cogently by Bernice Kaplan in her candidacy statement for election to the Ethics Committee: "The increasing size of the Association has led to greater heterogeneity of membership, a lack of a shared value system and a dispersal of the membership which vitiates social pressure as an enculturation mechanism" (Kaplan 1974: 29).

In contrast to anthropology in the United States and Canada, fewer native-born North American Blacks, Indians, Inuit, Chicanos, or Asians have entered the field of archaeology. To the extent that recruitment patterns are changing to incorporate a wider range of backgrounds, these changes are more evident in the United States than in Canada. It does seem that archaeology in North America offers little scope for acting out the concerns of the minority groups who have been most visible in academia since World War II, and this may be one factor in the differential recruitment patterns that have developed between archaeology and anthropology. It may also be that archaeology is simply twenty years behind anthropology in recruitment as in some other things. Some of the differences between archaeology and anthropology with regard to recruiting may be merely a matter of size of the two disciplines through the last forty years; but the rapid growth of archaeology has resulted in its becoming almost as large as its North American parent discipline, if relative membership sizes of the Society for American Archaeology and the American Anthropological Association are adequate guides.

It is not only the perspective of Blacks and Chicanos that has been missing from archaeological research. Some hint of what an archaeological world view might look like from an Inuit perspective was offered by the late Daniel Weeteluktuk, whose career was terminated in a tragic accident in the summer of 1982. The previous fall, he attended the annual conference of the Archaeology Association of the University of Calgary, where he presented a very emic view of archaeology in the north. Should archaeology begin regularly to recruit from Inuit and Native Americans, the "insider's" view of past cultures might become more evident. Alternatively, as ethnic groups become involved in identifying the significance of archaeological remains, as David E. Doyle (1982) has reported, another avenue of ethnic input into archaeological decision making is developing (*see also* Bielawski 1982).

Archaeologists seem generally to be less active with regard to professional affairs and contemporary social and political issues than their anthropological colleagues. Archaeologists do not identify with the same range of contemporary problems as deserving their action and concern as do other anthropologists, probably by virtue of the nature of their discipline, its subject matter, and the means of practicing it. If archaeologists are interested in activism, they can participate through their anthropological role. When issues concerning archaeology are at stake—the illicit traffic in antiquities, destruction of archaeological resources, or where legislation can effectively counteract unfortunate developments, archaeologists become very active indeed.

As a whole, archaeology in North America simply hasn't had its backup value system challenged and tested to anything like the extent evident in anthropology, perhaps partly because of the continuing homogeneity of the socio-economic backgrounds of people attracted to the field. The only groups in archaeology that parallel the visible sociological subgroups in anthropology are women's groups, and these appear to be somewhat less activist than their anthropological counterparts. Indeed, the Society for American Archaeology disbanded the Committee on the Status of Women in 1982, feeling that, within the Society itself, considerable progress had been made on issues of concern to women.

If one turns things around and looks at what attracts individuals to archaeology instead of considering where in society they come from, there are several factors to consider. Presumably the same range of factors Rowe (1965) identifies as necessary for the emergence of a modern discipline of anthropology must to some extent be duplicated in individuals who enter archaeology—that is, perception of cultural differences across time and space and a conviction that these are worth studying. Little seems to be known of how these attitudes develop, and yet this is the eye of the needle through which all potential anthropologists and archaeologists must pass.

Role expectations must play a part in recruitment and these, too, change over time. Concerns over professionalism in the sense of formal training of practitioners were very different in the 1930s. People who would be classed today as amateur or marginal were active, productive, and accepted as professionals then (cf. Davis 1979). The challenge of doing a great deal of fieldwork with scanty and often shaky support systems carried very different role expectations than do many contemporary archaeological activities. Economic constraints on acquiring a higher education were of a different magnitude in the 1930s than in the 1960s. Jane Kelley's father, W. C. Holden, ran archaeological field schools in the 1930s, accepting payments from students in the form of a $40 horse and other nonmonetary items. This barter level of acquiring field training is unlikely to be duplicated today. Before World War II, except for the Carnegie and similar situations, archaeology was likely to be regarded as exciting and interesting but not as a reasonable career choice if one was not independently affluent. At least through the late 1940s and early 1950s, a significant number of individuals entered archaeology without firm expectations of future employment. The perception of archaeologists of this older generation seems to be that there is a qualitative shift in expectations of students today, with greater emphasis being placed on employment opportunities in selecting archaeology as a career. Universities accept greater responsibility for placement of their graduates and consciously try to regulate the production of professional archaeologists in terms of the job market. Again, this may be largely a matter of the size of the discipline and pressures toward accountability; but whatever the reasons, the world of archaeology in the 1930s was qualitatively and quantitatively different—as was the society of which it was a part.

Views of archaeology that have long affected recruitment include what can be called the Boy Scout Syndrome, the image of the intrepid explorer of faraway places, and the feeling of the glamour of it all. The archaeologist in Roe's sample of anthropologists is quoted as saying:

> I suppose the aspect that appealed to me was that it combined both the rugged outdoor life, this side that I was always trying to be proficient in, with some intellectual content, and I couldn't just be a ditch digger but this was ditch digging with intellectual content and it was glamorous (Roe 1953:16).

The finding of untold riches of the past still shapes the public image of archaeology and supplies crucial role models affecting recruitment. Expeditionary archaeology of the sort described by Sir Leonard Woolley (1930) and Walter Taylor (1954) still exists. Dramatic finds such as the terra-cotta army of the First Emperor of Qin at Xian Yang, China, or the excavation of the Templo Mayor of Tenochtitlan (cf. Moctezuma 1980) receive wide publicity. In their concern with accountability to the public regarding the heavy public financing of archaeology, archaeologists actively perpetuate views of the field as exciting and dramatic. Such popular views of archaeology not only affect recruitment to the profession but also operate in the relationships between professionals, amateurs, and interested members of the public.

In an article entitled "The Archaeologist as Cowboy: The Consequences of Professional Stereotype," Woodall and Perricone (1981) discuss the legacy of earlier images of archaeologists in the contemporary context, concluding that:

> The ideological system created by archaeologists during and after the 1930s had adaptive value by attracting personnel to a low-paying and exacting profession. By endowing practitioners with the romantic macho complex a ready supply of recruits was insured, and thus the ideology spread rapidly through our ranks. Today, the posturing attending that outmoded ideology looks faintly ridiculous . . . (Woodall and Perricone 1981: 508).

The spectrum of role models, however, is multiplying and broadening. Nowhere is this more evident than in the additional dimensions offered by the advent of Cultural Resource Management, a topic to be discussed more fully later. Woodall and Perricone (Ibid.) offer evidence that "preservation archaeologists" regard themselves and are regarded by "archaeologists" as different (and vice versa), and that there is a relationship between these stereotypes and behavior. It is in this progressive differentiation of roles and role expectations rather than in major shifts in the socio-economic backgrounds from which archaeologists are recruited that archaeology is encountering the major loss of social homogeneity.

Training plays a significant role in producing individuals who reflect the views and attitudes of the archaeological sphere to which they belong. It has often been remarked, albeit only half seriously, that brilliant innovations are more likely to come from those who have not been subjected to the traditional training program of the field. In the Canadian and American context, it is possible to overemphasize the extent to which training forces researchers into a particular mold, this mold varying from place to place according to the theoretical or ideological views of the instructors. In the first place, if we take university training as our model, there are few North American departments that are as internally unified as the above picture suggests. It is probably more realistic to see most academic departments as operating within a fairly broad spectrum of contemporary archaeology, although this is affected by the age or youth of the faculty. Given university tenure and hiring systems and the effort that most departments of archaeology and anthropology make to maximize the breadth of expertise and diversity of their staffs, it is unlikely that any department would reflect a single approach. On the other hand, it is equally unlikely that students would be equally influenced by each member of a department; instead, it seems that they would tend to emulate the work of the better known or more dynamic individuals. If students read the journals and other literature, they have access to most of what is current in the field. Nevertheless, training is certainly essentially a conservative endeavor, and it is no less conservative when students absorb the newest of the new. That is, it is conservative in the sense of providing the students with a world view, a set of methods, a way of doing archaeology; and this inevitably raises the problem of balancing the values to be gained from practising routines against the potential productiveness of innovations of various kinds. In any case, an archaeologist's changing and adapting world view is presumably determined as much by what comes after his or her formal training as by the training itself.

Another issue currently facing North American archaeology is the question of the kind of training that is appropriate, given that most graduates of today will find employment outside of academia. Traditional academic training in the past has reflected the fact that most graduates would be employed in universities, museums, or closely related institutions. Just how the training of archaeologists should be altered to reflect this changed state of affairs, if at all, is beyond our scope here; but it is an issue of increasing importance in the field (Kelley 1979). A number of universities inaugurated "applied" M.A. programs during the late 1960s and early 1970s. Woodbury (1973b) reported such a program at the University of Arizona. In that case, some students opted to take degrees in the applied fields for a few years. In the tighter job market, students find that a broader background gives them better employment opportunities than a narrow specialization. Curriculum changes initiated in response to the shift in employment opportunities ended by enriching the mainstream programs. This evolution of specialized programs back into general ones seems not atypical of a number of universities, and it seems to have occurred in cultural anthropology programs as well as archaeological ones. Specialized or applied programs continue to do well at some universities such as the University of South Carolina, however.

To claim that most archaeologists are trained within a broad spectrum of contemporary archaeology with the variations of content and outlook remaining relatively minor is not, of course, to say that North American university departments of archaeology or anthropology are all on the same footing. On the contrary, significant variations in status occur, as Hurlbert's (1976) study of Ph.D. granting departments of anthropology in the United States and Canada shows. Applying exchange theory to data given in the 1973–1974 *Guide to Departments of Anthropology* and comparing her results with those of other relevant studies, she looks at the social structure of the discipline. Her findings suggest that there are elite and nonelite circles within North American anthropology. Elite departments tend to be older departments, to have produced the greatest number of Ph.D.'s, and to be highly ingrown. Elite departments often supply teaching faculty to non-elite ones, while the reverse is rarely seen. Of eighty Ph.D-granting universities listed, fourteen are classed as elite: Harvard, Chicago, Berkeley, Yale, Columbia, Cornell, Michigan, Pennsylvania, U.C.L.A., Indiana, Northwestern, Wisconsin-Madison, Stanford, and Arizona. Harvard draws 96% of its faculty from elite schools; 69.6% of its graduates teach in Ph.D. granting departments; and 34.6% of its faculty are Harvard products. Harvard is also the oldest department and has produced the most graduates. In contrast, the 1973–1974 guide showed that Southern Methodist University, with its six-year-old program, derived 85.7% of its faculty from elite schools but had none of its Ph.D.'s placed in elite schools.

In the past, especially in the expansionist 1960s, the mobility of professional archaeologists was fairly high, particularly at lower ranks, counteracting the tendency for internal stagnation to develop within particular departments. People came from different departments, many were fresh from training in graduate school, and they provided fresh perspectives. Now, however, mobility is greatly reduced, and for the first time, we face a major break in generational representation within departments. People who have jobs remain where they are. The average age of the faculty rises each year. Young people are not entering university employment at the lower ranks at anything like the rate of ten years ago. This potentially introduces a new locus for intellectual lag to occur, and if the trend continues (as it shows every sign of doing), more stable academic departments may very well become more insulated departments. This problem is presently common to most academic fields.

Empirical information about certain sociological aspects of anthropology and archaeology is found in reports emanating from the American Anthropological Association. In particular, reports on the status of women are of interest as we look at the context in which archaeology is practiced. Vance (1975) summarizes data taken from the *Guide to Departments of Anthropology 1974–75*. Of 3,279 academic anthropologists listed in the *Guide* for that year, twenty-two percent are women. Women cluster disproportionately in the lower ranks. While the number of women holding professorial rank increased over previous years in 1974–1975, the percentage of female professors remained roughly the same at thirteen percent, while thirty-two percent of the men in the sample held the rank of professor (Vance 1975:11). Given the Hurlbert study discussed above, it is notable that Vance finds that women receiving Ph.D.'s from fourteen "selected schools" (a list that largely overlaps but does not duplicate the fourteen "elite" departments in Hurlbert's study) achieve greater success than other women—but the edge is not great.

A later study by Forman (1978) is based on the 1977–1978 *Guide*, which lists 3,974 people of whom 24.8 are females. It was found in this study that women are underrepresented in the fourteen major departments, underrepresented at higher ranks, and do not advance through academic ranks as rapidly as their male colleagues (*see also* Forman 1977). Another study (Sanjek 1978) looks specifically at the position of women in twenty-two major departments (including Hurlbert's fourteen elite) for the period 1967–1976. Sanjek concludes that "the experience of women in the major departments of anthropology . . . has been one of less opportunity for employment and for achievement of senior rank than in all departments [in the guide] taken together" (Sanjek 1978:903). Women "disappeared" from the listings of major departments at about twice the rate of men over the nine-year study period; for those who could be identified as changing jobs to another department listed in the guide, men were more likely than women to move to another major department. Sanjek saw improvement in hiring practices during the study period as being due to affirmative action guidelines, the 1972 American Anthropological Association resolution on the status of women within the profession, and studies such as that by Vance referred to above. This is an example of the profession's identifying a sociological problem and taking pragmatic steps to deal with it.

One topic which a number of the reports on the status of women in anthropology have addressed is the representation of women within the several subdisciplines. Vance (1975) finds them to be "extremely underrepresented in archaeology." Forman (1978), using figures from the 1977–1978 *Guide*, reports that 26.29% of cultural/social anthropologists are female, 27.48% of linguists are female, but only 14.01% of archaeologists listed in the *Guide* are female. Such figures, which reflect academic employment, are at odds with the percentage of males and females doing dissertation work in archaeology according to the *Report of the 1984 Survey of*

Anthropology Ph.D.'s, in which 38.5% of the 1983–1984 archaeology specialists were women; for the 1981–1982 survey (as reported in 1982), males and females were evenly matched. These percentages suggest that more women specialize in archaeology in graduate school than are evident in employment figures.

A similar situation is seen in Yellen's 1983 study. Yellen, reporting on an analysis of grants submitted to the National Science Foundation over the last few years, found that the number and nature of dissertation research proposals was about the same for males and females. "Senior" grant proposals told a different story. A much lower percentage were from women; their success rate was lower; women tended to request smaller grants; and most of the proposals submitted by women involved analytical rather than fieldwork. As he points out, it is unclear how faithfully the National Science Foundation's granting picture reflects archaeology as a whole. Whether the selection process operating against women in applying for and receiving grants is temporary (i.e., the changing position of women in society hasn't yet had time to be reflected in senior academic standings) or whether there are other factors that will perpetuate this kind of statistic is not yet apparent.

Results of a preliminary survey questionnaire sent to women in archaeology are reported by Wildesen (1980). The overall picture emerging from the 107 responses (a 39 percent return rate) is not dissimilar from those of the studies cited above. Because the questionnaire collected qualitative as well as quantitative information, Wildesen is able to report that the women perceive themselves as less well off than their male counterparts in terms of salaries and prestige, especially in academia. The U.S. government hires more women than do other employers. In the U.S. Forest Service, twenty-four of eighty-two full-time archaeologists are women (twenty-nine percent) and, of the higher administrative positions, "3 of the 9 Regional Archaeologists, and that of the Chief Archaeologist, are held by women" (Wildesen 1980:8). There are several other studies (*see* Gero 1985; Conkey 1978, 1982; and Conkey and Spector 1984) that reflect that the differential representation of women in academia and in research is deeply rooted.

Gero (1983) argues that the wave of the future lies in careful analysis of existing collections from a problem-oriented perspective rather than in fieldwork. Women, she feels, already have a corner on the analytical market, and therefore have the potential for making the most significant contributions to archaeological theory and substantive explanations in the future—an optimistic interpretation of the picture portrayed by the statistics. She also notes that among South Americans, class outweighs sex as a determinant of achieving status as a professional archaeologist. All of the studies on the status of women in anthropology and archaeology lend support to the observation that the population of archaeologists is not a microcosm of the population at large in any country, and, further, that the roster of practicing professionals is not a faithful reflection of those entering archaeology at the graduate level.

Wobst and Keene (1983) express a view of archaeology that comes much closer to extreme sociological and even political views of science than those we have examined to this point. In a paper entitled "Archaeological Explanation as Political Economy," they argue that archaeologists, "like other normal folk," must "fit themselves into the structure of rewards which society provides for people and scientists" (1983:81). Archaeology has differentiation of social or niche space. Among the variables operating in social differentiation are: "(a) ease of access to financial and personnel resources; (b) degree of external control over one's position and one's resources; (c) ability to reproduce oneself professionally; (d) ability to introduce constraints upon the behavior of other practitioners; and (e) ability to maintain control over personnel, space, time, and resources" (1983:81). They illustrate niche adaptation on the part of archaeologists using the examples of origins research and the construction of regional typologies. In the former, the individual or individuals who control the perceived point of origin or are closest to it (whether the origin in question is agriculture, civilization, or early man) will enjoy greater independence from the behavior of other archaeologists, will influence other archaeologists, and will have a higher citation count. Similarly, what is accomplished diachronically by origins research is accomplished in political economic terms synchronically through spatial control of regional typologies. "This allows the owners of the richer data base to suck up formal variation from elsewhere, as part of the range of variation around their *own* base. In this way, they can co-opt the typological behavior of the greatest number of adjacent archaeological practicioners" (1983:84). They conclude that "certain methodological and theoretical approaches to archaeological data are popular, *not* because they help us in understanding change, but because they help to establish positions of high visibility, power, status, and social control within the discipline" (1983:86). These two forms of archaeological practices and behavior have survived in spite of changes in the major theoretical shifts in the discipline which might have been expected to render them obsolete.

If one accepts that the range of factors discussed above affects North American archaeology in real if (usually) only dimly perceived ways, it seems reasonable to systematically investigate such aspects in order to clarify their role. We need to be able to evaluate in realistic terms the claims that archaeology is dedicated to the maintenance of existing class structures, is the handmaiden of colonialism, or is only a Euro-centric way of seeing. Such claims have a great deal of historical support. The virtue of combining a scientific approach with sociological awareness is that these issues can, to an appreciable extent, be sorted out.

To this point our discussion has focused on archaeology in the United States and Canada, but this perspective needs to be broadened. Even without undertaking a socio-historical analysis of archaeology in Mexico (*see* Bernal 1980), Peru, England, France, or other countries, we can note that there are significant differences with respect to the factors discussed above. Archaeologists in Catholic or Islamic countries are obviously unlikely to be recruited from predominantly Protestant backgrounds, and social class is more important than sex with regard to professional roles in some countries (Gero 1983).

As the political makeup of the world alters, as more emerging nations take control of their own education systems and cultural resources, the pattern of archaeology on a global scale is perceptibly altered (cf. Miller 1980; Kense 1982; Daniel, ed., 1981). The socio-economic and national background of archaeologists is expanding rapidly—or it was until recently. Instead of a strong pattern of colonial or quasi-colonial archaeology in other parts of the world, the trend is toward more "in-house" archaeology. A great deal of Mexican archaeology has long been done by Mexicans, but even less is being done by outsiders now than formerly. In contrast, the Sudan has only recently begun to train its own archaeologists, who will assume progressively greater control over Sudanese archaeology in the future. Jerome Jacobson (1979) offers insights into recent developments in South Asian prehistory and protohistory, finding that the proportion of foreign archaeologists is very low in India, somewhat higher in Pakistan, and higher still (at the time he wrote) in Afghanistan. Sankalia, "the most productive prehistorian ever to have worked in South Asia" (Jacobson 1979: 469), is said to employ an emic approach in his studies. Jacobson feels:

> Indian archaeology today is characterized by scientific methods without *the scientific method*, descriptions of environments and traits without systems theory, and reconstructions of culture history without regard to culture process. These features . . . give the observer an impression not of Europe in 1860 but of United States archaeology of 1960 (Jacobson 1979:468).

A common pattern in emerging nations is to send the first generations of archaeologists abroad for training. The old colonial ties often determine where these archaeologists are trained, so that Sudanese archaeologists frequently go to England or to other Commonwealth countries such as Canada; but some go to Russia, and this means that alternative world views are built into the emerging loci of archaeology activity. By way of illustration, a recent graduate of the M.A. program at the University of Calgary also holds a degree from Leningrad. In the Sudan, he held the position of Senior Inspector of the Nile Province for the Department of Antiquities.

New national centers of archaeology respond to their own local cultures, and training and resource allocation are directly affected by economic and political factors. This may not be inherently different from the situation in North America, but because the forces shaping archaeology in these new contexts are different from those in North America or Europe, they are easier for a North American or European audience to perceive. Many of these countries are consciously rejecting colonial trappings and seeking to establish a distinctly noncolonial identity. Archaeology not infrequently assumes an additional role (absent for United States and Canadian archaeologists) in establishing links between a precolonial past and the modern national identity. This in turn affects the selection of research topics. West African archaeologists from Ghana or other countries are now tending to react against diffusionist ideas that attribute West African developments to strong influences from Egypt. Such ideas are increasingly rejected by other, foreign archaeologists working within the area as well. In this context, however, it seems that West Africans are rejecting the colonial flavor of those earlier views as well as the diffusionist principles they embody. Jacobson (1979) notes that diffusionist principles are beginning to be rejected in South Asian archaeology. For the reverse side of the coin, Trigger (1980) has argued that early views of American Indians as savages placed American archaeology in a Eurocentric mode of thought in which Indian remains were regarded more as objects than as subjects (*see also* Blakey 1983).

Prior to World War II there was a fairly patterned distribution of where European and North American archaeologists worked around the world (e.g., the French worked in Indochina, the English in India). The basic pattern of which nationalities do research elsewhere has not changed drastically, although the French, for example, are now active in the Indus Valley. The country whose archaeologists have most noticeably moved into new geographical areas is Japan, whose involvement in Peru since the late 1950s is an example of their expansion into new fields.

For several decades, European-derived archaeologists tended to regard the world as their laboratory. Although the attitude persists, the reality of gaining access to the archaeological resources of the world is forcing retrenchment with regard to where archaeologists from certain countries actually can work. It is more difficult to get permission to work in many nations than it was thirty years ago. When permits are obtained, the host country now usually imposes various conditions. The Mexican government levies a fifteen percent surcharge on funds expended in that country for foreign-controlled archaeological research. Since Canadian and American granting agencies do not customarily allow for this amount in the calculation of grants, the surcharge initially curtailed the involvement of foreign archaeologists from those countries. Now, however, the National Science Foundation in the United States allows the fifteen percent to be incorporated into the grant with the proviso that it be expended for facilities or services related to the funded research. Colombia also imposed financial constraints that affected the participation of foreign scholars in projects in that country for several years. Other changes facing today's researcher are constraints on export of archaeological material for either short- or long-term purposes. Political conditions and armed conflict in certain parts of the world are also causing many archaeologists to look for more compatible places to work, and, at least temporarily, effective access is lost to parts of Asia, Africa, and Latin America.

Although archaeology can be seen as derived from the European intellectual tradition, world political conditions have fostered the development of several archaeological centers of varying degrees of independence from European and American areas for several decades. For instance, Chinese archaeology, in the main, has developed in isolation since the 1930s. Only in the last few years have North American archaeologists begun to visit China. The groundwork for actual collaboration with Chinese scholars, both in China and in North America, has been laid now, and this should result in fuller intellectual exchanges in the future. A number of archaeologists from both the United States and Canada have visited the Soviet Union, and North American students have studied there, but it is difficult to think of any North American archaeologists who have seriously carried out major research there or of any Soviet scholars who have seriously worked in North America. When Okvladnikov collaborated with Laughlin in the Aleutians, it was a major event (Laughlin 1975). Clearly, differential interaction between archaeologists on a global scale affects access to primary data.

In considering decreasing access to global archaeological resources, one needn't go so far afield. Canadian Inuit and Indian Councils have more than once refused permission to non-native archaeologists to work on their land. These refusals have been of a temporary nature, and usually when communication among the parties has been fully established, mutually satisfactory arrangements have been made. The British Columbia government has successfully formalized relationships between archaeologists and Indian bands of that province by developing workable procedures for their interaction. Events of the past decade have nonetheless demonstrated that archaeological research is constrained by various factors at home as well as abroad. One need only think of the issue of reburial of prehistoric human skeletons in Iowa or California to realize how anthropological and archaeological research can be restructured by forces outside the discipline. This serves to underlie the changing relationship of archaeologists to their data base.

On a global scale, we see several major geographical centers of archaeological research tending to be self-contained although it is difficult to quantify their degrees of isolation. Soviet publications are more often translated for North American audiences than are Chinese reports, and, until now, there has been more interaction between Russians and North Americans than between Chinese and North Americans. Siberian archaeology is routinely scrutinized by both North American Arctic and Early Man specialists for culture history purposes. Semenov's (1964) book on functional analysis of lithics is a classic in the west as well as in the Soviet Union, and several western scholars have personally studied with him. Knorosov (1963) made significant leaps forward in deciphering Maya hieroglyphs, and other examples of Soviet archaeological contributions of great interest to archaeologists in other parts of the world could be listed. Still, archaeology as taught and practiced in the Soviet Union constitutes a different intellectual sphere from that in North America, with some profoundly different views (cf. Klejn 1977). Speaking of the Soviet Union, Soffer notes that during the late 1970s "one still had to prove that the nuclear family both existed and was a minimal social unit during the Paleolithic . . . Older evolutionary views still prevailed and are enshrined in all the textbooks" (1983:97). Recent interest in typology and the study of material remains replaces the previous view in which such pursuits were condemned as "thingization" (ibid). Soffer reports her own "culture shock" when she realized that "no one at present was working on questions of paleoeconomy" (1983:98).

The spectrum of archaeological beliefs, attitudes, methodologies, and problem orientations will undoubtedly be broadened by the continuing development of archaeology in new contexts on the global scale (cf. Glock 1985). To return to our spherical model, the global sphere will be larger than the North American sphere, with multiple and overlapping "cores." Newer emerging loci of archaeological work will not be faithful reproductions of the more established centers. Paradoxically, there will be greater attendant opportunities for increased degrees of isolation.

Throughout the 1960s and 1970s, North American archaeology became relatively aflluent. Not only were the employment opportunities greater and more competitive with other professions than previously, but the level of logistic support for research reached significant new plateaus. Many of the problems now occupying North American archaeologists depend on this support base in order to acquire or analyze certain classes of data. It is fairly routine for North Americans to have the financial resources to afford Carbon 14 assays, for example. This level of support is simply not duplicated in many nations. When a Sudanese student undertook his research in the Sudan for his Cambridge Ph.D. dissertation, he went from department to department at the University of Khartoum to collect his field and subsistence supplies. One of his cherished acquisitions was a jar of strawberry jam. His support base for a major piece of research was much closer to 1930s standards than to those of most projects operating today in or from North American bases, including foreign-run projects in the Sudan. He was able to do a great deal with very little.

In general, the different levels of support in various archaeological centers will inevitably have an impact on the kind of data that can be collected, the kinds of analysis that can be undertaken, and, consequently, the kinds of problems that can be addressed. The effects of a worldwide recession are already evident on some of the newer loci of archaeological activity. Both Kense (1982) and Tamplin (1981) have commented on the reduction of fledgling archaeological programs in Africa.

No doubt many additional cultural factors could be mentioned that influence the theory and practice of archaeology in many ways, but we believe we have made it clear that these influences are real and need to be taken into account. On the whole, archaeologists have paid less attention to these variables than have anthropologists (especially ethnographers), probably in part because the latter, dealing as they do with living human beings who respond to the researchers, have been forced to a realization that the cultural bias of investigators is a serious problem. The view gaining in popularity, however,—that archaeology is making sense of the past *in the present*—is moving us toward a greater awareness of the effects of our own cultural backgrounds on archaeological questions. Thus Mark Leone, in an article on the concept of time, opens with this statement:

> Archaeology exists within our society's concepts of time and space, and all our findings take on meaning as a function of these vectors, not independently of them. We rarely examine how past societies conceived these dimensions, and we never ask how our own definitions of time and space influence our interpretations of other societies (Leone 1978:25).

If we do not yet seriously assess such factors, perhaps we should—they have very real consequences with respect to the archaeological problems that are chosen for investigation, the world view presupposed, the archaeological methods selected, and the interpretation of the results achieved.

Archaeology's institutional affiliations

Analogous to the decentralization of archaeology on the world scale is that occurring in the United States and Canada with the massive increase in nonuniversity-based archaeological activity. Museums and universities have been the two traditional institutional bases for archaeology, and in the twentieth century, the university grew to be the more important of the two. Archaeologists are trained at universities, most jobs in archaeology have been university related, and until recently most archaeological research was conducted from a university base. Many archaeologists now pursue career patterns combining administrative and managerial skills with archaeology, however, in contrast to the previously more familiar pattern of combining teaching or museum work with archaeology.

Legislation in Canada and the United States has resulted in a lessening of the universities' domination of archaeology in these countries. The new statutory provisions recognize the value of archaeological resources as an important and nonreplenishable part of the cultural heritage and require the acquisition, study, and management of these resources by designates of the Canadian and American governments (federal, state, or provincial).

A variety of governmental agencies in both countries now either have archaeologists on staff or grant contracts to archaeologists, many of whom do not have any university affiliations. While most university departments have been in a period of slower or no growth, employment opportunities for archaeologists increased rapidly outside of academia until approximately 1980 or 1981, when the employment picture responded to the economic recession; and more companies are (or until recently were) being incorporated for the purposes of handling contracted archaeological research. Contract and agency-related research presently constitute the largest portion of the archaeological research carried out in the United States and Canada—and this research grew rapidly until the recession.

Such a major and rapid shift in organizational structure inevitably initiated a series of problems and adjustments at every level of the profession. The problems associated with a shrinking profession will be equally severe. Perhaps the most immediate reaction to rapid growth was widespread concern for the maintenance of professional standards, a concern largely responsible for the creation of the Society of Professional Archaeology in the United States. Other concerns are reflected in the monograph *Digging for Gold* (McDonald, ed. 1976) and in the symposium held at the Canadian Archaeological Association meetings in 1979 on the new professional training needs. Many of the papers in another symposium, "Directions in Archaeology: A Question of Goals" (Francis and Poplin, eds. 1982), also address these questions.

The concern over the maintenance of standards which was expressed at the beginning of the Cultural Resource Management (CRM) expansion was more or less intuitive. Now, more than a decade later, the concerns and debates are rooted in experience. There exists a widespread folklore about the disproportionate number of poor quality and largely descriptive reports being turned out in the CRM framework (cf. Longacre 1981). A long-term critic of CRM, Thomas F. King, is quite pessimistic about the state of CRM science, noting that it has "most often fulfilled the ugly destiny I prophesied for it in 1975" in the area of "little jobs" such as well-pad surveys (1984:25). He further feels CRM has attracted less well-qualified people or, when well-trained scholars have entered CRM work, they are often well trained in areas of marginal significance to conservation-related work, and they are often overwhelmed by "real-world conflicts" (Ibid.:26).

Dena Dincauze (1984) fixes an anthropological eye upon the "present discomforts" surrounding conservation archaeology:

My thesis is that archaeology and archaeologists have been, in a manner of speaking, colonized. That is, they have been co-opted into the capitalist political economy that surrounds them, and that co-option has been accomplished in ways that are strikingly like the colonization of northeastern North American native peoples a few centuries ago (Dincauze 1984:49).

Before 1970, archaeologists "produced goods and meanings intended for consumption within the community," rather like subsistence economies of native Northeasterners (Ibid.). Changes in archaeology since 1970s have resulted in archaeologists' becoming dependent on sources of power:

> archaeologists gave away what control we have over our values, our work lives, and our resources when we made a fateful bargain for the uses of external sources of power and money. No matter that we had little choice, if *anything* was to be saved. What matters is that we did not become fully equal partners in the bargain. We became dependents, dependent upon the bureaucrats at the interface (Ibid.:51).

Some responses to the perceived problems are likened to revitalization movements and cargo cults. The strategy that Dincauze and others (cf. Wildesen 1984) feel *"can* make a difference" is "effective, informed political action, directed over the heads of the bureaucrats to the lawmakers, who are, not incidentally, our 'public servants'" (Dincauze 1984:51).

Empirical studies are beginning to appear that document the nature of problems in conservation archaeology and suggest socio-economic-political reasons for these problems. Paynter (1983:25) looks at the production process in archaeology, noting some disquieting tendencies for archaeological labor to become deskilled, for management and scientific decisions to be made by the higher authorities within a firm or organization most distant from the "dirt and its contents". Zubrow (1984) reviews nearly 200 samples of contract work in New York and Colorado, finding, for example, that only twenty-three percent of the Colorado reports and forty-five percent of those from New York report artifacts found.

In a paper entitled "The Development of Least Effort Strategies in CRM: Competition for Scarce Resources in Massachusetts," Lacy and Hasenstab (1983) review more than ninety percent of the

1970–1979 survey reports submitted to the Massachusetts Historical Commission, evaluating and scoring each report on 140 variables in 5 categories: background research, technical documentation, assessment of project impacts, evaluation of resource significance, and recommendations for mitigation (1983:32). They specifically note that neither field methods nor research orientation were evaluated. We will not review their analysis in any detail here, but we draw attention to their claim to have identified a trend toward "standardized mediocrity"—a claim derived from the scores of each report reviewed and their graphing of the results in various formats. Four main points emerge from their analysis:

> (1) the bureaucratic process favors and rewards predictable, conservative formats and results which best serve client goals, i.e., non-archaeological or archaeologically neutral priorities;
> (2) the maturation of the contracting system has brought about the entrenchment of minimal standards for adequate performance;
> (3) the observed standardization was facilitated and compounded by a competitive bidding process which will tend to produce consistently minimal results;
> (4) an increasing number of people committed to contract-employment as their primary source of income, at a time when contracts were decreasing in number, inevitably led to stiffer competition for increasingly scarce resources in the form of contracts.
>
> Given these conditions, it is not surprising that the variability in report scores decreased to the point where information exchange reinforced or maintained an across-the-board mediocrity, paralleling minimal adequacy requirements, rather than encouraging improvement through innovation, scientific inquiry, or the pursuit of a more anthropological archaeology (Lacy and Hasenstab 1983:46).

Lacy and Hasenstab are entirely sympathetic toward the view expressed by other authors (Dincauze and Roberts 1978: Keene and MacDonald 1980; Lipe 1978) that there is a recognizable dichotomy between bureaucratic client needs and archaeological research goals (1983:32).

A chronic problem targeted by many commentators on the contemporary archaeological scene has to do with the enormous submerged literature spawned by CRM. Like an iceberg, nine-tenths of it is hidden from view. Indeed, some feel that this situation (with absence of peer review, grave difficulties in data retrieval, and subversion of the fundamental goal of the production of knowledge) constitutes one of the most serious problems facing archaeology today. In the first Plenary Address delivered in 1982 to the Society for American Archaeology, and published in 1983, Colin Renfrew addresses this problem insightfully, concluding:

> If the major outstanding defect in public archaeology today is this lack of an automatic commitment to good publication, it springs, I would argue, from the corresponding defect in the New Archaeology. The New Archaeology has never formulated a coherent policy for dealing with great masses of data: Information processing and data retrieval are not its highest priorities (Renfrew 1983:11).

Lipe, in discussing this "gray literature," argues that: "If this informtion is not accessible even to reasonably determined scholars, it might as well not have been obtained. Explicit recognition that information value provides the basic justification for CRM programs might lead to greater efforts by both agencies and research groups to see that reports are published" (Lipe 1984:6). He feels, however, that the "extreme variability of information return among CRM projects at any given level of scope and funding" is an even more serious issue (Ibid.:6).

A related concern focuses on curation of collections generated by CRM programs. As Mark Raab comments, "Many regard this topic as a ticking bomb" (1984:17). John Speth is concerned about:

> the thousands of collections generated over the last decade by CRM work. Where are all of these materials today? How accessible are they for research? Are they properly catalogued and curated? We have developed guidelines and laws to protect, preserve, and when necessary salvage archaeological sites, but we have paid woefully little attention to the equally vital issues of publication and the future accessibility and integrity of collections once they have been removed from the ground (Speth 1983:649).

It is not, of course, that CRM archaeology cannot be as problem oriented, thorough, rigorous, explanatory, or inferential as any other kind of archaeological endeavor (cf. Schiffer and House 1977; Hester 1981; McGimsey and Davis 1977; McGimsey 1981). The time constraints, the vested interest of the people in bureaucratic settings, the competition for contracts, the profit-making syndrome, and other factors seem inevitably to set up a different set of tensions and demands that must affect the archaeology done in that context.

Scale undoubtedly affects how problem-oriented and problem-solving many CRM projects can be. In one extreme, Hester (1981: 495) quite rightly notes that "a 1,000-foot sewage line" does not provide "the forum for explicit method and theory posturing." (Hester's article cited here is a response to Longacre's [1981] critical evaluation of CRM publications cited earlier.) On the other hand, one might hope that even 1,000-foot sewage lines and all other small projects within a region or locality might collectively provide information relevant to both method and theory. On the plus side, as most critics acknowledge, large-scale projects are usually much more successful in terms of employing viable research designs (cf. King 1984). The large projects are on a scale unlikely to be undertaken in more traditional research contexts. For example, extremely useful and comprehensive data bases have been created in the overviews (cf. Cordell 1979; Tainter and Gillio 1980) which also have significant components of method and theory. The state of archaeological knowledge has been significantly altered as a result of massive CRM programs such as the American Bottoms Project, and other examples could be given.

The changed affiliation of archaeological research in the United States and Canada has structural implications of enormous importance for future developments in the discipline. The decentralization is somewhat analogous to scientific specialization: some of the same sociological tendencies toward segmentation operate. Very real structural reasons underlie the trend of agency-related archaeology to move away from the academic. Interaction and communication patterns shift, and these changes carry with them certain problems. The large and growing literature consisting of reports filed in contract firm and agency offices, receiving less than full peer review, is worrisome. An even greater problem is the incipient attitude that contract archaeology might become a different kind of archaeology—a "service industry." To allow a schism to develop along these lines would not be in the best interests of the discipline, for not only would the intent behind the enabling legislation be subverted, but the rest of archaeology would be cut off from much of its data base and/or the new context would produce a data base that becomes irrelevant to the questions raised by the discipline. It would appear that the method and theory of archaeology in at least the United States and Canada needs to operate in the new context, and this new context with its conservation and management mandates is not just academic archaeology transplanted—it has new dimensions. In sociological terms, this is the New Archaeology of the late 1970s and the 1980s (cf. Wildesen 1984). Our concern for its direction is not idiosyncratic. At the Society for American Archaeology meetings in 1979, Fred Wendorf (1979: 641–643) directed his presidential remarks to his concern that archaeology was in danger of becoming a service industry "like beauticians and plumbers." If a president of the Society publicly expresses such sentiments, the concern is probably quite widespread.

If we espoused a strictly Kuhnian view of the sociology of science, we might see progressive segmentation along these structural lines as inevitable. But American anthropologists have demonstrated their awareness of the effects of the organizational structure in their own professional association upon the nature of that association, and, as a result, have carried out two structural reorganizations aimed at maintaining the AAA as the umbrella society for all the subdisciplines. Although some structural factors are moving archaeology toward segmentation, these tendencies can be consciously counteracted once the fundamental importance of the factors is recognized. As archaeologists understand the problems, "the current muddling through can be replaced with a sane plan for perpetuating the profession" (Rogge 1976: 839). It is only reasonable to expect more change in at least the near future, and probably over a much longer time, as archaeology comes to grips with the many and diverse organizational issues it faces.

Up to this point we have discussed the CRM developments of the past decade as a fairly stable state of affairs. But, as Don Fowler reminds us, "the whole federal preservation system is fragile— somewhat akin to a house of cards built on sand—to mangle a metaphor" (1984:47). Continuation of conservation archaeology depends on: "The supporting legal structure remaining *workably* intact . . . It's being eroded—exponentially. We can either watch it go—or we can do as we did 10 years ago, in standard American fashion: go do something about it" (Ibid.:48).

The technology of archaeology

Although it may seem strange to discuss the technology of archaeology in a chapter devoted to the social context of the discipline—the technology is, after all, an integral part of scientific methodology, there are social and political aspects relating to the acquisition and utilization of technology, and it is primarily these aspects we address here. The introduction of new techniques and technologies to archaeology has itself been responsible for radical changes in the field. Thirty to forty years ago Carbon-14 dating, statistical sampling procedures, pollen analysis, X-ray diffraction, and neutron activation, to give just a few examples, were unfamiliar to archaeologists. Now, these are well-known throughout the discipline and their use has significantly altered not only archaeologists' views of specific issues, such as the temporal scale of human history, but those technological factors have affected the selection of research problems and even archaeologists' perception of what constitutes data. Notwithstanding these significant developments, archaeologists have not yet even begun to tap the potential of the techniques and technologies recently introduced to their discipline. This underusage has been attributed to the organizational structure of the discipline.

When Struever (1968) advocated a reorganization of the discipline, he did not refer to the kind of reorganization brought about by the introduction of CRM as discussed in the previous section. Rather, he argued that the traditional university departmental structure does not adequately serve contemporary research needs of long-term, multi-disciplinary, multi-investigator projects, nor does it provide adequate access to specialized technical facilities and expertise. In some views of science, increased technological sophistication should virtually always be advantageous, and if such views are accepted, it would be reasonable for archaeology to strengthen its technological repertoire. This would mean moving the available technical aids from the ideal and esoteric into routine usage, though many problems would occur in achieving this transition.

Many sciences (chemistry and physics, for example) bring technology into their disciplines and their laboratories in a way that archaeology has not yet accomplished. Nor has archaeology developed a strong coterie of archaeological technicians. No doubt archaeology could profit enormously from establishing more internal control over its technology, but even if technicians were trained, there would remain the problem of placing the necessary specialists within the existing organizational structure and making their services and the necessary facilities accessible as needed.

Aside from sending samples off to be analyzed for, say, Carbon 14 or Neutron Activation, archaeologists often rely on the good will of colleagues in other disciplines for technical assistance. As useful as this is, it has limits. Trying to explain archaeological problems to nonarchaeologists is sometimes as difficult as understanding the results. Here, as in all borrowing, a failure to understand the process and its limitations carries unfortunate consequences.

Exploration for potentially useful technical aids is quite unsystematic in archaeology, proceeding as it does largely within the framework of personal interaction. As a result, archaeology faces not only the problem of improved access to techniques known to be suitable but also the additional problem of effectively monitoring techniques across the worlds of science and technology for other aids that will be both useful and within the limits of archaeological resources to implement.

The monitoring process between science and technology has been discussed by Price and Bass (1969), who find that historical and sociological studies indicate that the worlds of science and technology are relatively independent of each other. Being interested in the dynamic interface, they describe different relationships between the two. Since most technical innovations in archaeology do not come directly from the world of technology but from other sciences, archaeology tends to exhibit what those authors designate as an "indirect coupling" relationship to technology in which there is a lack of direct dialogue between originators and users. Such a relationship contrasts with direct participation and a two-way partnership. Archaeology also lacks the "gatekeeper" system described by Price and Bass in which individuals are assigned monitoring responsibility (Ibid.: 806).

Interestingly enough, the impetus of New Archaeology to make archaeology more scientific tended to emphasize epistemological matters; the role of technology in scientific methodology was played down. Other archaeologists, perhaps more numerous in Europe, South Africa, and Australia, tend to see future breakthroughs in understanding and the acquisition of knowledge as being much more directly tied to technical advances and the internalization of technology within the discipline. Some archaeologists, such as Nikolaas von der Merve of Capetown University, are deeply involved in the physics and chemistry of certain analytical procedures, preferring to do their own analytical work rather than using other scientists or technicians as technological brokers (cf. van der Merve 1982).

It seems unlikely that archaeology could or needs to achieve the technological sophistication of chemistry. But it does need to make technology more widely and routinely available and to monitor outside technology in a more systematic way. The more that archaeological technology is effectively located within the discipline, the more successful the results will be. The two major obstacles appear to be resources and the organizational structure of the discipline.

Communication and interaction among archaeologists

As is true of other disciplines, archaeology does not rely exclusively on formal communication to keep its members in touch with one another and apprised of developments in the profession. In fact, personal contact between archaeologists has been a vital aspect of communication in the discipline. When the archaeological population was smaller, people working in a particular area—for example, the southwestern United States—or attending even national conferences were able to meet personally with a large percentage of their colleagues, and a great deal of information was exchanged in this way. Woodbury traces the history of several regional conferences, noting common elements in their founding and development:

> They are begun to meet a strongly felt need for the archaeologists in a specific and sometimes sharply defined region to exchange information and ideas on a face-to-face basis, with a minimum of formality . . . Nearly all regional conferences seem to grow gradually and to reach a point at which more formality and organization are needed . . . As conferences have grown in size, there has been some tendency for new, more specialized, sub-regional conferences to develop (Woodbury 1985:442–443).

The overall size of the profession affects communication patterns, and archaeology has grown tremendously in recent years.

For example, according to figures published in *American Antiquity*, the Society for American Archaeology had approximately 1,900 individual members in 1968, 4,257 in 1974, and 5,829 in 1982 (cf. membership figures in *American Antiquity*, vol. 34: 221–239; 39:674 and 47:473). In comparison with the established sciences, archaeology is still a small discipline, but as these figures show, one of its major professional societies more than doubled its membership in five or six years and tripled it in twelve years. In 1981 and 1982, however, the Society lost members at the rate of two or three percent per year.

Even though informal communication continues—and many archaeologists have even enlarged their network of personal contacts with the expansion of the discipline, person-to-person communication necessarily plays a smaller role in the discipline-wide dissemination of information. The reversed trend toward a shrinking population would have to proceed much farther than it seriously has to date to restructure communication lines along, say, the scale of the 1940s. All fluctuations of any magnitude will cause some restructuring of both the lines of communication networks and the content of the communication.

There are at least two aspects of all disciplines, even those with very large populations, to which informal communication is essential: the development and the operation of the small groups that inevitably form within them. One form that such groups take, and it is a form whose membership usually does not exceed 100 persons, has been called the "invisible college" by Price (1963). Griffith and Mullins (1972) have identified the same phenomenon as Price, which they call alternatively "invisible colleges" or "coherent social groups," and which they claim are the loci of creativity in science. Schneider summarizes these authors' ideas in this way: "Essentially their thesis is that groups which are in the process of formulating a radical conceptual breakthrough form a revolutionary society or an elitist ingroup characterized by intense democracy and by barriers to communication to the outside" (Schneider 1977: 17).

One of the most striking aspects of this work in "invisible colleges" is the extent to which it seems to confirm some of Kuhn's sociological claims. For example, in recommending that an extensive research program be undertaken to study the structure of what he calls "scientific communities," Kuhn has said that:

The analytic unit [for the research] would be the practitioners of a given specialty, men bound together by common elements in their education and apprenticeship, aware of each other's work, and characterized by the relative fullness of their professional communication and the relative unanimity of their professional judgment. In the mature sciences the members of such communities would ordinarily see themselves and be seen by others as the men exclusively responsible for a given subject matter and a given set of goals, including the training of their successors . . . Typical communities, at least on the contemporary scientific scene, may consist of a hundred members, sometimes significantly fewer . . .

Groups like these should, I suggest, be regarded as units which produce scientific knowledge (Kuhn 1970c:253).

In many places Kuhn has discussed what he sees as one of the disadvantages of specialization, namely the obstacle outsiders encounter to understanding what goes on in scientific communities. He goes on to say that:

The programme just outlined . . . highlights what has perhaps not been clear before, the extent to which I regard scientific knowledge as intrinsically a product of a congeries of specialists' communities. Sir Karl [Popper] sees "a great danger in . . . specialization" . . . but with respect to [that danger] . . . the battle has clearly been lost from the start. Not that one might not wish for good reasons to oppose specialization and even succeed in doing so, but that the effort would necessarily be to oppose science as well (Ibid.: 254–255).

It does seem as if "invisible colleges" and Kuhn's scientific communities share many of the same features.

Unfortunately (if we believe Kuhn's claim that they are the most productive units in science), "invisible colleges," or Kuhnian scientific communities, are not easily recognized in archaeology. Rogge has written that: "There do seem to have been some great-teacher-and-his-students groups, but the only formalized group that resembles an invisible college in anthropology, that I am aware of, is the Southwestern Anthropological Research Group (SARG), which was established only in 1971" (1976: 838). There may be more approximations of "invisible colleges" in archaeology than Rogge admits. Some have seen the participants in the New Archaeology during its early years at the universities of Michigan and Chicago as at least pale reflections of an "invisible college."

It is not uncommon for archaeologists to work in groups that approximate lower level and sometimes short-term "invisible colleges." An example of a small, informal, dynamic group that has been highly productive can be called the Mayan glyph group; it is composed of Floyd Lounsbury, Linda Schele, Peter Mathews, and David Kelley, whose initial interaction occurred in the context of a Dumbarton Oaks Mini-Conference. Located now at Yale, Texas, Harvard, and Calgary respectively, they have face-to-face contact only for short periods at widely separated intervals. A degree of overlapping interests and expertise forms the core of their endeavors. At the same time, each has perceptions, methodological approaches, and knowledge that is unique within the group. All have sturdy enough egos to survive the free-for-alls that inevitably develop as they pound away at some idea one of them has thrown out; and there is a remarkable lack of territoriality within this group. The result has been a collapsing of the time it would normally take for ideas to be rigorously investigated. Had the ideas been presented in individual publications to which others slowly reacted, the advances which have been compressed into a few years would have taken twenty or thirty years to evaluate and revise. No joint publications have yet emerged, but each has taken the results of their collaboration into his or her own work. Maya decipherment, partly as a result of the consensus reached by this group, is now approaching a period of "normal" science which some of the participants view as a mixed blessing.

Archaeologists previously tended to undertake one-person research, even though interdisciplinary and multi-investigator research is generally acknowledged to be more compatible with contemporary research objectives. Even the one-person style of research, however, causes archaeologists regularly to engage in fieldwork which, given its intensive labour requirements, is usually a group effort. Some long-term field schools with a substantial duplication or overlap of staff from year to year come to resemble "coherent social groups" or "invisible colleges." Notable here is the Vernon, Arizona, field school under the direction of Paul S. Martin, where Michael Schiffer, Fred Plog, Ezra Zubrow, William Longacre, James Hill, and others interacted intensively. Other long-term southwestern field schools exhibiting similar characteristics at different times include those at Chaco, Point of Pines, and Grasshopper. Carol Gifford and Elizabeth Morris conclude that one important function of the southwestern field schools is that: "The close encounters fostered by the field settings meant that those continuing in archaeology were well acquainted with significant numbers of their colleagues and interaction continued at regional and national meetings" (Gifford and Morris 1985:410). The importance of something like "invisible colleges" may be underestimated in archaeology because they are seldom recognized or acknowledged, because the participants become scattered, and because new research claims the participants' attention.

Great changes have occurred in formal modes of communication among archaeologists with the growth of the profession and increased specialization in the field. One (not surprising) result has been a proliferation and increased specialization of journals and other publication outlets. While at present a few publications with wide circulations still act as a central forum for the profession, there are enough smaller ones that individual archaeologists are unable to monitor the entire range of publications available. Also, description of basic data formerly received more printed space than it now does in primary journals (cf. Erasmus and Smith 1967; Dyson 1985); instead, interpretive and methodological matters have become progressively more important.

Derek Price draws an interesting distinction between "Big Science" and "Little Science" which is especially relevant to sociological questions about what kind of science archaeology might be. One of the differences between Big and little Science has to do with the role of scientific papers. These, he argues, originated because there were too many books for scientists to read in the seventeenth century. Scientific papers are closely associated with little Science, as they allow scientists to stake out scientific property and to establish their prestige. With the transition to "Big Science," this is drastically transformed:

> we tend now to communicate person to person instead of paper to paper. In the most active areas we diffuse knowledge through collaboration. Through select groups we seek prestige and the recognition of ourselves by our peers as approved and worthy colleagues. We publish for the small group, forcing the pace as fast as it will go in a process that will force it harder yet. Only secondarily, with the inertia born of tradition, do we publish for the world at large (Price 1963:91).

By and large, the role of scientific papers in archaeology falls toward the Little Science end of the scale, significantly tempered, however, by collaborations, small working conferences, and other prepublication interactions and communication networks.

Another characteristic of Big Science is group research and joint authorship. Although archaeology has moved firmly toward multi-investigator projects, Rogge (1976:837) estimates that anthropologists are sixty to eighty years behind chemists on this score (*see also* Roe 1972). On the other hand, archaeologists, like other scientists, enjoy the benefits of Price's "affluent scientific commuter" syndrome, which is characteristic of "Big Science." As one would expect, archaeology appears to be in a transitional phase between Big and Little Science, using Price's criteria (which, of course, have little to do with disciplinary size).

Archaeologists have wholeheartedly embraced one mode of communication that has features singled out by Kuhn, Price, and other sociologists of science as effective—the small working conference. Gatherings of this type constitute some of the most productive events in current archaeology. Recent examples are the Dumbarton Oaks Mini-Conferences and those held at the School of American research, but early Pecos Conferences and numerous other examples can be cited from earlier decades. They seem to be so useful partly because they tend to be quite limited in the size and scope of issues addressed. Many of these conferences result in books or monographs which are more narrowly focused than, say, festschrifts or journals, and which often include valuable discussions of the main papers (cf. Culbert, ed. 1973; Adams, ed. 1977; Renfrew, Rowlands and Segraves, eds., 1982). The number of people who can interact effectively at the conferences appears to be well below the maximum of one hundred suggested by Price for "invisible colleges." Conferences of as few as three to five people can be successful, but with more than fifteen or twenty, participation by all in attendance, which is highly desirable, is unlikely. The School of American Research now sets a limit of twelve.

One difficulty in organizing such working conferences is finding the degree of commonality of interest among participants that will maximize productivity. As Kuhn makes clear, an extreme diversity in attitudes and knowledge base is inefficient—too much time has to be spent on laying the foundations for discussion. On the other hand, if participants are all identically inclined and equally well informed they have little, if anything, to gain from one another. The most desirable level of diversity is no doubt impossible to establish—the question must be settled on a conference-by-conference basis.

Between informal communication and the more formal professional literature, there is a fairly small class of writings containing more personalized accounts of archaeology and its practitioners. One sort includes such works as Woodbury's archaeological biography of Alfred F. Kidder (Woodbury 1973), the Listers' (1968) book on Earl Morris, and Brunhouse's (1973) account of early Mayanists. In a different context, recent "Women in Science" books offer certain insights (cf. Williams 1981). A related genre is Flannery's (1976) parable of "The Real Mesoamerican Archaeologist, the Great Synthesizer and the Skeptical Graduate Student," or his more recent (1982) parable of "The Golden Marshalltown." Insights into archaeologists' personal development are found in, for example, each of the chapters in *Archaeological Researches in Retrospect* edited by Gordon Willey (1974), the vignettes between the chapters of Binford's (1972) *An Archaeological Perspective*, and Paul S. Martin's (1971) article "The Revolution in Archaeology." Personalized information combined with commentary on some current topic characterizes the "Profile of an Anthropologist" column in the *Anthropology Newsletter*. Some of these feature archaeologists; the one about Hester Davis (vol. 22, no. 1, p. 7) gives some biographical information coupled with her concerns about "Cultural Resource Management: Crisis in the 80's?" These publications are obviously not so much the products of archaeological inquiry as they are commentaries on that inquiry. As such, when well done, they are invaluable contributions to this history and sociology of archaeology, but they are not primarily reports of archaeological progress, as are the more formal scientific papers. To the Kuhnian, this point is reinforced by the fact that such publications are easily read by, and often even written for, the public at large. This is not to say, of course, that similarly personalized publications are not written about and by practitioners of the established sciences as well—for they are, and often. Howard Gruber and Paul Barrett's (1974) psycho-historical account of Darwin, *Darwin on Man*, and Karl Popper's (1976) intellectual autobiography, *An Unended Quest*, are but two examples.

Communication and interaction patterns in archaeology do not seem to conform exactly to any single sociological model of science, and this is not surprising. Thus, in Kuhn's view, increasing specialization (which is taken to be the distinguishing feature of the advanced sciences) is reflected in scientific social groups and communication patterns. While specialization is highly visible in archaeology at some levels, it is counteracted by the equally strong trends toward interdisciplinary studies and attempts at integration—things Kuhn sees as impeding the development of "normal" science. Nor can archaeology be placed firmly in Price's Big Science/Little Science dichotomy; rather it exhibits characteristics of both.

Still, much can be learned from the sociological accounts of communication and interaction patterns in other sciences. Not the least of the lessons is that these patterns fundamentally affect the way science is done. These patterns can be analyzed, and when their nature is better understood archaeology will have taken another step from "muddling along" to more sane planning. The creation of stronger communication and interaction networks between academia and the rest of archaeology will be the major line of defense in counteracting the structural factors favoring segregation along structural lines.

The role of a sociology of science[1]

As we have noted in earlier chapters and in Hanen and Kelley (1983b), philosophers of science traditionally wanted to draw a sharp line between the philosophy and the sociology of science. The task of philosophy of science was to engage in analysis of the fundamental concepts of science, to understand the logic of science, and to develop criteria of acceptability and rationality in science.

The fundamental notion was that of justification, and the question to be addressed was whether a given theory was true or at least justified or warranted. Virtually all of the philosophical work to which the New Archaeology attended, at least at first, was in the tradition emerging from the tenets of logical positivism that had its roots in the Vienna Circle of the 1920s and 1930s. But concomitantly with the growing interest of archaeologists in philosophy of science, that discipline was itself undergoing major changes, with greater attention being paid to historical and social studies of science. It increasingly came to be realized that the logic of science could not be divorced from its history, and that an account of the structure of science needs to be informed by an understanding of the way the field has developed. Such an account can still take two quite different forms, however; it can remain within the tradition of viewing science as an essentially rational enterprise, or it can ignore considerations of rationality in favor of a simply descriptive account. Newton-Smith has described the former sort of approach this way:

> The rational model of Popper, Lakatos and Laudan of the scientific enterprise embodies a normative account of the factors that ought to govern theory choice. While these rationalists offer considerably different accounts, they are united by a belief in the importance of articulating how one ought to decide which of a number of rival theories is most likely to be the best relative to a given body of evidence (1981:237).

Thus the model is rational, progressive, and normative in the sense that it is meant to tell us what is good science, which theories are better than others, and what constitutes scientific progress. Newton-Smith continues:

> Rationalists hold that theory transition in the case of mature sciences like physics is by and large change from the decent to the even better. A rationalist concedes that there may be occasions on which the change was not progressive as judged by his latter-day lights. It is on those occasions that a sociological or psychological explanation of the change is appropriate. External non-scientific factors are to be brought into play when and only when we have deviations from the norms implicit in the rational model (Ibid.: 237–238).

Not surprisingly, the controversy that is implicit in this way of putting the rationalist point has indeed developed—a controversy that pits the rationalists against the advocates of the so-called strong program in the sociology of scientific knowledge, much as in the 1960s we had the debate between the Popperians and Kuhn on very similar issues (*see* Lakatos and Musgrave, eds. 1970).

The central claim of the proponents of the strong program in sociology of science is that explanation should be symmetrical in the sense that the same type of explanation is appropriate for all transitions, whether or not we regard them as rational. Belief and changes in belief are to be explained causally, without reference to truth or falsity, for belief is just one natural phenomenon among many, and none is inherently better or more rational or even closer to the truth than any other. Beliefs are not true or false in virtue of something independent of ourselves; and the notion of reasons for beliefs that would be reasons for anyone, regardless of his or her own inclinations, social position, or ideology makes no sense. The position is thus relativistic through and through, and there is no real distinction between propaganda and rational argument.

This sort of extreme view is not new; indeed, most of its critical force is already present in the work of Feyerabend (1975). Nor is it without strong appeal, for its relativism seems to go along with an antiauthoritarianism suitable for the times, and the history of attempts to specify what should count as true or rational or even as good evidence is not a history of success piled upon triumph. Add to this the fact that the advocates of the strong program are seeking explanations of beliefs and actions in terms of causal laws linking the social and cognitive worlds and theories that will lead to correct predictions about the beliefs of particular individuals. This is a longstanding goal in social science, and thus enhances the appeal of the program; but unfortunately the goal is not less Utopian in 1984 than it was in 1924 or even earlier, for we still have nothing even remotely like a general theory for the explanation of human belief and action.

Furthermore, tempting though the extreme relativist position is, as an account of science it is ultimately self-defeating. In the first place, the strong program requires that we be in a position to identify beliefs, but this cannot be done without some assessment of the truth or falsity of the beliefs and some overlay of theory, however low level. In the second place, if the sociologists want to claim, as they do, that their own theories are the correct ones, they must admit the legitimacy of the notion of truth. As Newton-Smith says: "the legitimacy of the scientific enterprise is dependent on the legitimacy of the concepts of truth and rationality. If there is to be a scientific sociology of knowledge its practitioners will have to learn to live with this fact" (1981:250).

So the strong program is too strong and ultimately incapable of explaining the growth of scientific knowledge. But the rationalist program is equally too strong in the other direction, and the more useful path lies somewhere in between. There may be several hypotheses or theories at a given time that fit the facts, and the choice among them may have to be made on pragmatic or other grounds that are not purely logical. Nevertheless, this goes no way toward establishing the anarchist conclusion that any belief is as good as any other, or that the notions of objectivity and truth are entirely dispensable.

What implications, if any, does this theoretical debate between rationalists and proponents of the strong program have for the sort of social or political study within archaeology that we have been looking at in this chapter? The temptingly short answer is "none," for of course the more knowledge we can have of the socio-eco-nomic-political workings of the field or the ideology underlying its research programs the better, regardless of the truth or falsity of some philosophical theory at the macro level. But in a deeper sense there *is* a connection, for it is important that the theoretical under-pinnings of the detailed studies be recognized and made explicit, and that we be clear about the view of the world and of science in particular that is implicit in the problems we study and the ways in which we study them. At the very least, we should not want our research to presuppose an account of the scientific enterprise that is untenable through being too strong either in the direction of absolutism or of relativism.

What is at stake, ultimately, is the objectivity of the discipline. The old view of objectivity, according to which all social science was suspect because it was not amenable to complete control of variables, has been seen to be untenable for quite some time, both in natural science and in social science. But the next step—that of recognizing the issues that are pervaded by ideological and other human considerations—is only now being taken. The awareness produced by this sort of study both enriches the field and opens the way to a greater objectivity than had earlier been possible. Once we come to understand the socio-political and ideological factors af-fecting the discipline, we are in a position to take the next step of evaluating the factors involved with a view to selection, on care-fully justified intellectual and moral grounds, of directions to be pursued. But is is important to see that what makes this sort of move possible is a commitment to a view of archaeological research that, however much it is conditioned by the realities of science as a social enterprise, is nevertheless prepared at least to attempt to understand the phenomena in terms that are not purely relativistic, but that allow some accounts to be seen as mistaken and not merely different.

Notes

1. Most of Chapter 4, *The role of a sociology of science*, is taken from Hanen and Kelley 1983b.

5 / Explanation

N ew Archaeologists adopted the concept of explanation together with the idea of deductive testing of hypotheses as
the touchstone both of what distinguished their work from the
"old" archaeology and of what promised to make their work scientific. It is not surprising that early proponents of the New Archaeology turned to the work of Hempel. Although there had been many
contributions to the logical positivist philosophy of science, the
classic analyses of the two fundamental concepts of confirmation
and explanation were produced by Hempel (1945) and Hempel
and Oppenheim (1948). These accounts were in part a codification
of work done by many philosophers and in part an original extension of these by Hempel himself. In any case, both papers were
extremely thorough treatments and came to be the starting points
for any further discussion and development of these concepts
within philosophy of science itself. In addition, they were very
clearly written and so were natural sources for persons in other
disciplines, such as archaeology, who were interested in philosophy of science. They represented a ready-made account of what
was thought to constitute good science and, because these writings
were often referred to as "methodological," it was natural that they
also should have been taken to represent prescriptions for doing
good science (even though Hempel and others were careful to
disclaim this intent for their work). It would seem that the discov-
ery/justification distinction was not fully appreciated among archaeologists who were in fact searching for ways to make their field
more scientific and to ensure that their results were objective,

165

testable, and capable of being built upon by other researchers. Much of the concern within the field had become focused on the fact that data collection seemed to be leading to an accumulation of facts without very much in the way of integrated theory. As the method of traditional archaeology was usually described as inductive, the work of Hempel, in which the notion of a deductive method appeared prominently, had a natural appeal. This deep discontent with the traditional archaeology and the desire to provide a sharp contrast with it, coupled with the perceived high status of science—especially in North America—and a concomitant desire for archaeology to become more scientific, led archaeologists to accounts of the nature of science that were most obviously found in the writings of philosophers of science. Hempel was perhaps the most accessible of these philosophers, and thus the focus of much of the attention to methodology in the early New Archaeology came to be on the concept of explanation, and specifically on the deductive model.

Rationality and the positivist background

The disputes about both the nature of rationality in science and the extent to which science is a rational enterprise bring us back to one of the themes of the early chapters of this book. The empiricist philosophy of science of the early logical positivists was intended to provide what was termed a "rational reconstruction" of science—a characterization of the scientific enterprise that was both a more-or-less accurate description of how that enterprise was actually carried out and a normative account of how it should be done if it was to be consistent and rational. The vehicle for providing this reconstruction was an analysis of the crucial terms and the concepts to which they referred—terms such as "explanation," "confirmation," "law," "theory," and the like. The analysis to be carried out was variously called "logical analysis," "conceptual analysis," or "explication"[1] and was developed, often with a high degree of technical detail, through the 1930s, '40s, and '50s. Since that time this mode of doing philosophy of science has given way to other interests, sometimes called "post-empiricism" or "postpositivism." Much confusion has ensued concerning these labels, as well as concerning the term "realism," which has come to be applied to one of the main views of the nature of scientific theories in current literature. These confusions will need to be sorted out as we proceed, but for the moment we must at least note that for archaeologists to embrace realism as the answer to their foundational questions (as there is evidence some are inclined to do [cf. Gibbon 1983 and 1984]) is no more likely to succeed than was the wholesale adoption of a particular model of science in the early days of the New Archaeology. Putting the move toward realism aside for the moment, we turn to a consideration of models of explanation.

The deductive model

Not surprisingly, the original Hempel-Oppenheim (1948) account of scientific explanation has many historical antecedents. Aristotle held a similar view (though his notion of lawlikeness would not have coincided with Hempel's), and Hempel and Oppenheim themselves cite Mill (1843), Jevons (1920), Ducasse (1925), Popper (1935 and 1950), Hull (1943), Feigl (1945), and Hospers (1946). But any serious examination of the notion of explanation must start from the 1948 paper and the highly developed view found in Hempel's later writings.

With the stress on scientific explanation and specifically on the work of Hempel and Oppenheim, what was at first seldom noticed in the archaeological literature was that the Hempel-Oppenheim model had been the subject of severe criticism on several fronts. Interestingly enough, the criticisms that were advanced shed considerable light on the problem of applying the model to archaeology, but any reasonable assessment of the possibilities for applying the standard model of scientific explanation in archaeology (or any other field) required a clearer understanding of the model itself and the uses to which it was ordinarily put than was at first available in the archaeological literature. Indeed, with all of the stress on the original Hempel-Oppenheim deductive-nomological (D-N) model, little attention was initially paid to the fact that there were several standard models, not one. The others were statistical models, which yielded probabilistic or inductive explanations, and these models, too, were due to Hempel (1965; 1962).

The form of the D-N model is well known. Such explanations are conceived by Hempel (1966: 51) as deductive arguments whose conclusion is the explanandum sentence, E, and whose premiss-set, the explanans, consists of general laws, $L_1, L_2. . . , L_r$ and of other statements, $C_1, C_2 . . . , C_k$ which make assertions about particular facts. The form of such arguments . . . can be represented by the following schema:

$L_1, L_2, . . . , L_r$ Explanans sentences
$C_1, C_2, . . . , C_k$

E   Explanandum sentence

The explanandum sentence E is a description of the empirical phenomenon to be explained, and this phenomenon is to be explained by subsuming it under general laws.

Let us look at an illustration of the D-N model offered by Hempel —an explanation of Périer's finding in the Puy-de-Dôme experiment—that the length of the mercury column in a Toricelli barometer decreased with increasing altitude. Hempel offers the following formulation and account for it:

(a) At any location, the pressure that the mercury column in the closed branch of the Torricelli apparatus exerts upon the mercury below equals the pressure exerted on the surface of the mercury in the open vessel by the column of air above it.
(b) The pressures exerted by the columns of mercury and the air are proportional to their weights; and the shorter the columns, the smaller their weights.
(c) As Périer carried the apparatus to the top of the mountain, the column of air above the open vessel became steadily shorter.
(d) (Therefore,) the mercury column in the closed vessel grew steadily shorter during the ascent.

Thus formulated, the explanation is an argument to the effect that the phenomenon to be explained, as described by the sentence (d), is *just what is to be expected* in view of the explanatory facts cited in (a), (b), and (c); and that, indeed (d) follows *deductively* from the explanatory statements. The latter are of two kinds; (a) and (b) have the character of general laws expressing uniform empirical connections; whereas (c) describes certain particular facts. Thus, the shortening of the mercury column is here explained by showing that it occurred in accordance with certain laws of nature, as a result of certain particular circumstances. The explanation fits the phenomenon to be explained into a pattern of uniformities and shows that its occurrence *was to be expected*, given the specified laws and the pertinent particular circumstances (Hempel 1966: 50, emphasis added).

One can, of course, produce much less exalted examples. Hempel discusses the statement: "The slush on the sidewalk remained liquid during the frost because it had been sprinkled with salt" (Ibid.: 52). Here, though no laws are explicitly mentioned, at least one is tacitly presupposed: that the freezing point of water is lowered whenever salt is dissolved in it. The explanation is thus elliptical but, when filled out, can be seen to fit the D-N model.

Innumerable variations are possible, but the fundamental point is that a D-N explanation is to be viewed as a deductive argument of the form, roughly,

All F are G
x is F
—————
x is G

Besides having correct deductive form, explanations were required by Hempel and Oppenheim to meet three further conditions of adequacy: that the explanans sentences be true, that they have empirical content, and that they contain general laws that are actually required for the derivation of the explanandum. All of these conditions raised serious difficulties, and indeed much of the progress of detailed discussion of the D-N model was through modification of these conditions toward a more adequate formulation. More recently, Hempel proposed two very general requirements for scientific explanations—a requirement of explanatory relevance and a requirement of testability—but the problem is to characterize precisely what would count as meeting or failing to meet these requirements. We shall return to this point later.

The striking similarity between Hempel's account of explanation and other influential contemporary accounts is worth noting explicitly. In the fall of 1934 Karl Popper published *Logik der Forschung* (imprinted 1935), in which he put forward a model of causal explanation virtually identical to Hempel's D-N model. There Popper claimed that to give a causal explanation of an event was to deduce a statement describing that event—in Hempel's terminology, "the explanandum"—from a set of premises, "the explanans," consisting of initial conditions and at least one general law. Popper used the example of the breaking of a thread with a tensile strength w when a weight greater than w is placed on it. The fact that both Hempel and Popper demanded that an adequate explanation of an event could have served as a prediction of that event is a further similarity between their views on scientific explanation.

Popper's model of explanation is put forward as a model of *causal* explanation. There has been much dispute in the philosophical literature about the relation between causes and scientific laws, and it would take us too far afield to go into this dispute in any detail; however, the issue is an important one in the philosophy of science and it should be briefly addressed.

Basically, the question at stake is this: does saying that A caused B mean anything more than saying that whenever A occurs B occurs? The view that causation is really nothing more than correlation or constant conjunction, and thus that the question posed should be answered in the negative, is usually associated with David Hume, the eighteenth-century philosopher who is credited with first having put it forward (*see* Hume, 1955 edition). Since D-N explanations express correlations—if the explanans is true, then the explanandum must be true—in a Humean approach to causation, causal explanations are just D-N explanations and vice versa.

Of course, acceptance of the Humean view has hardly been universal. Many philosophers have taken the position that we mean much more by "cause" than Hume suggested and that a D-N explanation which invokes a mere correlation as its general law cannot be causal (cf. Reid, 1969, especially Essay 4; Ewing, 1934, Section 3). Indeed, some would go further, arguing that mere coincidence is not illuminating and that because we want explanations for increased understanding, D-N "explanations" which do not invoke causal generalizations are not genuine explanations at all.

In 1953, R.B. Braithwaite published *Scientific Explanation*, in which he offered a characterization of explanation much like the accounts of explanation offered by Hempel and Popper. According to Braithwaite, when a person asks why a particular picture fell to the floor at noon yesterday

> what he is requesting is the specification of a preceding or simultaneous event which, in conjunction with certain unspecified cause

factors of the nature of permanent conditions, is nomically sufficient to determine the occurrence of the event to be explained (the explicandum-event) in accordance with a causal law. . . . (Braithwaite 1953: 320).

In this model an explanation is a deductively valid argument; if the premises (the explicans) of the argument are true, then the conclusion (the explicandum) must be true. Or, as Braithwaite said, in an explanation "the existence of the explicans-event ensures the existence of the explicandum-event" (Ibid.: 321).

Obviously, Braithwaite's characterization of explanation is very similar to the others we have discussed, but there are differences between his views and those of Hempel and Popper. For example, although Braithwaite views deductive-nomological explanation as the paradigm or central case of scientific explanation, it seems clear that he thinks it is not the only kind of scientifically acceptable explanation. Of course, in Hempel's account, not only D-N explanations are acceptable: deductive-statistical and inductive-statistical explanations are as well. But for Hempel, the explanandum of all scientific explanations must follow from the explanans according either to a deterministic or to a statistical law, and the law involved—if the explanation is not to be elliptical or a mere explanation-sketch—must be *explicitly* mentioned. Neither of these conditions is necessary in Braithwaite's view. First, according to Braithwaite reference to laws must be made in scientific explanation, but this reference may be either explicit or implicit; and second, although usually the explanandum follows according to a law from the explanans, this, Braithwaite argues, is not always the case. In most instances, he says, the person posing the "Why?" question "is taken to be asking for a *sufficient condition* for the explanandum-event" (Ibid.: 321). But, Braithwaite continues,

the "Why?" question is sometimes a request for a necessary condition for the explicandum-event; it then asks for the specification of an event which is such that, had it not occurred, the explicandum-event would also not have occurred. In this case it is the explicans-event which is nomically determined by the explicandum-event instead of the other way round (Ibid.: 321).

Braithwaite calls the answer to this sort of "Why?" question a teleological explanation, an explanation "expressed by the sentence 'In order that g' " (Ibid.: 320). So, "if a physiologist is asked why the heart beats, and replies 'To circulate the blood round the body' or . . . (in terms of an ultimate biological end) 'In order that the body may continue to live,' he will be giving in each case a teleological explanation of the action in terms of the goal or end of the action" (Ibid.: 323). As Hempel makes abundantly clear, the logical structure of such accounts, which involves reversal of the logical relationship that usually obtains between explanans and explanandum, does not fit his model of explanation (cf. Hempel, 1965: 308–314).

Another difference between the views of Braithwaite and Hempel is that Braithwaite is not, in general, as concerned as Hempel to distinguish pragmatic from formal considerations in his account of scientific explanation. Thus, Braithwaite can endorse the view that

> Any proper answer to a "Why?" question may be said to be an explanation of a sort. . . .
> What is demanded in a "Why?" question is intellectual satisfaction of one kind or another, and this can be provided, partially or completely, in different ways (Braithwaite 1953: 319).

Hempel, of course, could not accept such a view. It is not that Hempel wants to deny the importance of the role that pragmatic aspects play in explanation. Indeed, in his article, "Aspects of Scientific Explanation," Hempel wrote:

> Very broadly speaking, to explain something to a person is to make it plain and intelligible to him, to make him understand it. Thus construed, the word "explanation" and its cognates are *pragmatic* terms: their use requires reference to the persons involved in the process of explaining (1965: 425).

But while Hempel does not want to belittle the significance of pragmatic aspects of explanation, he does want to maintain that consideration of them is not the business of the logic of science. Hempel supports this position by suggesting that the goals of the philosophy of science should be similar in certain important respects to those of scientific research. He writes:

> scientific research seeks to account for empirical phenomena by means of laws and theories which are objective in the sense that

their empirical implications and their evidential support are independent of what particular individuals happen to test or to apply them; and the explanations, as well as the predictions, based upon such laws and theories are meant to be objective in an analogous sense. This ideal intent suggests the problem of constructing a nonpragmatic concept of scientific explanation—a concept which is abstracted, as it were, from the pragmatic one, and which does not require relativization with respect to questioning individuals any more than does the concept of mathematical proof. It is this nonpragmatic conception of explanation which the covering-law models are meant to explicate (Ibid.: 426).

The question for archaeology, of course, is whether any form of the covering-law model is applicable to archaeology; but in order to discuss this matter properly it is necessary also to outline the other standard model of explanation developed by Hempel.

The statistical model

The inductive-statistical (I-S) model of explanation differs from the D-N model in that the laws employed are probabilistic or statistical rather than universal in form and the connection between explanans and explanandum is inductive or probabilistic rather than deductive in nature. There are also, we should note, explanations that can best be termed *de*ductive-statistical (D-S). These "involve the deduction of a statement in the form of a statistical law from an explanans that contains indispensably at least one law or theoretical principle of statistical form" (Hempel 1965: 381). Hempel classes these with statistical explanations on the grounds that they make use of statistical laws. The explanatory inference they contain is deductive, however, which differentiates them rather sharply from inductive-statistical explanations. We consider this difference to be more important than the one between universal and statistical laws, and so we are inclined to class D-S explanations with D-N ones rather than with I-S ones. As these explanations have relatively few applications in science, they are of minor interest for our purposes. We shall say no more about them.

A schematization of the *in*ductive-statistical model can best be given in relation to another example of Hempel's. We might explain why a particular patient, John Jones, recovered from a streptococcus infection by saying that Jones was given penicillin and most cases of streptococcus infection clear up on administration of penicillin. This argument is schematized by Hempel thus:

p(R, S.P) is close to 1
Sj . Pj

_____[makes practically certain (very likely)]

Rj

where p=probability, j=John Jones, R=recovery, S=streptococcus infection, P=penicillin is administered. The symbol "." stands for "and," and the double line separating premises from conclusion signifies the relation of inductive support (Ibid.: 383).

The important difference between I-S explanations and D-N or D-S ones lies in the relation between explanans and explanandum: in the I-S case it is inductive, whereas in the other cases it is deductive. The difference between deductive and inductive inference is extremely important, but it is not that which is often supposed. Thus, the standard textbook characterization of deduction as proceeding from general to particular and induction as proceeding from particular to general is incorrect, for the difference has nothing to do with generality or particularity.[2] It is rather that, in a correct deduction, the truth of the premises guarantees the truth of the conclusion; in acceptable inductive inference, all the premises may be true and the conclusion yet be false.

This point is extremely fundamental and not particularly controversial, yet failure to appreciate it was certainly at the root of some of the unprofitable dispute about explanation in archaeology. For one thing, on the face of it one would have thought that if there are laws in archaeology at all, they are likely to be statistical rather than universal in form. One would have thought, therefore, that a model of explanation appropriate to archaeology would involve statistical rules, and that if one of Hempel's models were adequate it was more likely to be the I-S than the D-N model. Yet the archaeologists dealing with explanation as a deductive phenomenon in the early days of the New Archaeology seemed to be unaware of the existence of the I-S model, or of the possibility that one might have scientific explanation without having *deduction* of statements describing what is to be explained from statements of universal laws.

The question to be asked, then, was whether, assuming that archaeologists do engage in some form of explanatory activity, explanations in archaeology fit either the D-N or the I-S model reasonably closely. To do this it was necessary to establish, among other things, what would count as a reasonable fit, bearing in mind that it may be the case that archaeological explanation, if there is such a thing, fits neither model.

Some fundamental objections

It would surely be unreasonable to expect the fit for archaeology to be any better than it is for the physical sciences. The literature of philosophy of science abounds in demonstrations that the original Hempel-Oppenheim model and various fairly obvious revisions of it will not do (cf. Eberle, Kaplan, and Montague 1961; D. Kaplan 1961; Kim 1963; Ackermann 1965; Omer 1970; Morgan 1970). Broadly speaking, the difficulty has been that models so far proposed, beginning with the original D-N model, have turned out to be either trivially satisfiable (in the sense that virtually any theory explains virtually any particular fact) or unsatisfiable. The definition of a potential explanans that Hempel gave when he originally proposed the D-N model was one source of difficulty, for that definition can be shown to allow the explanation of any particular fact by itself (cf. Hempel 1965: 293).

The result that the original D-N model of explanation was trivially satisfiable was totally unsatisfactory, and it was one which Hempel had gone to some lengths to avoid when formulating the definition of a potential explanans. Hempel's original definition did make impossible the straightforward explanation of any fact by itself. There is no need here to go into the details of how this was done, but the following, obviously inadequate, explanation was inadmissible on Hempel's account:

All swans are white.
It rained today.
_____

It rained today.

However, the following explanation did satisfy his requirements:

> It is true of all swans both that they are white and it rained today.
> It rained today or it did not rain today.
>
> ---
>
> It rained today.

And that this explanation was acceptable on his account trivializes his model. For since "It is true of all swans both that they are white and it rained today" is logically equivalent to "It rained today," this explanation, it turns out, is nothing more than a thinly disguised self-explanation. To get around this difficulty, various additional requirements for explanations were proposed (*see* Kim 1963), but it was not at all clear that these requirements could be justified on anything other than the question-begging grounds that they saved Hempel's model from trivialization.

Another source of difficulty was Hempel's definition of the concept of derivative law, a definition which requires that a derivative law be deducible from a set of fundamental laws. Ernest Nagel (1961: 58) revealed this concept to be overly restrictive by showing that Galileo's and Kepler's laws of planetary motion "cannot be derived from the fundamental Newtonian laws of mechanics and of gravitation alone," which is to say that Galileo's and Kepler's laws cannot be explained by Hempel's model (cf. Hempel 1965: 291). Since one would have supposed that these laws are perfectly good scientific laws, and yet they turn out not to be explainable on Hempel's model of the explanation of general laws, that model appears to be as good as unsatisfiable. Modifications were attempted, however, which required that we have an adequate characterization of lawlikeness. Clearly such a characterization would be very helpful; indeed, the concept also plays a fundamental role in areas of scientific methodology other than those concerned with explanation. As we will argue in the next chapter, however, it is doubtful that an adequate *formal* characterization of lawlikeness is possible.

As a result of these and other difficulties with Hempel's D-N model of explanation, there remained considerable doubt about what might constitute an acceptable formal model of explanation, even in physics. And it is of central importance to Hempel's ac count that the notion of explanation be characterizable in formal terms.

There are numerous other objections to the Hempelian models besides the cluster cited above. It is not difficult to find cases that satisfy all of Hempel's conditions of adequacy but which cannot be regarded as genuine explanations. Kyburg (1965) has offered the following deductive example: "This sample of table salt dissolves in water, for it has had a dissolving spell cast upon it, and all samples of table salt that have had dissolving spells cast upon them dissolve in water." And Salmon gives several inductive examples, among them: "John Jones experienced significant remission of his neurotic symptoms, for he underwent extensive psychoanalytic treatment, and a substantial percentage of those who undergo psychoanalytic treatment experience significant remission of neurotic symptoms" (Salmon 1971: 33). As Salmon says of this example:

> The problem with [it] is the substantial spontaneous remission rate for neurotic symptoms of individuals who undergo no psychotherapy of any kind. Before we accept [the example] as having any explanatory value whatever, we must know whether the remission rate for psychoanalytic patients is any greater than the spontaneous remission rate (1971: 33).

The main trouble pointed up by these examples is, as Salmon puts it, that "the 'explanatory facts' adduced are irrelevant to the explanandum event despite the fact that the explanandum follows (deductively or inductively) from the explanans" (Ibid.: 36). That is, the salt will dissolve regardless of the spell, and there is some evidence that there is a substantial spontaneous remission rate for neurotic symptoms of people who do not undergo psychoanalysis. Very roughly, what seems to be needed is that the probability of the explanandum event in question should be substantially greater relative to the explanatory facts cited than its prior probability, and this is not the case with the above examples. Salmon actually regards the increase in probability as a by-product of a more fundamental feature, but the rough characterization will suffice for our purposes.

Another group of examples involves the distinctions among "cause," "symptom," and "explanation." The fact that a given patient has measles is not explained by the prior occurrence of

Koplik spots (plus the relevant law) even though the explanation has the standard D-N form:

All people who have Koplik spots develop measles.
Patient *a* has Koplik spots.

Patient *a* will develop measles.

and the patient's getting measles could have been predicted from his having spots. The difficulty here seems to be that both the spots and the measles are caused by and thus explained by, something else: the spots-measles correlation is just that, and not a causal law (Hempel 1965: 374–375). This suggests that for genuine explanation we require more than mere correlations: we require causal laws. As with the remission of Jones' neurotic symptons, what we are looking for is the *correct* explanation of the phenomenon in question, and for this we need causally relevant factors. If this is so, we shall have to offer an account of the difference between mere correlation and causal law—a notoriously difficult task.

We are not, of course, suggesting that correlations have no place in science. Indeed, correlations unquestionably are necessary, for they provide a basis from which the search for causal laws can progress and, since they are often reliable for predictive purposes, they have a use in their own right. Further, as with many other procedures that might be characterized as "low level" in highly developed sciences such as physics, the identification of correlations is still especially critical in archaeology where general laws are scarce and even correlations are few.

Bromberger's flagpole example provides another "explanation" in which the allegedly explanatory facts are irrelevant to the explanandum event even though the explanandum follows from the explanans. The position of the sun and the height of a flagpole (together with the relevant laws) explain the length of the shadow cast by the flagpole; we can equally (deductively) infer the height of the flagpole from the length of the shadow and the position of the sun, but we do not say the length of the shadow explains the height of the flagpole. Again, the difficulty seems to have to do with causation: the flagpole "produces" the shadow and not vice versa (Salmon 1971: 71–72).

What all of these examples show is that satisfaction of the D-N model cannot be regarded as a sufficient condition for there being an adequate explanation. Hempel's own later introduction of a "requirement of explanatory relevance" seems to acknowledge some of the difficulties and to point a way to their solution. But the characterization of explanatory relevance is by no means an easy matter.

Is satisfaction of the D-N (or similar) model, then, a necessary but not sufficient condition of an adequate account of explanation? Against this suggestion, many purported counterexamples have been offered. Functional explanations in biology or genetic explanations in history (where we explain the phenomenon in question by showing how it came to be—by giving the chain of events that led to it) are alleged not to fit the model. An example of the former might be the explanation of the presence of hemoglobin in the blood by saying that it has the function of transporting oxygen from the lungs to other parts of the body. When this case is appropriately expanded, it is seen not to fit the D-N pattern, for in order to turn this into a valid deductive argument, the statement of the facts to be explained must function as a premise rather than as the conclusion of the argument:

> We are told . . . that it is the function of hemoglobin to transport oxygen from the lungs to the cells in the various parts of the organism. This fact is taken to provide an explanation for the blood's hemoglobin content. Trouble arises when we try to fit such explanations to schemata . . . for deductive explanation, for we always seem to end up with a necessary condition where we want a sufficient condition. Consider the following:
>
>> Hemoglobin transports oxygen from the lungs to the other parts of the body (in an animal with a normally functioning circulatory system).
>> This animal has hemoglobin in its blood.
>> _____
>> This animal has oxygen transported from its lungs to the other parts of its body.
>
> Here we have a valid deductive argument, but unfortunately the explanandum is a premise and the conclusion is part of the explanans. This will never do. Let us interchange the second premise and the conclusion, so that the explanandum becomes the conclusion and the explanans consists of all of the premises, as follows:
>
>> Hemoglobin transports oxygen from the lungs to the other parts of the body.

> This animal has oxygen transported from its lungs to other parts of its body.

> This animal has hemoglobin in its blood.

> Now we get an obviously invalid argument. In order to have a standard deductive explanation of the Hempel variety, the particular explanatory fact cited in the explanans must, in the presence of the general laws in the explanans, be a sufficient condition for the explanandum event. In the hemoglobin case the explanatory fact is a necessary condition of the explanandum event, and that is the typical situation with functional explanations (Salmon 1971: 60).

Yet despite these difficulties, many people would insist that the above example (or something identical in form) does represent an adequate explanation. As we have already discussed, Braithwaite argues that explanations of the form illustrated by the above example are perfectly acceptable explanations. Further, in the case of genetic explanations, it is not obvious that there is an appeal to general laws, and while this is often alleged to be a problem, there are numerous disciplines (for example, archaeology and history) in which such explanations are considered useful.

We are now in a position to offer a preliminary assessment of the Hempelian models of explanation: however well the models may appear to capture what we mean by scientific explanation, they face the two parallel problems of the existence of (1) arguments that fit the Hempelian schema but are not explanations, and (2) explanations that appear to be satisfactory though they do not fit the models. And, it should be noted, these are difficulties with the models that are largely independent of problems of applying the models to archaeology.

Further objections

In the present section we shall sketch very briefly some further objections to the Hempelian models to show where the philosophical debate was focused, to illustrate how Hempel tried to accommodate his critics' points to his theory, and also to indicate where some of the remaining controversies and open questions lay. Many of the disagreements between Hempel and his critics centered on whether or not various common kinds of explanation fit the Hempelian model. Dispositional explanation is one of these, which was "held to defy a covering-law analysis" (cf. Hempel 1965: 457). To attribute a dispositional characteristic—say, solubility or brittleness—to an object is to say that if certain events were to occur then that object would behave in the specified manner. Thus, to say that a pane of glass has the dispositional property of brittleness is to say that if a rock was hurled at the pane of glass, under ordinary circumstances, the glass would shatter. That we can dispositionally explain the shattering of a windowpane on being struck by a stone by saying that the glass in the pane was brittle has been taken to show that dispositional explanations succeed without invoking general laws and thus the Hempelian models of explanation will not do. Hempel's response to this objection is illustrative, as we shall see, of the approach he often took to meeting criticisms of his views on explanation: he points out that the explanation presupposes the general law that glass (under normal conditions) is brittle (or breaks when struck sharply with a stone) (Ibid.: 458). There are serious philosophical difficulties in explicating the nature of disposition statements, but we can ignore these for our purposes.

Dispositional explanation is used in archaeology, although there is nearly always the problem of our having insufficient or inadequate data for formulating statements ascribing dispositions to individuals or groups. The scarcity of relevant data not only restricts the plausibility of such formulations, but it also renders them vulnerable to falsification. For example, the concept of the "peaceful Maya" was deeply entrenched among Mayanists for several decades, although it is exceedingly difficult to trace the development of this idea. Beginning with the discovery of the Bonampak murals in 1946, the view that the Maya were disposed toward peaceful activities was challenged, and most archaeologists now believe that this earlier view was false (cf. Webster 1977).

There are some rather direct analogues in archaeology to the classic "glass will break when struck with a rock" case in the philosophical literature—for example, with regard to lithic production. On the other hand, it is often difficult to sort nonfalsified dispositional explanations or arguments from other kinds of explanations such as genetic or evolutionary arguments. We suggest, however, that Mark Cohen's (1977) arguments presented in his book *The Food Crisis in Prehistory* can in some respects be characterized as dispositional in nature. Indeed, a twofold sequence of dispositions can be identified. In the first instance, humans were disposed toward hunting and gathering, preferring to acquire at least part of their food from high trophic levels in the food chain. In the second instance, humans, as a consequence of a disposition to population growth, were "forced to eat more and more unpalatable foods, and in particular to concentrate on foods of low trophic level and high density," because of population pressure.

It is perhaps a moot point whether stress imperatives can be viewed as dispositional, but there are certainly dispositional elements in the Cohen example, as in many others that could be cited. The Cohen cases suggest not only that dispositional explanations sometimes occur in archaeology, but also that human dispositions attributed to hominids can change over time as conditions change. These examples thus seem to illustrate the claim that philosophical arguments about dispositional explanation may be more relevant to archaeological concerns than one might assume at first glance.

Just as dispositional explanation has been used as a counterexample to Hempelian models, the applicability of the covering-law model to history has been the subject of a different and more far-reaching series of philosophical debates. A number of the questions raised in this context may have some relevance to archaeology because of its culture-history component. If this is the case, these debates should have significance for the archaeological assessment of explanatory models. William Dray is the philosopher of history who has been most actively engaged in this issue. He claims that the D-N model fails for (a) "explaining-what" or "explanation-by-concept," (b) "explaining how-possibly," (c) genetic explanations, and (d) rational explanation, all of which are used not just by historians but by investigators in other disciplines as well.

(a) Explanation-by-concept is provided where we ask, in effect, "What was it that happened in this case?" and the reply is "It was a so-and-so," as in "It was not merely an economic change that was beginning; it was a social revolution." This explanation, according to Dray, "explains what happened as a social revolution" (Dray 1959: 403). It is "explanation by means of a general concept rather than a general law; for the explanation is given by finding a satisfactory *classification* of what seems to require explanation" (Ibid.: 404)—by explaining-as.

Hempel's reply is that not all cases of classification will count as explanatory. In the case cited by Dray, the events that are allegedly explained as a social revolution could "be truthfully but unilluminatingly classified also as changes involving more than 1,000 persons and affecting an area of over 100 square miles. If there is explanatory significance to characterizing x, y, and z collectively as a Q, it is because the characterization implies that the particular case fits into, or conforms to, some general pattern that is characteristic of Q" (Hempel 1965: 454). All explanation may be classification, in that all explanation may involve the subsumption of events under a general concept; but not all classifying is explaining, as Hempel's counterexample shows. An *explanatory* subsumption of events under some general concept always, then, according to Hempel, relies on corresponding general hypotheses, and is thus a case of subsumption under general laws as in the D-N model.

Recognition of a similarity between two things, which is a prerequisite to classification, may often be a matter of having an appropriate familiar term that applies to both. But the similarities that are explanatory would seem to be the ones that depend upon general features that we should expect to be repeated in similar circumstances. In order for the categorization of an A as a B (an economic change as a revolution, for example,) to be explanatory, we should expect that, were we to come across another A similar in relevant respects to the first, then it too would be illuminatingly classifiable as a B. Not just any similarities will do. As Hempel might put it, *relevant* similarities need to be lawlike.[3]

Dray seems to assume that any answer to a "What happened?" question is an explanation. This assumption is surely false, as we have seen, but this does not end the matter, for we still need to ask how we can characterize the answers to "What happened?" questions that *are* explanatory. Hempel claims that if an answer is explanatory then it can be shown to have, at least implicitly, the logical structure of his covering law model. Not all answers with this structure are explanatory, however, as we noted earlier and as Hempel seems to concede in discussing one of Dray's examples of "explanation-by-concept" (Hempel 1965: 455).

Whether classifications in archaeology are explanatory is a question that has bothered at least some archaeologists. For example, in their article, "The Nature of Archaeological Explanation," Fritz and Plog (1970) address this issue, and the view they express there seems to be that archaeological classifications are explanations *because* they conform to Hempel's covering-law model. Thus, they would argue that the following archaeological classification, which Michael Levin (1973) puts forward in his critique of the Fritz and Plog position, was explanatory because it has the logical structure of the D-N model:

C is a bifacially flaked cobble.
All bifacially flaked cobbles were once choppers.

C was once a chopper.

It should be clear from our discussion of the debate between Hempel and Dray that mere conformity to the D-N model does not guarantee explanatory worth. But in considering the classification in question we are immediately met with another difficulty, for what, if anything, does the above explain? Surely not why C was a chopper. Instead, if it is explanatory of anything, it would seem to be of why we should call C, the bifacially flaked cobble in question, a chopper. Or, as Levin says of the explanatory value of such archaeological classifications:

> [The generalization "All bifacially flaked cobbles were once choppers"] does not serve as an explanation of why object C is a chopper. This object is not a chopper *because* it is bifacially flaked and found at site so-and-so. Rather, the generalization here serves to give me *a good reason to believe* that C is a chopper. It explains, not: why C is a chopper; nor: why there is a chopper here; but: why it is reasonable for me to believe that C is a chopper (Levin 1973: 391).

Levin refers to generalizations such as that in the sample archaeological classification we are considering—"All bifacially flaked cobbles were once choppers"—as "reason-giving generalizations" and calls "the use of reason-giving generalizations for the ascription of classificatory predicates the first degree of nomological involvement" (Ibid.: 391). He then contrasts this use of laws in archaeology with another:

> The second degree of nomological involvement is attained by attempts causally to explain the occurrence of the data itself: Why is there a chopper here? At this point, laws connecting the structure of the object, ends attainable by objects of such structure, and the intent of the vanished society to attain these ends, become germane (Ibid.: 391–392).

Levin then argues that there is a close connection between the two degrees of nomological involvement because "the explanatory laws at the second degree also serve *to explain and justify the general-*

izations at the first degree" (Ibid.: 392). According to Levin, reason-giving generalizations have little, if any, explanatory force, while genuine explanations in archaeology involve the use of cause-giving generalizations.

Levin thus makes the following reasonable claims:

(1) not all archaeological classifications are in any interesting way explanatory,
(2) those that are explanatory attempt to give a causal account of the archaeological record and
(3) the reason-giving and cause-giving generalizations are closely related.

He suggests, however, that archaeological classifications generally involve the reason-giving variety of law and as a result are not explanatory. "It is at least arguable," he writes, "that the generalizations to which archaeologists are committed by retrodictive classificatory inferences are largely of the reason-giving, rather than the cause-giving, sort" (Ibid.: 39). This may ultimately be misguided because it presupposes a clear distinction between reasons and causes, a view which is probably false and one which we discuss later in this chapter in connection with Harré and Secord's account of explanation. A better position seems to be that classifications do not *uniformly* succeed or fail as explanations. Instead, we suggest that the explanatory worth of classifications varies: a classification has explanatory value to the extent that it reveals why a certain state of affairs came to pass. In this account it would appear that most archaeological classifications are only weakly, if at all, explanatory. This may be because most archaeological explanations involve generalizations that look more like reasons than causes. We would nevertheless want to insist that reasons and causes shade into one another and, thus, explanatory worth is not an all-or-nothing matter, but one of degree.

We have suggested that a classification will count as explanatory to the extent that it reveals why a certain state of affairs came to pass. That is, being told that A is a B is explanatory when seeing A as a B is helpful to an understanding of A. But this will not be the case when, for example, saying something is a B amounts to no more than giving A a new name. Thus, labelling a primitive sculpture of a woman in a museum "Human Figure" tells us nothing more about the object than would simply assigning it an exhibit number. In the same way, someone completely unfamiliar with the game of chess gains no insight into how he or she is to use a piece by being told it is a bishop. The player needs to know what rules govern the movement of the piece—not its name.

Myriad other examples could be given of classifications that lack explanatory force, but it is much more difficult to say what will (in the sense of what should) be helpful to understanding in any particular case. As a first approximation, however, we suggest that a relevant similarity between A and B should count as illuminating, or aiding understanding, and hence can be regarded as explanatory. And if this is so, we are here again directly confronted with the problem of characterizing relevance, for it is surely not obvious which similarities are relevant ones in particular contexts.

Another difficulty concerns the concept of "similar," for it is trivially true that any two objects or events are both similar to and different from one another in some respects. So the difference between trivial and significant similarity is one of degree, and one category or class shades into another. (As Watson Smith [1952] argues in "When is a Kiva?," pit houses grade into non-pit houses and into semisubterranean structures with other designations.) Thus, the specification of similarity, like that of relevance, must be carried out from a particular point of view, and how we categorize any A will depend upon the similarities and differences we see between A and other things. If we see A as more like a B than a C, for example, then we will, presumably, classify A as a B, rather than as a C. Whether the archaeologist's explaining his find, for example, as a B has explanatory force will depend upon whether B, the category he has placed it in, has explanatory relevance for archaeology. Thus, Emil Haury's (1958) classification of the archaeological remains at Point of Pines as evidence of a *migration* has at least potential explanatory force, because the concept of population movement carries along with it much that is of interest to archaeology. This is true notwithstanding the skepticism with which many archaeologists view interpretations which involve population movement. Haury's classification may be controversial, but if it is accurate, we would suggest that it would most certainly have explanatory worth. Other archaeologists hold contrary views (*see* Martin and Plog 1973).

While it is exceedingly difficult to say just what should count as explanatory, it does seem clear that the explanatory force of a comparative classification is greater than that of a noncomparative one: to be told that A should be categorized as a B because it is more like a B than a C or a D is to be told more than just that A is a B. Thus, to take the example from Dray we discussed earlier, our understanding is greater having been told not just that the event in question was a social revolution, but, as well, that it was not merely the beginning of an economic change. This is in part, it seems, because the term "social revolution" encompasses a broad range of phenomena, and that range is narrowed with the imposition of the caveat that the social revolution in question was not just an economic change.

While we have discussed the difficulties with classification and categorization from the theoretical point of view, the practical consequences of these problems should not be overlooked. Inevitably, modes of classification influence ongoing research: the labels and categories we use dictate how we view the data at hand. For example, to classify a certain object as a projectile point rather than as a cutting tool may determine to a considerable extent how we see other objects found in the site or similar objects found elsewhere. Given that archaeologists are often faced with limited data on which to base their conclusions—a problem we discuss at some length in Chapter 4—this can easily result in a compounding of errors brought about by judgments based on expectations of finding a particular kind of object, pattern of residence, or what have you.

(b) To explain "how possibly" something could have occurred is required when "what we know seems to rule out the possibility of the occurrence which is to be explained" (Dray 1957: 161). In these cases the occurrence to be explained seems to us highly improbable or even impossible, so that it is not clear how citing general laws could provide any sort of explanation. Such puzzling facts, one supposes, would not be uncommon in archaeology. For example, the D-shaped kiva—unique among Point of Pines kivas in southern Arizona but known in the northern part of the state—that Emil Haury undertook to explain could be seen as just this sort of phenomenon (cf. Haury 1958a; *see also* Chapter 7).

According to Hempel, what comes into play in explaining "how possibly" is the pragmatic aspect of explanation, as distinct from the logical. We shall need "to ascertain the empirical assumptions underlying the question and then to show either that some of these are false or else that the questioner was mistaken in thinking that those assumptions warranted his belief that x could not have occurred" (Hempel 1965: 428). A third possibility is that

> all of the questioner's relevant assumptions might be true, and his belief that they make the occurrence of x very improbable may be correct. In that event, the perplexity expressed by the questioner's "how could it possibly have happened?" may be resolved by broadening the questioner's total evidence, i.e., by calling to his attention certain further facts whose addition to those previously taken into account will render the occurrence of x less improbable (Ibid.: 429).

The assumption on Hempel's part seems to be that if the event actually occurred and is explainable, it cannot have been quite as improbable as we supposed, and that a pragmatically acceptable explanation consists in showing this to be so. But, in Hempel's view, a logically correct explanation of the events in question is still simply a D-N account.

(c) Genetic explanations are, it is often claimed, widely used in history as well as in applications of social evolutionary thought, which occur quite frequently in contemporary archaeological, anthropological, and other social scientific literature (cf. Dunnell 1980). A genetic explanation "presents the phenomenon under study as the final stage of a developmental sequence, and accordingly accounts for the phenomenon by describing the successive stages of that sequence" (Hempel 1965: 447). But, of course, Hempel insists that more is involved than simply describing the successive stages of the sequence. The stages have to be connected in such a way that the outcome is seen to proceed inevitably (or with a high degree of likelihood) from the various stages. The difference between explaining the events in question by simply citing a general law and explaining them by reference to the various stages is that in the latter case we have, in effect, an account of why the general law holds. The reason, for example, why an oil leak in a properly built car generally leads to engine failure is that "such a leak causes the oil to drain out, the cylinders and pistons to be deprived of lubrication thus leading to frictional heating and expansion of the pistons and cylinder walls so that the metals lock tightly and the engine stops" (Hempel 1965: 452, in which Hempel was using an example from Dray 1957).

In history, such explanations are particularly apt because we lack general laws that are nontrivial. Instead we can provide additional data—some of it explanatory and some of it descriptive—that will link up the various stages and thus provide an understanding of the phenomena in question.

(d) A similar problem arises for rational explanation or explanation in terms of reasons, also widely held to defy analysis by means of the covering-law models. Again, Dray can be taken as Hempel's chief opponent. In this type of explanation, according to Dray, we need to know what considerations convinced the agent that he should act as he did and also we need to show that what was done was the thing to have done for the reasons given. In other words, the reasons given have to be *good* reasons (Hempel 1965: 470). According to Dray, such explanations invoke a principle of action, such as "When in a situation of the type $C_1 \ldots C_n$ the thing to do is x" (Dray 1957: 132). It is not an appeal to a general empirical law. Rather, an element of appraisal is involved, permitting us to say that what was done was, indeed, the appropriate thing to have done in the circumstances.

Hempel counters with the following considerations. First, it is highly unlikely that we can single out a particular course of action as *the* thing to do in the circumstances, for this depends on many factors, including our appraisal of the likelihoods of the different possible outcomes. Second, to show that a certain thing is the appropriate thing to do in particular circumstances is still not to show that the agent in question is a rational agent and thus disposed to do the appropriate thing. Hempel restores the covering-law model of explanation for rational action by substituting for Dray's normative principle of action a general-law statement. Thus, we explain A's having done x by saying that he was a rational agent in situation C and that in a situation of type C any rational agent will do x. But this explanation is incomplete because it does not address itself to the question of what counts as rational. Presumably, the question about A's behavior that we wanted answered by an explanation in terms of reasons was "Why was what A did rational?" Thus, to explain why what A did was rational in the way that Hempel suggests does not really get us anywhere. That there is this problem with casting rational explanation in the D-N mould should certainly have concerned those archaeologists who advocated adoption of the Hempelian models of explanation and at the same time approached their work from some version of the "man as an actor" or "man as a decision maker" point of view.

The problem illustrated in the above discussion of rational explanation is a general one. We can no more explain why what A did
was moral (or legal) by saying that A was a moral (or legal) agent in
situation C and that in a situation of type C any moral (or legal)
agent will do what A did, than we could explain why what he did
was rational. Further, we can hardly explain why a piece of wire
conducts electricity by saying that the piece of wire is a conducting
agent with a current passing through it and whenever a current
passes through a conducting agent, the conducting agent conducts
electricity. Far better, for example, would be an explanation that
related the molecular structure of copper to a general theory of
electricity. The point, of course, is that what we are looking for in
explanations of all these states of affairs are good reasons for why
the states of affairs obtained.

There is no special reason for thinking that explanation in science is a different enterprise from explanation in other disciplines.
In particular, generalizations probably play an essential role in all
explanation, not just the scientific. But even if this is so, it is highly
likely that in some types of explanation the generalizations referred to will not be empirical. It is still an open question whether
the logical structure of these explanations is deductive-nomological, but it is at least clear that a full explanation of rational action
requires some account of what it is to be rational. Hempel, in fact,
realized such an account was necessary and attempted to incorporate one into his theory of rational explanation (*see* Hempel 1965:
463–487).

This Hempel-Dray controversy about the nature of historical
explanation has numerous antecedents, of course. Perhaps the
most important figure who requires mention in this connection is
R.G. Collingwood (*see especially* 1922, 1939, 1940, 1946). Collingwood is often described as the major proponent of the view that
history is unique or *sui generis*, that history is a matter of recreating
past experience, and that causality has a different function than it
has elsewhere, especially in science. The crucial distinction between science and history is stated by Collingwood in this way:

> When a scientist asks "Why did that piece of litmus paper turn
> pink?" he means "On what kinds of occasions do pieces of litmus
> paper turn pink?" When an historian asks "Why did Brutus stab
> Caesar?" he means "What did Brutus think, which made him decide

to stab Caesar?" The cause of the event, for him, means the thought in the mind of the person by whose agency the event came about (1946: 214–215).

This is clearly, at least in part, a matter of explanation by reasons or motives or intentions; and according to Collingwood there is no special inferential problem about knowing the motives or intentions of others, for we can rethink them within our own minds by recreating the experience of the agent. Clearly, this view is attractive because it mediates between the respects in which we think of motives as both like and unlike causes; but it is also disturbing because of its being couched in overly metaphysical language which leaves us with the uneasy feeling that we have not really understood the process involved.

Even more interesting for our purposes is Collingwood's discussion of whether history and science are different kinds of knowledge. Collingwood views history as a science of the past based on evidence. He argues against the view that history is nonscientific because it is concerned with the particular whereas natural science is concerned with the universal; both of these are false abstractions:

> The scientist generalises, certainly: but generalisation is subordinate to his real work as a scientist, the interpretation of individual fact. But the historian does not remain at a level of thought below generalisation: he generalises too and with exactly the same kind of purpose. Such generalisations as charters, medieval scripts, types of handwriting characteristic of the early fourteenth century, guild institutions, and so forth, go to the interpretation of a scrap of parchment which fits into its place as a link in the history of a town precisely as fossils, Jurassic fauna, shells peculiar to the Portland beds, and so on, are the concepts through which a geologist works out the geological history of a valley (1922: 449).

In a continuation of this passage, Collingwood argues for the importance of the interpretation of individual fact, the reconstruction of historical narrative. Archaeologists may have forgotten this in their pursuit of the scientific, but both science and history need to be seen as processes rather than as products, in which case the alleged differences of method and of logic disappear (cf. Spaulding 1968). The elements of generalization and individualization, far from being characteristic of science and history respectively, are essential to both. Data require interpretation in archaeology (which Collingwood classes, with paleography, numismatics, and such, among the sciences that constitute, or at least contribute to, history) as much as they do in any science, or indeed, in philosophy. The antecedents of some of Dray's views in Collingwood are thus clear, even if couched in rather different, and indeed Kantian, language.

From the preceding discussion of Dray's objections to the Hempelian view of explanation, three important differences between Dray's and Hempel's approaches emerge. We touched upon one of these differences, the greater reluctance on Hempel's part to include pragmatic features in an account of explanation, in connection with Dray's views on explaining "how-possibly," and we will discuss this aspect of Hempel's view in greater detail later when the work of Scriven and Achinstein on explanation is considered. Another difference is their disagreement as to whether improbable events can be explained—Hempel saying that they can't and Dray that they can, which also arose in the discussion of "how-possibly." In the next section, where we look in some depth at Salmon's account of explanation, Hempel's position on this matter will again be considered, for the issue is a major one dividing Salmon and Hempel.

The third, and most significant, of the differences between Hempel and Dray concerns the role of general statements in explanation, a difference apparent in their accounts of explanation by concept, genetic explanation, and rational explanation (or explanation by reasons). The issue here is a somewhat complex one, for it has both formal and substantive aspects. When Dray argues that general statements are not always part of adequate explanations, sometimes what he is saying is that they need not *appear* in explanation, i.e., need not be made explicit, which is a claim about the formal character of explanations. On the other hand, sometimes what he is saying is that they need not play any part at all in explanation, which is a claim of substance about the content of adequate explanations. With respect to the former claim, Hempel replies that all complete explanations involve explicitly general statements. This raises the question of what standard of completeness is appropriate, a question which will be considered in greater detail later in this chapter in connection with Achinstein's views. With respect to the latter, Hempel, of course, argues that without general statements a putative explanation is completely devoid of explanatory force, and this controversy involves an issue that is a very deep and central one in that part of the philosophy of science most relevant to archaeology where, as in history, general laws are not easily discovered.

One of the most tenacious critics of Hempel's account has been Michael Scriven (*see* 1958; 1959a; 1959b; 1962; and 1963), and his criticisms have ranged rather widely. He has argued, for example, that one can often offer perfectly acceptable explanations that one cannot justify by reference to laws. One can explain, for example, how the carpet was damaged by saying: "I did it, by knocking over the ink." Scriven claims that such explanations require no knowledge of the relevant laws of physics and make no appeal to them.

Hempel's reply is, of course, that the kind of explanatory statement Scriven envisages "claims by implication that the antecedent circumstances invoked were of a kind which generally yields effects of the sort to be explained. Indeed, it is just this implicit claim of covering uniform connections which distinguishes the causal attribution here made from a sequential narrative" (Hempel 1965: 360–361). According to Hempel, an explanatory statement of the form "q because p" implies the existence of certain covering laws which ground the explanation. And this reinstates the covering-law model.

Scriven has objected that the Hempelian account of explanation fails to consider a great many ordinary uses of "explain," such as explaining the rules of a game, explaining the meaning of a difficult literary passage, etc. In all these cases there is some basis that is understood, and the explanation relates what is to be explained to that basis. Hempel takes the position that these considerations relate to the pragmatic rather than the logical aspects of explanation and it is only the latter with which his models are concerned. Nevertheless, Scriven's objection may be important for the question of applications to archaeology, for it may be that pragmatic features are as crucial as logical ones in that context. For example, it seems clear that the standard demanded of explanations in archaeology varies with the novelty of the phenomenon being explained (*see* Chapter 4 for related points). Thus, archaeologists don't have to give complete explanations of Carbon 14 dating methods now that an understanding of that technique is virtually universal among archaeologists. In contrast, an archaeological explanation involving specific and unfamiliar concepts from systems theory would have to be more detailed. (For further disussion of the pragmatic aspects of explanation, see the next section where Achinstein's work is considered.)

Another point on which Scriven and others have taken issue with Hempel is the matter of whether every explanation is also potentially a prediction. According to Hempel, since "in a fully stated D-N explanation of a particular event the explanans logically implies the explanandum, we may say that the explanatory argument might have been used for a deductive prediction of the explanandum-event *if* the laws and the particular facts adduced in its explanans had been known and taken into account at a suitable earlier time" (Hempel 1965: 366). But further, Hempel holds that the thesis of the symmetry between explanation and prediction is supported also by a more general principle which, he thinks, constitutes a general condition of adequacy for any acceptable explanation. The condition is this:

> Any rationally acceptable answer to the question "Why did event x occur?" must offer information which shows that x was to be ex-pected—if not definitely as in the case of D-N explanation, then at least with reasonable probability. Thus, the explanatory informa-tion must provide good grounds for believing that x did in fact occur; otherwise that information would give us no adequate reason for saying: "That explains it—that does show why x occurred." And an explanatory account that satisfies this condition constitutes, of course, a potential prediction in the sense that it could have served to predict the occurrence of x (deductively or with more or less high probability) if the information contained in the explanans had been available at a suitable earlier time (Hempel 1965: 367–368).

Scriven offers the following counterexample. We can explain a certain patient's having paresis by pointing out that he previously had syphilis, and the only cause of paresis is syphilis. The paresis could not have been predicted, however, for only a small propor-tion of syphilis patients develop paresis. Hempel's counterclaim is that we do not have an explanation here at all: he writes "precisely because paresis is such a rare sequel of syphilis, prior syphilitic infection surely cannot by itself provide an adequate explanation for it" (Hempel 1965: 369). In respect of the state of knowledge Scriven describes, on Hempel's view we do not have an acceptable explanation of a person's developing paresis. If in the future we come to find out that some, let us say, chemical imbalance is present in those syphilitics who develop paresis and absent from those who don't, then having had syphilis *and* this particular chemical imbalance might adequately explain the paresis and at the same time provide a basis for predicting cases of it.

It may be that we are simply faced here with a conflict of intuitions, Scriven holding that we do have an adequate explanation in this case, and Hempel that we do not. But there are other arguments to be presented on Scriven's side, and these relate to some of the relevance conditions we mentioned earlier. As a matter of fact, it has been forcefully argued that there is a whole class of cases in which it is incorrect to think of showing that the event in question was to be expected as a desideratum of an adequate explanation. We discuss this claim in the next section where we deal with Salmon's treatment of Scriven's paresis example.

Some alternative models

As the discussion of explanation developed in the literature, the various Hempelian models seemed to crystallize into alternative views. There are quite a number of these accounts, including, we would suggest, Dray's criticisms of Hempel, since his criticisms, unlike those of Scriven perhaps, seem to amount to a clear alternative. In this section, however, we deal with just three that seem to point in particularly interesting directions.

(1) The first is a view characterized by its main proponent, Wesley Salmon, as a statistical-relevance (S-R) view. Salmon's work (*see* Salmon 1971) having become familiar to archaeologists over the last number of years, this may be the development of greatest interest in archaeological circles. Earlier in this chapter we noted some examples (deductive and inductive) of purported explanations which fit the Hempelian models but which could not be regarded as genuine explanations because the explanatory facts adduced are irrelevant to the explanandum event. There is no essential disagreement between Hempel and his critics, including Salmon, regarding these examples, all of which can be dealt with by a version of the requirement of explanatory relevance developed by Hempel, namely his requirement of maximal specificity. According to Hempel, the idea of the requirement of maximal

specificity "comes to this: In formulating or appraising an I-S explanation, we should take into account all that information provided by K which is of potential *explanatory* relevance to the explanandum event; i.e., all pertinent statistical laws, and such particular facts as might be connected, by the statistical laws, with the explanandum event" (1965: 400–401). K is the set of all statements accepted in a given "knowledge situation." Salmon points out a difficulty with Hempel's requirement, and offers a reformulation, but we need not enter into the technicalities here. Suffice it to say that the differences between Hempel and Salmon on this issue are only differences of detail.

The important divergence comes with other cases. Scriven's paresis example is one such case. Scriven regards it as a genuine counterexample to the Hempelian view, but does not offer an alternative account of explanation that would incorporate this and similar cases. Salmon remedies this lack. Roughly, he disagrees with Hempel on the point that explanations show the event in question "was to be expected." According to Hempel this means, for the statistical case, that the event to be explained must be shown by the explanation to have been highly probable. Salmon, on the other hand, maintains that when the probability of an event, independent of the explanatory facts cited in explanation of it, is very low, it is not necessary that its probability relative to those facts be made high in order to have an inductive explanation. The probability of a person's contracting paresis is very low, and even though the probability that a person with untreated latent syphilis will contract paresis is still low, it is considerably higher than the probability of paresis among people in general. Thus, in Salmon's view, untreated latent syphilis is explanatory of paresis (Salmon 1971: 56–57).

To counter the claim that this is really just a partial explanation which we believe will be filled out when we acquire further knowledge, Salmon cites an example involving radioactive decay where, he argues, there is less reason to suppose that further knowledge will yield a more complete explanation:

> Suppose we had a metallic substance in which one of the atoms experienced radioactive decay within a particular, small time period, say one minute. For purposes of the example, let us suppose that only one such decay occurred within that time period. When asked why that particular atom decayed, we might reply that the

substance is actually an alloy of two metals, one radioactive (for example, uranium 238) and one stable (for example, lead 206). Since the half-life of U^{238} is 4.5×10^9 years, the probability of a given uranium atom's decaying in an interval of one minute is not large, yet there is explanatory relevance in pointing out that the atom that did decay was a U^{238} atom (Salmon 1971: 57).

Since we have no reason to think that we will ever be able to distinguish among U^{238} atoms on the basis of radioactive decay, Salmon concludes that "it is not necessary to suppose that examples such as the paresis one can derive their explanatory value solely from the conviction that they are partial explanations which can someday be made into full explanations by means of further knowledge" (Ibid.).

Salmon believes that his account will also serve to take care of Bromberger's flagpole example and similar cases. To do this he capitalizes on the temporal asymmetry of causation. The "flagpole produces the shadow in a sense in which the shadow certainly does not produce the flagpole" (Ibid.: 72). The intuitive idea is that explanations can also involve an element of temporal asymmetry, even where it may at first sight appear that the events involved are simultaneous. Salmon's account is also able to sustain a thoroughgoing adherence to the thesis of the symmetry of explanation and prediction.

In Hempel's account, an adequate D-N explanation shows that the explanandum event was to be expected and an adequate statistical explanation shows that its occurrence was highly probable. Thus, an explanation can be treated as an argument in which the explanandum can be inferred from the explanans. This aspect of Hempel's theory gives rise to the fundamental difference between his and Salmon's accounts: Salmon does not regard explanations as arguments or inferences because, in his view, there are insuperable problems with conceiving of inductive logic (under which Salmon subsumes deductive logic as a limiting case) as being concerned with inductive arguments that have premises and conclusions. Behind this lies the worry that inductive logic does not contain rules for the detachment of premises from conclusions, a view which Salmon illustrates with the famous lottery paradox:

> If inductive logic contains rules of inference which enable us to draw conclusions from premises—much as in deductive logic—then

there is presumably some number r which constitutes a lower bound for acceptance. Accordingly, any hypothesis h whose probability on the total available relevant evidence is greater than or equal to r can be accepted on the basis of that evidence. (Of course, h might subsequently have to be rejected on the basis of further evidence.) The problem is to select an appropriate value for r. It seems that no value is satisfactory, for no matter how large r is, provided it is less than one, we can construct a fair lottery with a sufficient number of tickets to be able to say for each ticket that [it] will not win, because the probability of its not winning is greater than r. From this we can conclude that no ticket will win, which contradicts the stipulation that this is a fair lottery—no lottery can be considered fair if there is *no* winning ticket (Salmon 1971: 77).

Further, even if the difficulty revealed by the lottery paradox could be circumvented, there would still be problems with saying that the conclusion detached from the statistical explanation had any particular probability. "For," as Hempel has written of this matter, "the credibility of an empirical assertion—in sharp contrast to its truth or falsity—depends on the available evidence" (Hempel 1965: 62). Thus: "If terms such as 'almost certainly', 'probably', and 'with probability r' are to express the force of the inductive statistical arguments we have been considering then they must be understood, not as qualifiers of single statements [i.e., detached conclusions], but as representing relations between statements" (Ibid.: 60).

Even in Salmon's view there are statistical cases where the inferential model of explanation seems correct, but these are where the probability of the event to be explained is very high—indeed, so close to 1 as to make no real difference in any decision problem (Jeffrey 1971: 20–21).

Now, it should not be thought that either Salmon or Jeffrey would deny that statistical explanations usually support, or involve, statistical inferences. Their view is rather that, in other than the cases of very high probability (i.e., the "beautiful cases"), the inferential aspects of those explanations should not be identified with the explanatory. As Jeffrey says:

[W]here the strength of the inference is more modest [than a probability of nearly 1], I think it simply wrong to view the inference as an explanation and to identify the strength of the inference with the strength of the explanation. To explain the phenomenon that there was at least one head in two tosses of a coin, I would point out that the process is stochastic with probability 1/2 of heads on each toss, and with different tosses independent of one another. I would give the same explanation if matters turned out differently: if, improbably, there had been no head on either toss. The difference between the two cases would be entirely with the gloss: in the first case one would point out that the probable happened, while in the second one would point out that the improbable happened. But the strength of the explanation would be the same in either case (Jeffrey 1971: 27).

And the strength of the explanation in each of the cases is the same because the process involved, i.e., the coin tossing, is governed by a probabilistic law. In fact, each series of heads and tails that can be generated by the tossing of a coin would be explained in the same way as any other, even though the probabilities of each would not necessarily be the same. As Jeffrey says, "[W]e shall know all there is to know about the why of it and the how, when we know that the process which yielded the [particular series] is a random one and when we know the probabilistic *law* governing the process. The knowledge that the process was random answers the question, 'Why?'—the answer is 'By chance'" (Ibid.: 24).

Salmon's S-R account of explanation gets around these difficulties quite well by not casting explanations in the form of arguments. It thus fits in nicely with a view of inductive logic that does not depend upon there being inductive inferences, and we shall see in Chapter 6 that such a view allows us to integrate our conceptions of explanation and confirmation into a plausible general account of the logic of science.

It should be noted that Salmon has not claimed that his S-R model characterizes the only legitimate kind of explanation (*see especially* his "Postscript" 1971: 108–110; *see also* Salmon 1975a). As we have mentioned above, however, he does hold that deductive explanation is accounted for by his model: because the explanans of a deductively valid explanation on the explanandum has a probability of 1, Salmon treats deductive explanation as a limiting case of statistical explanation. It is also interesting to note that he classes the hypothetico-deductive method as inductive because one aspect of the method involves confirmation and, therefore, inductive logic.

(2) Peter Achinstein (1968 and 1971) offers another account of explanation that has some advantages over the received view. These are mainly that it makes pragmatic and contextual features an integral part of the explanation rather than trying to separate them from the "logical" features—a task which, as we have seen in our discussion of Scriven, for example, is extremely hard to carry out (*see* Hempel 1965: 425–433). Achinstein holds that what will count as an explanation or as an attempt to explain something depends, in part, upon the sort of person to whom one is doing the explaining. The point is to render something understandable to a person or persons in a particular situation, and the material cited for this purpose will vary with the persons and situation. The supposition is that the persons to whom we are explaining either do not already understand what is being explained or, if they do, can be treated as though they do not. The concept of understanding is thus basic to Achinstein's notion of explanation, and what is to count as someone's having understood that which is being explained would, of course, have to be spelled out.

The question also arises as to criteria for deciding whether an explanation is a good one, and here Achinstein offers several considerations that can be used; explanations, he suggests, can be evaluated on grounds of relevance, correctness, depth, completeness, unification, and manner of presentation. In keeping with the emphasis he puts on the pragmatic aspects of explanation, Achinstein argues that even in science—which, on the face of it, presents us with more obviously standardized situations than we find in everyday cases where we need to offer explanations, it is still important to relativize the notion of explanation to the situation.

Besides insisting that pragmatic and contextual features are essential to an adequate account of explanation, Achinstein disagrees with the Hempelians in another fundamental respect. He argues that it is not necessary for a satisfactory explanation to contain general statements of laws, that laws are often not invoked, and that even when they are, there is no reason to suppose that they must be considered part of the explanation itself. For example, a "physicist may explain why a certain track was produced in the cloud chamber by saying that an alpha particle passed through the chamber" (1968: 102). Not surprisingly, Hempelians reply to this suggestion by arguing that if all that is implicit in such explanations was made explicit, then general statements would be revealed as integral to such completed explanations. Achinstein's rejoinder shows, once again, the importance he places on the pragmatic features of explanation:

> The problem with this reply is that whether an explanation can be judged to be complete depends in part on the knowledge and concerns of those in the situation in which it is given. In order to provide an answer to Q the explainer may need to provide answers to the questions related to Q, and the completeness of his explanation will be determined in part by the extent to which it does this. But what questions these are depends on the situation. Completeness is not something that can be judged solely by examining the explanation and the Q (Ibid.: 102–103).

To support his claims about the importance of contextual considerations in scientific explanation, and against the need for explicit appeal to laws, Achinstein cites examples that he thinks we would all accept as paradigms of scientific explanation from such physicists as James Clerk Maxwell and J.J. Thomson. What is important about this is that, if Achinstein is right, then satisfactory explanations, even in the physical sciences, need not follow the rigid patterns we have been led to believe. This would remove one obstacle to the acceptance of explanations in archaeology, where laws often are not explicitly cited. But we should remember that Hempel's reply would very likely be that general laws, even in this case, are implicit in our explanations.

(3) In *The Explanation of Social Behaviour* (1972), R. Harré and P.F. Secord put forward an account of explanation that they think is a clear and preferable alternative to the positivistic behaviorism that has been adhered to by most psychologists, and indeed most social scientists, during the first five or six decades of the twentieth century. Though a number of different accounts could have been chosen as examples of this sort of alternative, the Harré and Secord view is of particular interest for archaeologists for two reasons. First, that view incorporates some principles of systems theory, a field which has captured the attention of a number of archaeologists; and, second, some anthropologists think that the Harré and Secord account may provide a basis from which an adequate model of explanation in anthropology may be found (cf. Hanson 1974). Since advances in anthropology are often relevant to, and often eventually adopted by archaeology, the serious interest on the part

of anthropologists in Harré and Secord's "realist, nonpositivism" gives archaeologists reason to expect that the Harré and Secord account may prove useful to their discipline as well. In addition, the underlying realism in the Harré and Secord account is of interest both because of the central place scientific realism has come to occupy in a philosophical account of theories and because some archaeologists have put foward a version of this view as a correction of the earlier positivist-empiricist underpinnings of the New Archaeology (cf. Gibbon 1983 and 1984). The methodological rigidity of the positivist framework, together with its questionable applicability to "real" archaeological problems seem to have motivated the turn to realism among some archaeologists. As well, it may be that the renewed importance of ethnographic analogy, and the emergence of ethnoarchaeology, have also encouraged the appeal to realism through a linkage of the "real" ethnography with archaeological evidence. In this section we confine ourselves to a discussion of the development of realism in one of the more obvious social science contexts. Scientific realism and other related developments are discussed more fully later in this chapter.

While Harré and Secord may have an exaggerated view of the extent to which their account of explanation differs from, for example, Hempel's, there is no doubt that their account is significantly different from the behaviorism that emerged from positivism. That positivism, as we have already discussed, was distinguished by its followers' attempts to formulate a criterion of cognitive significance. What lay behind their efforts was the view that only propositions that could be verified through observation had meaning. Thus, logical positivism, as we have already noted, was a form of empiricism, a philosophical movement essentially concerned with issues of observability.

Behaviorism was, as well, a product of the empiricist tradition. While the positivists dealt with questions about the meaningfulness of language, however, the behaviorists had more epistemological and metaphysical concerns: if putative entities cannot be observed then their existence and our ability to know about them are in doubt. As a result, scientifically respectable conclusions can hardly be drawn about them and such entities are not the proper subject of scientific study. For example, in the behavioristic model, thoughts, because they are not observable, cannot be the subject of legitimate social scientific investigation.[4] If we want to understand a human being's actions, then we must focus solely on his or her observable behavior (and, under certain circumstances, his or her physiology). An account which relies upon the individual's thoughts simply will not do.

The connection between positivism and behaviorism, then, is a close one, but not perhaps as close as it has often been thought to be. For the sake of clarity, at least, we need to note that one *could* be a behaviorist without espousing logical positivism, and, at least in theory, the converse is true as well.

In putting forward their alternative to behaviorism, Harré and Secord argue that social scientists, and social psychologists in particular, are studying the wrong entities. Instead of studying human beings, as they do now, as if they were only "object[s] responding to the push and pull of forces exerted by the environment," they should, as well, be studying people as if they were "agent[s] directing [their] own behaviour" (1972: 8). Harré and Secord argue that the behaviorists' mechanistic model of human beings is inadequate. They claim that the most "unique feature" of human beings is their "capacity to monitor the control of [their] own actions" (Ibid.: 6), and that because behaviorism cannot take this feature into account, it is an inadequate theory for social science.

It follows from their insistence on human beings' having the "capacity to monitor the control of [their] own actions" that some causes of human action are not external to human beings. While the position is a controversial one, some philosophers have argued that reasons are phenomena internal to human beings which can and often do serve as causes of human action (cf. Davidson 1963: 685). Following in this tradition, Harré and Secord explicitly adopt the view that, while not all reasons are causes, some are. They use the example of a person seen to cross the street, enter a building, and return a few minutes later. If asked why he did this, he may reply with a reason:

> that he went to buy cigarettes. [T]his may direct us to quite a different structuring of reality from that revealed by the application of the concept of cause, in its mechanistic signification. It directs us to consider such items as the plan according to which the man acted, or the impression he was trying to create and so on (1972: 40).

While their view that reasons can function as causes may be contentious, nowhere do Harré and Secord deny the received view that one of the major goals of science is causal explanation. They do want to distinguish between the causal "mechanisms" of the natural and social sciences, however. They say that "It is the self-monitored following of rules and plans [by human beings] that we believe to be the social scientific analogue of the working of generative causal mechanisms in the processes which produce the non-random patterns studied by natural scientists" (Ibid.: 12). This issue of the difference between the natural and the social sciences is one to which we shall return shortly.

What is essentially different about their view of the explanation of human behavior from that of the positivists is that Harré and Secord are committed to the view that there are such things as intentions, reasons, and expectations, and it is in this respect that their view is appropriately called realism. The behaviorists, on the other hand, at least wanted to avoid a positive commitment to such entities, and some behaviorists went further, claiming that these things simply didn't exist. The issue of whether scientific realism is a better theory than positivism, and, indeed, whether these are the only available alternatives, is one we shall address in a moment. The problem is complex, and is related to the question in the social sciences as to whether such things as intentions, reasons, or expectations exist, which is part of a larger debate concerning the general status of theoretical terms and especially whether those terms refer to real entities or not. The literature on this controversy is vast (*see* Carnap 1970; Hempel 1970; Scheffler 1970; Maxwell 1970; Achinstein 1970; Sellars 1970; Nagel 1961, Chapter 6; and Suppe 1977).

One query that remains is whether the realist and the positivist views differ not only regarding what there is in the world for science to investigate, but also regarding what counts as scientific investigation. It is unquestionably true that the data which in Harré and Secord's view are relevant, for example, to confirmation of theories about human behavior will be different from those which were relevant in the behaviorist view. As Harré and Secord say: "[t]he things that people say about themselves and other people should be taken seriously as reports of data relevant to phenomena that *really exist* and which are *relevant* to the explanation of behaviour. This contrasts with the mistaken view that the statements themselves are the phenomena" (1972: 7).

Further, they say that "There is no simple route to discovering [causal mechanisms]. Some may be simply inspected; others are only quasi-accessible" (Ibid.: 5–6). Thus, explanation in physical science is meant to be seen as complex and fraught with many of the same difficulties as explanation in social science, especially once we accept the fact that these explanations are meant in terms of real entities actually existing in the world. In neither case can we know for certain that our explanations are correct or immune from revision.

Harré and Secord are, in general, very eager to emphasize the novelty of their account of the scientific investigation and explanation of human behavior, and it is clear that in a number of respects their account is significantly different from the received views. In other respects, however, the differences are much harder to discern. We have already mentioned their acceptance of the widely held view that causal explanation is the primary goal of science. There are many other similarities. For example, the positivists were ultimately concerned with ensuring that their theories, laws, and explanations had cognitive significance. They were concerned to show, that is, that evidence could count for or against their theories, laws, and explanations. Harré and Secord are concerned in part with precisely the same thing. And there is nothing in their view that suggests a logic of confirmation substantially different from that of the positivists. Nor is there anything in their view that suggests a substantially different logic of explanation from that of the positivists.

The main difference between the Harré and Secord approach and the earlier standard one, then, stems from their acceptance of realism, and the main focus is on their rejection of behaviorism— the positivists' philosophy of mind. They believe that the behaviorist reduction of "complex actions to simple, independent behavioural elements, capable of independent explanation, is impossible. . . . It is complex and deliberate actions, unified through their contributions to the meaning of the total act, that constitute the true subject matter of human social behaviour" (Ibid.: 40). Harré and Secord are recommending that social scientists accept as existing—and study—different entities from those which they have been studying through most of this century, and this constitutes a large part of their appeal to archaeologists. But, it appears, the results of these studies should be evaluated from a point of view that is not substantially different from that held by philosophers in the positivist tradition.

Social psychology, of course, is not the only area in which this sort of point has lately been made. Though the behaviorist tradition was strongest in psychology of all the social and behavioral sciences, there have recently been numerous criticisms of the standard models within psychology. Examples of the criticisms levelled include: (a) the artifactual nature of experimentation as a means to assess issues in human functioning; (b) the lack of generalizability due to subject and experimenter bias; (c) excessive emphasis on manipulation and not enough on criterion development; (d) the absence of identified laws of human interaction; (e) the sterility and lack of applicability of theories in psychology; (f) the fallacious attempts to explain the dynamic functioning of human beings using only static models and techniques; and (g) the effects that social structures exert on researchers. Many psychologists have acknowledged that these criticisms are accurate, at least in part, and a number have been attempting a radical reorientation of approach in studying psychological phenomena and social interaction.

In related fields, sociologists and anthropologists have long used "subjective" data from informants in constructing pictures of the workings of institutions and societies, and this sort of method has come to be recognized as more respectable in many areas that share important features with archaeology. For example, we have recently seen a much wider acceptance of the value of results of certain kinds of interviewing and psycho-history in such fields as ethnology, anthropology, and history.[5] Thus, there are signs that the narrowness of methodological approach that went along with the earlier positivist metaphysics is coming to be replaced with a more tolerant attitude to possible means of acquiring knowledge—though, one hopes, with no diminution of intellectual rigor.

Criticisms and alternatives—a summary

As our discussion of Hempel's critics has shown, some of the main tenets of the Hempelian account of explanation have been vigorously attacked. The requirement, for example, that a law must be part of any adequate explanation has been called into question by many, including Scriven, who claims rather straightforwardly that many perfectly satisfactory explanations do not involve laws. The requirement also has been challenged by Dray, in the context of philosophy of history. His arguments must be seen as motivated in part by the fact that general laws in history, as in archaeology, are difficult to come by. Dray is committed to the view that it is the job of historians to explain the significant human events of the past. But if historians are to explain, and if there are not—or, more importantly, cannot be—historical universal generalizations, then an alternative to the Hempelian account must be found. Thus, the historian's, or archaeologist's, search for an account of explanation that does not require laws is understandable. Nonetheless, we fear that it is fundamentally misguided. While there are (as we already have pointed out and will point out again below) some weaknesses in the Hempelian account of explanation, Hempel's claim that laws are always, at least implicitly, part of explanations seems sound. To show how or why something, x, occurred would seem to require making clear what brought x about, i.e., what caused x. Even if correct, this doesn't mean that only causal explanations—those involving universal, or deterministic, laws—will do. Sometimes, in the words of Jeffrey (1971: 24), "we shall know all there is to know about the why of it and the how" when we know that the process which resulted in x was a probabilistic one, and in these cases statistical generalizations will have to suffice. But the notion underlying explanation may perhaps be said to be an illumination of the *process* which yielded the explanandum event. And since processes would seem, by their very nature, to be rule governed, it is very difficult to see how explanation could be achieved without the invocation of laws of one kind or the other.

Another central aspect of the Hempelian account which has been widely challenged is Hempel's insistence that a philosophical account of explanation should be concerned only with the logical features of explanation, not with the contextual or pragmatic. We have primarily considered the views of Scriven and Achinstein on this point, and their claim that extralogical features could be, and probably will have to be, part of any adequate account of explanation seems to us correct and the Hempelian position on this issue arbitrary. If explication of the concept of explanation is the object of the exercise and an approach which has addressed itself strictly to logical considerations has not succeeded, it is not readily apparent why further attempts at that explication should not try to incorporate extralogical features. We shall have more to say on the pragmatics of explanation in the next section.

A third element of the Hempelian account which has been disputed is the claim that properly construed explanations show that the events being explained were to be expected, or at least that their occurrence was highly probable. Salmon has been Hempel's main critic on this issue, and his views are compelling; it does seem as if we both want, and can have, explanations of improbable events and, therefore, that Hempel is wrong on this point. A related matter, and one which is central to Salmon's statistical-relevance model of explanation, is whether explanations should in general be characterized as arguments, or inferences. Salmon argues that they should not be—that the relationship between an explanatory sentence and the statement of which it is the explanation is not that of premise to a conclusion, but instead one of relevance. And while Salmon's account seems preferable to Hempel's on these points, it raises the problem of characterizing relevance.

We have also considered quite a different sort of attack on the positivistic approach to explanation—that made by Harré and Secord. The chief distinction between the positivistic account and that of Harré and Secord concerns what sort of entities social scientists should study. Motivated by worries about their unobservable nature, the positivists claimed that such things as reasons and beliefs, even if they existed, could not properly be the objects of scientific investigation. Harré and Secord, on the other hand, argue that only by studying such entities will social scientists be able to explain the aspects of human behavior in which we are most interested. And whether or not it is correct to say, as Harré and Secord do, that a reason can ever function as a cause, it does seem to us that the most intriguing features of human beings are, probably unnecessarily, put beyond the scope of science in the positivists' view. Thus, Harré and Secord's recommendations clearly point in a useful direction, whether or not realism can ultimately be sustained.

In addition to the Harré and Secord position, we considered two other accounts of explanation—those of Salmon and of Achinstein. An interesting feature of all three of these alternatives to the Hempelian account is that they try to break loose from the narrow confines of an account simply of the logical structure of explanation. In so doing, not only are considerations which Hempel regarded as "merely" pragmatic introduced as central features of explanation, but also concepts such as relevance, of which we as yet have no adequate logical account, are relied upon. In addition, in some cases entities often regarded with suspicion are accepted as part of the data base. This shift encourages us to view as real things many entities (motives, desires, communities) that were earlier seen as merely hypothetical, or even as inherently nonverifiable and, therefore, beyond science. Whether giving an account of more than the logic of explanation is necessary for the explanatory enterprise remains to be seen. But at the very least, it is clearly an important part of the current search for more adequate models of explanation, and it appears that it may have much greater significance than that. As we suggested in Chapter 1 in respect to both the positivists' search for a criterion of meaningfulness and Hempel's account of explanation, the assumption of formalizability may be disastrously inappropriate.

Recent developments

As we noted earlier, the various criticisms of the Hempelian model of explanation have been crystallizing over a number of years into alternative views; but the alternatives that have come to occupy the field at present are more than just reactions to the D-N model. Indeed, it is not at all uncommon to view the covering-law models of explanation as merely one manifestation of the positivist philosophy of science, the central tenet of which was a particular view of scientific theories, often referred to as the Received View (Putnam 1962).[6] According to this view, scientific theories are to be construed "as axiomatic calculi which are given a partial observational interpretation by means of correspondence rules" (Suppe 1977: 3). This Received View of Theories was seen also to encompass the standard positivist accounts of explanation, intertheoretic reduction, induction, and confirmation, and its final and most sophisticated version was thought to be owing to Carnap and Hempel (*see* Suppe 1977: 50–53).

By 1977, Suppe was able to declare quite confidently that:

(1) Positivistic philosophy of science has gone into near total eclipse;
(2) the more extreme *Weltanshauungen* views of Feyerabend,

Hanson, and Kuhn no longer are serious contenders for becoming a replacement analysis; and (3) philosophy of science is coalescing around a new movement or approach which espouses a hard-nosed metaphysical and epistemological realism that focuses much of its attention on "rationality in the growth of scientific knowledge" and proceeds by the examination of historical and contemporary examples of actual scientific practice (Suppe 1977: 618).

The first two of these we have already had occasion to see, in part explicitly and in part implicitly. The first has been treated in connection with the Hempelian models of explanation, and will be discussed again in the next chapter where we deal with the nature of scientific inference. The second, however, requires a bit more explicit treatment here.

Suppe refers to the *extreme Weltanschauungen* analyses of Feyerabend, Hanson, and Kuhn, intending thereby to distinguish the views of these writers from those of such writers as Toulmin, Lakatos, and Shapere, with which they nevertheless share some features. The fundamental criticisms of the extreme views turn primarily on the charges

> that their positions lead to an extreme subjective idealism which is incompatible with the objectivity of science; that their doctrines on meaning, incommensurability, and the theory-ladenness of observation make impossible any rational basis for comparing the empirical adequacy of competing theories; and that it debases knowledge into the joint prejudices of the members of a scientific discipline who accept the same theory (Suppe 1977: 636).

Feyerabend is certainly the most extreme of these writers, and his refusal for the most part to engage in the usual exercise of scholarly and critical debate has led to his views' being ill understood and often caricatured. But it is not a caricature to note that "his view makes the rejection of a global theory on the basis of observation a fundamentally irrational process" (Suppe 1977: 640), and this seems a serious shortcoming. Further, it is unclear what is the relation between the acceptability or truth of a theory and its correspondence (or lack of correspondence) with reality in Feyerabend's scheme, and this, too, seems at best counterintuitive. Feyerabend advocates proliferating hypotheses inconsistent either with well-established theories or with well-established facts, and proceeding counterinductively as well as inductively (Feyerabend 1975), in order that truth and knowledge will emerge in some dialectical fashion; but again, the means for achieving this dialectical resolution are just not clear and cannot be discerned from Feyerabend's writings.

Kuhn, as we have seen earlier, presents us with a very different case, for he has engaged in lengthy exchanges with and responses to his critics, often resulting in considerable clarification of and alteration in his position. To be sure, he has often insisted that his critics have misinterpreted him or that their disagreements with him are differences over detail rather than substantial disagreements. Still, there remain substantial difficulties with Kuhn's position. For one thing, as we noted in some detail in Chapters 3 and 4, the alleged oscillation between normal and revolutionary science is not clear. Indeed, as D.L. Hull (1975: 397) has maintained:

> The cyclic pattern which he describes for science is much too simplistic. In fact, rarely can any of the stages be found exemplified very clearly in the course of science. The periods which he had previously described as pre-paradigm contained paradigms not that different from those of normal science. Nor does normal science alternate with revolutionary science; both are taking place all the time. Sometimes a revolution occurs without any preceding state of crisis. In short, the most one can say for Kuhn's three stages of scientific development is that they are abstract, ideal types which certain episodes in science occasionally approximate, just the sort of abstract rational reconstructions which Kuhn found so objectionable in the work of philosophers such as Hempel.

In fact, Kuhn's views have developed in a direction that makes it appear that he is moving closer to a kind of neo-positivist stance on such matters as the interpretation of terms in theories. In addition, there are the problems of the short shrift given to the role of rationality in the growth of scientific knowledge and the sociological reduction of scientific knowledge to the beliefs of members of scientific disciplines.

Interestingly enough, one of the complaints commonly heard about Kuhn's work is that it is unclear concerning the role of discovering how the world really is in the growth of scientific knowledge. This complaint comes primarily from a perspective known as scientific realism, which has come to be the dominant metaphysic in current philosophy of science. As Suppe tells us:

> contemporary work in philosophy of science increasingly sub-scribes to the position that it is a central aim of science to come to knowledge of how the world *really* is, that correspondence between theories and reality is a central aim of science as an epistemic enterprise and crucial to whatever objectivity scientific knowledge enjoys—in sharp repudiation of the "sociological" views of knowl-edge found in the more extreme Weltanschauungen analyses while acknowledging the defects of positivistic and earlier empiricist treat-ments. This has led to an emphatic belief that an adequate philoso-phy of science must embrace a "hard-nosed" metaphysical and epistemological realism wherein how the world *is* plays a decisive role in the epistemic efforts and achievements of science (Suppe 1977:649).

One of the concomitants of this sort of emphasis has been an intensification of the phenomenon we noticed earlier—a very close scrutiny of actual scientific practice, both historical and contempo-rary. Discussions of explanation, confirmation, or theories reflect the actual usage of these concepts in fairly sophisticated scientific reasoning, including the basic assumption of most working scien-tists that science yields knowledge descriptive of how the world really is. The use of historical material is intended not merely as an illustration of philosophical views, but rather as evidence for them, in the same sense that the philosophical theory of the rationality of science is supported by an examination of actual scientific practice which does, it is claimed, yield knowledge much of the time.

A number of authors have developed this rationality approach in some detail. Lakatos (1970), for example, picks up where Popper leaves off, developing the idea of a scientific research program which does not collapse in the face of refutations. Instead, a re-search program, for Lakatos, is "an objective, 'third world' [objec-tive] reconstruction of Kuhn's sociopsychological concept of para-digm" (Ibid.: 179). Research programs are defined in terms of problem shifts which may be "progressive" or "degenerating" and which are generated in accordance with methodological rules com-prising a "negative heuristic" that tells us which research paths to avoid and a "positive heuristic" that tells us which paths to pursue. Lakatos provides extensive historical illustrations of these ideas which "provide at least a *partial* account of ways in which theoretical advance in science *could* have occurred under at least *some* circumstances" (Suppe 1977:670). But the role of rationality in the growth of scientific knowledge in general remains obscure.

Toulmin is another philosopher who has pursued rationality, but in a manner quite different from Lakatos. Rationality, for Toulmin, is an attribute of human activities and enterprises (*see* 1972: 43–134). That is:

> rationality in science is an attribute of the *collective* use, evaluation, criticism, and alteration of concepts. And rationality in the growth of scientific concepts is to be understood in terms of the *process* whereby scientific disciplines evaluate and change their concepts— which process Toulmin sees as being evolutionary in nature (Suppe 1977: 672).

Toulmin develops his evolutionary model in some detail, but he does not explain just how this leads to knowledge or accounts for the success of science in providing knowledge of the world. Thus, we do not yet have an adequate account of the role of rationality in the growth of science.

A brief consideration of the views of Dudley Shapere leads us back to the central theme of this section, namely scientific realism. Shapere has been one of the most assiduous investigators of actual scientific practice as a means of discerning patterns of reasoning that can be described as justified and as exhibiting the rationality of scientific development. One of his purposes is to "steer a middle ground between the excesses of logical positivism and the *Weltanschauungen* analyses" (Suppe 1977: 684), and he believes that the reasoning involved can be formulated without reference to psychological or sociological factors. Not only scientific development that occurs "relatively continuously" but also the introduction of radical new hypotheses can be seen to be rational and to involve justified patterns of reasoning. But rationality alone does not provide objectivity:

> The point of the rational evaluation of knowledge claims by science is to determine whether the claims do or do not express facts *about* the world—to determine whether what the claims assert is descriptive of how the world actually is. Thus part of the objectivity of science on Shapere's view consists in the items in domains being

putative *facts*—that is, on adopting some realistic view of truth wherein the truth of scientific claims depends on whether what the claims assert corresponds with the way the world is (Ibid.: 702).

Suppe believes that Shapere's work provides the most promising direction to be followed in developing an account of rationality and objectivity in science. More importantly, he argues that actual scientific practice presupposes realism and that this is at odds with the positivist Received View. This position—that realism supplants positivism and is the only viable alternative to it—seems to be in the process of being adopted by a number of archaeologists and by the philosophers to whose work they turn (*see* Gibbon 1983 and 1984, D. Miller 1982, Wylie 1981, Bhaskar 1978, Harré 1979). The interest in scientific realism among archaeologists is not necessarily a hopeful sign, especially if realism is somehow viewed as a model for *doing* archaeology; for, as we noted at the beginning of this chapter, scientific realism is no more able to provide a model for doing archaeology than was the earlier positivism. But even assuming that views on the relation between science and the world are, as they should be, seen as accounts that are meant to promote our understanding of the nature of science rather than to direct our practice of it, there are alternatives to scientific realism that are extremely appealing and that make it clear that realism is not the only possibility. One of the most interesting alternatives to realism is the one developed by Bas van Fraassen in his *The Scientific Image* (1980).

Van Fraassen develops accounts of the relation of a theory to the world, of scientific explanation, and of probability within physical theory. Only the first two subjects will concern us here, the second primarily in this chapter and the first in Chapter 6, though it is important to note that van Fraassen believes the three accounts "need each other for mutual support" (1980: vii). He develops a nonpositivist "constructive" empiricism—constructive in the sense that "scientific activity is one of construction rather than discovery: construction of models that must be adequate to the phenomena, and not discovery of truth concerning the unobservable" (Ibid.: 5). Much of the exposition of van Fraassen's constructive empiricism is devoted to an account of inference to the best explanation, belief, and theory acceptance, and we shall discuss these matters in Chapter 6; but some of them must at least be introduced here.

For van Fraassen, "to accept a theory is (for us) to believe that it is empirically adequate—that what the theory says about what is observable (by us) is true" (1980:18). He believes, then, that a distinction can be drawn between what is observable and what is not, even though he agrees with the realists (and virtually everyone else) that the positivist reconstruction of *language* into a theoretical and nontheoretical part is untenable. It is not philosophical theory but, rather, science itself which "delineates, at least to some extent, the observable parts of the world it describes" (Ibid.: 59). This does not, however, mean that science is confined to the observable; clearly theories go beyond the data and are underdetermined by them. Van Fraassen says:

> My view is that physical theories do indeed describe much more than what is observable, but that what matters is empirical adequacy, and not the truth or falsity of how they go beyond the observable phenomena. And the precise definition of empirical adequacy, because it relates the theory to the *actual* phenomena . . . does not collapse into the notion of truth (Ibid.: 64).

The important point, though, is that, even though a claim of empirical adequacy is much weaker than a claim of truth, both clearly go beyond the evidence, or beyond what we can *know* at any given time. Theory is involved in both cases, and the anti-realist's claim is just that his account provides a better and less metaphysical picture than does that of the realist. If this is so, then there is no particular reason why archaeologists must adopt a realist framework for their own work.

It should be noted that inference to the best explanation is often thought to lead to scientific realism. If this is so, how can we espouse the former and remain skeptical about the latter? The answer is that the rule of inference to the best explanation needs to be broadly enough interpreted so that it is taken to license "the acceptance of a theory or hypothesis, on the basis of its explanatory success" (Ibid.: 20). The crucial question then becomes: what is the set of rival hypotheses from which we are to choose the best? The realist and the anti-realist have different answers to this question.

The range of hypotheses for the anti-realist includes only those which attribute empirical adequacy to the theories under consideration and not also those that claim that every entity postulated by the theory, whether observable or not, exists. The anti-realist is concerned to insist that regularities in nature may or may not "have an explanation in terms of unobservable facts 'behind the phenomena'" (Ibid.: 24) and it is inappropriate to assume otherwise. Thus it seems to us that agnosticism about realism is the appropriate stance for archaeology for two reasons: first, from the point of view of accomplishing the scientific tasks at hand, it makes no difference whether one is a realist or not; and second, there are sound theoretical arguments that lead to skepticism about realism. The insistence on realism on the part of some archaeologists thus seems to be a red herring.

The anti-realist position, contrary to the view of some of its critics, is certainly meant to allow us to draw conclusions that go beyond the evidence, and thus the charge that it leads to a self-defeating scepticism is unwarranted. The choice among a range of available hypotheses or theories is made, according to van Fraassen, on the basis of how well the hypothesis explains the available evidence. This is essentially a matter of inference to the best explanation, which we discuss further in Chapter 6. When this evaluation of the available evidence is decisive, then we accept the hypothesis or theory as empirically adequate; no claim is made about its truth. When the hypothesis is about only what is observable, then empirical adequacy and truth come to the same thing; but still, we go beyond the evidence available in the sense that we draw conclusions about the nature of the observable phenomena.

Van Fraassen argues that his constructive empiricism makes better sense than does realism, and without the "inflationary metaphysics" of scientific activity. What he means by this is that scientific experimentation can be seen in the empiricist model both as testing for the empirical adequacy of the theory as developed to that point and as filling in the blanks in the developing theory, taking it to a farther point in its construction, or to completion. Experimentation can thus be construed not just as a means of discovery, but also as a means of continuing the construction of theories that are empirically adequate: "On the one hand, theory is a factor in experimental design; on the other, experimentation is a factor in theory construction" (Ibid.: 77). In both cases, van Fraassen's argument is that there are plausible alternatives to the realist explanation of the influence of theories on experimental testing as deriving from a belief that the theories are true. The alternative he puts forward, constructive empiricism, takes seriously the idea that we are all totally immersed in the scientific world picture of our time—but this is an argument not for the truth, but for the empirical adequacy of that world picture. Objectivity, and even observability, become intrascientific matters, and it is unnecessary to embrace ontological commitments about the existence (apart from the theory) of the various entities to which one refers.

It is often said that theory choice in science must be based on many features other than mere empirical adequacy: strength, simplicity, scope, elegance, unification, and explanation. As van Fraassen says: "Theory acceptance has a pragmatic dimension" (Ibid.: 88), and the features mentioned above enter into the evaluation of theories along this pragmatic dimension. The most interesting of these features, for our purposes, is explanation, and the question to be asked is what—if explanation is a pragmatic (that is, person-and-context-related) feature of theories—it contributes to the rationality of science. Van Fraassen argues that the features of an explanation relevant to the rationality of science include those that make it empirically adequate, empirically strong, and so on. The presence of such features is not sufficient to ensure that we have a good explanation; but they are the features that connect with our common ideas of what gives a theory epistemic value, even if it is the pragmatic features that add the dimension necessary for a satisfactory explanation. Both the language of theory appraisal and the language of the use of theories to explain phenomena are viewed by van Fraassen as radically context dependent. Whether a theory is explanatory depends upon the speaker, the audience contemplated, and various features of the context; and explanation is not a preeminent virtue of theories in the sense of being preferred over other virtues when there is a conflict. In this sense, empirical adequacy is the preeminent virtue, and explanation is not even always required if further explanation would require extending the theory in ways that do not yield further empirical consequences. But just what is explanation?

The answer requires attention to just the sort of historical development of the concept of explanation that we have offered in the earlier sections of this chapter. As a kind of summary of what has emerged from that development, van Fraassen suggests that "Explanation of why an event happens consists (typically) in an exhibition of salient factors in the part of the causal net formed by lines 'leading up to' that event" (Ibid.: 124). But what these salient factors will be depends upon the context—the interests of the persons explaining and being explained to, and various related matters. As Hanson puts it:

> There are as many causes of x as there are explanations of x. Consider how the cause of death might have been set out by a physician as "multiple haemorrhage," by the barrister as "negligence on the part of the driver," by a carriage-builder as "a defect in the brake-block construction," by a civic planner as "the presence of tall shrubbery at that turning" (N.R. Hanson 1958: 54).

These different explanations are competing in a sense, though not in the sense that only one of them can be true. They are alternatives that set up different *contrast classes* or spheres of interest. Thus, in van Fraassen's discussion of the paresis example:

> If a mother asks why her eldest son, a pillar of the community, mayor of his town, and best beloved of all her sons, has this dread disease, we answer: because he had latent untreated syphilis. But if that question is asked about this same person, immediately after a discussion of the fact that everyone in his country club has a history of untreated syphilis, *there is no answer*. The reason for the difference is that in the first case the contrast class is the mother's sons, and in the second, the members of the country club (van Fraassen, 1980: 128).

This matter of contrast classes is important in several respects. For one thing, the selection of the appropriate contrast class eliminates various alternative hypotheses that might, in other circumstances, explain the event in question; and for another, the contrast classes are themselves determined by context. Just which alternatives are to be considered in the first place can make a considerable difference to the adequacy of the explanation for its purpose, and it is clear that a considerable variety of interests can determine the selection of a range of relevant factors for explaining some phenomenon. Van Fraassen develops a theory of explanations as answers to why-questions based on the importance of these two context-dependent notions of contrast classes and relevance relations. The technical details are beyond our scope here, but it should be noted that the theory seems to account adequately for various problematic features that we have noticed earlier, including the asymmetries of explanation. Explanations are not, like descriptions, merely relations between theory and fact, and it is the failure to perceive this that, according to van Fraassen, has led so many theories of explanation astray. Instead, explanation should be viewed as a three-term relation between theory, fact, and context. The information requested by means of the why-question varies from context to context, as does the background information we use to evaluate how well the answer replies to the question. Thus, scientific explanation should be viewed not as pure science but as an application of science. It is a way of answering specific questions in specific contexts—of providing information in those contexts. What makes this process of value for science is "that the search for explanation is *ipso facto* a search for empirically adequate, empirically strong theories" (Ibid.: 156–157). Van Fraassen's account of explanation thus fits into his anti-realism or constructive empiricism and seems to provide some confirmation for our earlier claim that neither is it essential for theoretically minded archaeologists to embrace scientific realism, nor does that view provide a panacea for resolving all the problems archaeologists have encountered in attempting to find a suitable theoretical framework for their discipline. We develop this point further in Chapter 6 where we discuss inference to the best explanation.

Explanation in archaeology

We have been suggesting that many of the difficulties with explanation encountered in archaeology are not unique to that discipline and that there is no reason to think full-fledged explanation in archaeology will turn out to be substantially different from explanation in the more developed sciences. It is not that history or archaeology or geology fails to fit a model which applies unproblematically to, say, quantum physics. The differences are as likely to result from difficulties of application to less precise sciences, difficulties of establishing even statistical laws, and so forth. It is not really that the social sciences are so much more complex than the physical sciences, as is often claimed. Rather, problems in social science involve a large number of dimensions or variables most of which, for a variety of reasons, cannot be controlled, so that a multi-variate or multi-dimensional analysis is required.

Nonetheless, it would be hard to deny that there are special problems associated with disciplines like archaeology (and history and geology) where laws are simply not forthcoming in the way that they are in physics. As Achinstein (1971:97) says: "Given an explanation which does not explicitly invoke a law, if all the assumptions made in such an explanation were themselves to be explained or justified by appeal to more basic assumptions, and similarly for these in turn, then at some point it is very likely that laws would be invoked." In the case of the disciplines cited above, we may simply not have discovered any laws that could be invoked. Are our explanations for that reason inadequate or incomplete? In a sense perhaps they are, and this is probably part of the reason that there is so much more fundamental disagreement about method in archaeology than in physics.

LeRoy Johnson takes New Archaeologists to task for applying the hypothetico-deductive method where it is inappropriate and for improperly using generalizations from other disciplines that are unverified or have been abandoned. Also, he says, "several 'new' archaeologists have inferred the existence of specific social institutions from data which could hardly be so specifically interpreted"—in particular, "the inference of post-nuptial residence norms, or lineal descent groups from archaeological evidence about the patterned distribution of artifacts and site features which might also reflect only the existence of localized task groups of a non-corporate sort or else slight differences in the ages of assemblages which archaeology is hard put to detect" (1972: 369).

The example that seems to recur in the literature is Longacre's explanation of the stylistic distribution of pottery at the Carter Ranch site on the basis of matrilocal residence groups:

> It was postulated by Longacre [in his publications of 1966 and 1968] that the pottery styles of one matrilocal residence unit shared and passed on by the unit's female potters, would differ somewhat from pottery style shared by related females in another residence unit in the same village. If so, this would explain localized patterns of ceramic styles at archaeologic sites like Carter Ranch in Arizona (Ibid.: 370).

There are obvious problems with this explanation. One is simply that there are many other equally plausible hypotheses that would account for the data as well, and none of these alternatives has been eliminated. Another is that ethnographic evidence runs counter to the hypothesis, for

> Hopi and Hopi-Tewa women teach styles and pottery craft to their daughters-in-law in many instances, not to their daughters. And these daughters-in-law usually reside elsewhere. At other times, pottery is manufactured, and style and techniques transmitted, in the context of simple neighborhood noncorporate task groups (Ibid.: 370).

Johnson's main objection to arguments like Longacre's is more far-reaching, however. He fears that there has been at least covert acceptance of

> the nineteenth century notions (1) that descent groupings in societies are usually corporate or face-to-face groups, and (2) that residence ideals are actually practiced in a consistent way. These are invalidated generalizations confounded by a large corpus of social-scientific literature. Descent groupings are often only categories of people specified for defining potential marriage mates, etc., and have no clear corporateness which would influence the archaeological record in a measurable way. Residence ideals are often only vaguely approximated statistically on the ground, and several forms of post-nuptial residence are usually common in one society (Ibid.: 370).

What Johnson could be construed as saying here is that the underlying generalizations to which Longacre's account appeals are incorrect and thus cannot possibly be construed as laws. No doubt he is right about this, and, further, it seems clear that Longacre's account doesn't fit the D-N model as had been claimed. As Charles Morgan (1973) notes, Longacre provides us with "a plausible argument from the hypothesized residence rule to a certain pattern of remains. If such a pattern is in fact found (as it was) then it should strengthen the plausibility of the presence of the hypothesized residence rule (which it did)" (Ibid.: 271).

The form of the argument that Longacre puts forward seems crudely to be:

$$\frac{\text{If P then Q}}{\text{Q}}$$
$$\text{P,}$$

and this is a clear instance of the deductively invalid form known as affirming the consequent. That reasoning in the above way is deductively fallacious should be obvious. It might be, for example, that if nomadic tribes had raided the Anasazi then the Anasazi would have abandoned their cliff dwellings, but the fact that the Anasazi did abandon their cliff dwellings does not show that such groups raided them. Alternative explanations of the abandonment of the Southwest plateau are available (cf. Martin and Plog 1973: 318–333). Perhaps, for example, severe drought forced the Anasazi to migrate. The above argument form certainly does not fit any of the Hempelian models for explanation, nor does it look like a particularly promising candidate for inductive validity, for it would indeed be odd if an invalid deductive argument form were automatically a valid inductive one. It is, nevertheless, a mode of reasoning often used in science and can be used to illustrate the extremely close connection between explanation and confirmation, a connection we will discuss in detail in Chapter 6.

The argument we are considering actually says something like "If P were the case it would explain or account for Q," or "If P had been the case, we would expect to find Q, and since we did find Q, it seems likely that P was the case." Whether or not this is plausible depends on many contextual factors, such as whether there are alternatives to P that are equally likely, and so on. But this argument is not explanatory in the Hempelian sense at all. Rather, P serves as a reasonable hypothesis to account for the facts. Whether we accept P or not (consider it confirmed or corroborated) depends on whether it has survived tests, whether alternative hypotheses have been eliminated, and so on, but note that in archaeology we frequently lack sufficient information to provide explanations in any of the Hempelian patterns. In the pottery-residence example we are inferring a hypothesis or perhaps making a conjecture to account for some puzzling facts, but we have no law of residence patterns and we do not know that the residence pattern in a particular site was an instance of such a law, even a statistical one. What is in question here is P itself, and we need to ask whether it is likely or not. Q counts as evidence for P in some circumstances, and what is really at issue is whether or not P is confirmed to some degree.

Although Longacre's account of the Carter Ranch site may, in substance, be mistaken, it will be clear from the discussion of van Fraassen's model of explanation that the reasoning involved in the Longacre example seems to us considerably more relevant for archaeology in its current state than much of what had earlier been so taken by New Archaeologists. This has to do partly with the stage the discipline has reached and partly with its nature. The overemphasis on D-N explanation in the literature has unfortunately led to certain distortions. No doubt explanation is an important goal of the scientific enterprise, but it is not the only goal, or even the central one.

Charles Morgan, for example, has insisted that the "major goal of scientific endeavour is the acquisition of knowledge about the world" (1973:273). This includes not only an effort to find the "facts," but also an effort to find general principles or laws, and the relationship between the facts and the laws. The search for explanations is a heuristic device for finding out about the world. According to Morgan:

> Scientists use the consideration of explanations as a heuristic device in at least three ways. First, an inability to explain indicates a gap in our knowledge, and thus it indicates a hole for science to fill. Hence asking "Can I explain this" is one way of determining how complete my knowledge is, and thus how successful my scientific endeavour has been. Inability to explain acts as a pointer to indicate areas where more investigation is needed. Second, the search for explanations in some cases serves as a discovery device. Frequently we know what kinds of principles could explain certain puzzling facts, and this knowledge helps cut down the range of hypotheses we must consider . . . Third, in some cases the ability of a given hypothesis to explain increases our confidence in the hypothesis. That is, explanation plays a role in the confirmation or evaluation of hypotheses . . . Thus, far from being an end in itself, explanation is a means to an end—the end being the acquisition of information about the world (1973: 273–274).

The information, or knowledge, we are after can be provided in various ways, depending, as we have noted, on many contextual factors. If we encounter puzzling facts we try to explain them, and these attempts require well-confirmed hypotheses or, preferably, established laws. But often the best we can do is to provide what might be an explanation, and we need to ask whether the explanation is correct. This raises the general question of when we should say that a hypothesis is confirmed by certain evidence (or confirmed to a certain degree), and this is a very vexing problem in philosophy of science. It also raises the question of how we come to establish or confirm scientific laws, and so, with the connection between confirmation and explanation presenting itself once more, we turn to confirmation.

Notes

1. The concept of "explication" is particularly associated with the work of Rudolf Carnap. *See* his "Testability and Meaning," 1936.

2. This characterization of the deductive-inductive difference in terms of generality and particularity was common not only in standard textbooks where scientific method was discussed, but also in some of the best treatments of these matters in the New Archaeology. *See* Watson, Le Blanc, and Redman 1971.

3. Our remarks on explaining-as and on classification are suggested by some of the literature on seeing-as. *See* Wittgenstein 1968, pp. 193–208; E.H. Gombrich 1961, *passim*; J. Bruner 1957, 123–52. *See also* Karl Popper 1962, 42–48; Nelson Goodman 1965, 82, 122; and Nelson Goodman 1972b, 437–446.

4. Of course, if thoughts should turn out to be observable—perhaps, as a materialist would argue, because they are identical with certain brain states—then the behaviorist's objections to them would be met.

5. One area in which this broadening has been occurring is women's studies. For some relatively early examples, *see*: Oakley 1974; *Frontiers, A Journal of Women's Studies* 2 (Summer 1977); Kanner 1977.

6. The Suppe volume, *The Structure of Scientific Theories* (1977), provides an excellent account of the issues involved in the development and subsequent demise of the positivist philosophy of science. The central section consists of the proceedings of a symposium on the structure of scientific theories held in Urbana, Illinois, in March 1969. This is supplemented by a detailed introduction outlining the background and development of the Received View, criticisms of and alternatives to it, and an Afterword which attempts to provide a perspective on philosophy of science as of 1977. The much abbreviated account in the text owes a great deal to Suppe's insights. Suppe's volume also contains an excellent bibliography.

6 / Inference in Archaeology

T he concept of confirmation is surprisingly little discussed in
recent archaeological literature. There are probably a number
of reasons for this, not least the fact that so much of the attention of
the New Archaeologists has been devoted to the concept of explanation, and especially to the deductive model. Along with the
emphasis on deductive method, there was a concomitant, even if
temporary, denigration of the role of inductive inference in archaeology. Inductive methods were considered retrograde—a simple
collecting of data without related hypothesizing or theorizing. As
ideas that arose from the early New Archaeology were developed,
the hypothetico-deductive method came to be discussed as part of
appropriate methodology for archaeology; but it took some time
before this was seen as a method of confirmation.

It is well known that some proponents of the hypothetico-deductive method—especially Popper and some of his followers—
were staunch anti-inductivists who thought of induction in overly
simplistic terms. They argued, quite correctly, that induction could
never establish the truth of a scientific hypothesis, whereas the
hypothetico-deductive method, given certain assumptions, could
prove that a particular hypothesis was false. This suggested to
them that we should not attempt to confirm hypotheses but only to
disconfirm them. But then hypotheses which withstand attempts
at disconfirmation or falsification are taken by Popper to be, to
some extent, corroborated.

It is clear that this notion of corroboration, as much as more
traditional notions of confirmation, involves nondeductive infer-

ence. If science is to progress, we must be able to accept certain hypotheses, even if only provisionally; and surely it is appropriate that hypotheses worthy of acceptance should be those that are confirmed or corroborated to some appropriate degree. Our hope in science is that confirmed hypotheses will have explanatory power, and this is impossible if those hypotheses are justified solely by deductive reasoning.

As we discussed in some detail in Chapter 2, inductive reasoning is necessary for inferences that go beyond the observed data. Such inferences are essential, for hypotheses that are no more than summaries of observed facts have no explanatory value. As Michael Levin writes:

> [A] hypothesis, to be explanatory, must go beyond the data it explains; it must have implications that the data themselves do not have. This means that the "explanation" of some data must be more than just a repetition of the data in other words. Thus, a hypothesis cannot only be a "summary" of the facts, but cannot be reducible to the data in any finite specifiable way (Levin 1973:394).

As Skyrms says: "the advantage which inductively strong arguments have over deductively valid ones" is this—inductively strong arguments afford us "the possibility of discovery and prediction of new facts on the basis of old ones" (Skyrms 1975:8–9). The great weakness of much of the early literature on scientific method in archaeology was that explanation, and especially deductive explanation, was overemphasized, with a consequent underemphasis of statistical models, and of the reciprocal process of confirmation.

It is not clear why this should be so, particularly in view of the fact that the classic paper on confirmation is also by Hempel (1945), and appeared within a few years of his original paper (with Oppenheim) on explanation (1948). It must be admitted, however, that typically the relation between explanation and confirmation has not been seen in a clear light even by philosophers. Hempel, of course, recognized that one of the fundamental concepts required in his account of scientific explanation—that of a scientific law or lawlike sentence—was itself in need of analysis. He thought that analysis might be achieved, at least in part, through an adequate account of confirmation, for a scientific law, roughly speaking, is a hypothesis that has been confirmed. The reciprocity between explanation and prediction has been clearly noticed and much debated in the literature, although we do not have any thorough account of the similarities and differences between confirmation and explanation. Such an exploration cannot be undertaken within the scope of this book; nevertheless, it is essential to point out that the interesting questions about scientific methodology in archaeology relate to both explanation and confirmation, that the two are much more closely related than is usually recognized, and that both depend in fundamental ways on certain more basic concepts.

Laws and lawlikeness

If one read only the literature on explanation, one might be led to suppose that the notion of what a law is (or what a lawlike statement is) is unproblematic. This, however, is not so, for there are at least two very serious problems connected with lawlikeness. First, we need to distinguish lawlike statements from accidental generalizations. We need, that is, to be able to say which of the many regularities that have been observed up to now are such that we expect them to continue in the same pattern of regularity in the future. Second, we must characterize the relation between evidence and hypothesis that enables us to say that a particular hypothesis is well confirmed. Roughly speaking, we can say that well-confirmed hypotheses are lawlike hypotheses, so that the problem of characterizing lawlikeness can be treated as part of the problem of characterizing confirmation. Viewed in this light, one can begin to see the interrelatedness of the problems of confirmation and explanation. For explanatory purposes we are looking for laws, or at least for generalizations that may be lawlike; lawlike hypotheses are usually thought to be those that are confirmable, i.e., capable of receiving support from their positive instances. In fact, Hempel and Oppenheim, aware of this interrelatedness, discussed at some length the problem of characterizing lawlikeness in their original paper on explanation. But they were unable to provide a satisfactory definition of lawlikeness, and it has since become apparent that a purely formal characterization is unlikely to be forthcoming (Goodman 1965; Stegmüller 1976; Sneed 1971).

With respect to whether there are or can be laws in archaeology, archaeologists hold widely divergent views. In the one extreme are those who believe that laws have not been and will not be discovered in archaeology. LeRoy Johnson seems to take this position when he says that "Archaeological data are only amenable to direct treatment with empirical hypotheses, not laws or theories" (Johnson 1972:268). An intermediate view is that laws are desirable—indeed, some archaeologists argue that the formulation of laws is the basic goal of the discipline—but that few have yet been found. In the other extreme are archaeologists who have claimed that not only can there be archaeological laws but, in fact, that some are currently available (Schiffer 1976:4–9; Fritz and Plog 1970:405–412).

It seems that what divides archaeologists on this issue are their different conceptions of what a scientific law might be. In particular, they disagree over whether a statement must be known to be true in order to be a law. Archaeologists such as Johnson apparently hold that such statements must be true and, thus, because there are no such general statements in archaeology, there are no archaeological laws. Other archaeologists, including those sometimes referred to as "processualists," have adopted a less strict requirement. As Flannery (1967:122) writes: "The process theorists assume that 'truth' is just the best current hypothesis, and that *whatever* they believe now will ultimately be proven wrong." For Flannery, however, process theorists are to be distinguished from law-seeking theorists, so that the reference to truth in relation to processualists probably tells us little about the truth requirement for laws. What Flannery should probably be construed as saying is this: the process theorists assume that they may have to reject some hypotheses in the future which now seem very likely to be true, and which have, therefore, been tentatively accepted.

Archaeologists also disagree over whether laws can be statistical. In LeRoy Johnson's view, for example, they cannot be; laws, it would appear, must be universal (1972:367–369, passim), and since, according to Johnson, there aren't even any plausible universal law *candidates* in archaeology, his reason for claiming that laws do not play a role in that discipline is clear. What Johnson does think appropriate in archaeology, as we have already mentioned, are what he calls "empirical hypotheses" or "empirical generalizations," and these are often statistical. He writes that "Thus behavioral disciplines [of which he claims archaeology is one] best explain and interpret their data by seeking *empirical generalizations*, not nomothetic regularities. . . . Generalizations contain no references to theories or laws, but are often determined in part statistically" (Johnson 1972:367). On the other hand, for archaeologists such as Schiffer (1976:4) the term "law" covers statistical as well as universal generalizations. But we need not pursue this controversy further, as it is obvious from our discussion of explanation in Chapter 5 that statistical laws must occupy a central place in any science.

The debate in archaeology over whether laws must be true is more interesting. According to Hempel and Oppenheim (Hempel 1965:265), the first requirement of laws is truth. This immediately occasions the difficulty that a given empirical generalization can never be definitely known to be a law, since we cannot know it to be true. Only generalizations that go beyond the data on which they are based are of interest in science, and because they do go beyond that data they are always vulnerable to refutation. Hempel and Oppenheim accept this apparently odd consequence because the alternative (to say that laws are merely highly confirmed generalizations) would have the consequence that a given statement might be a law at one time and not at another. Such a relativization of a statement's lawfulness

> does not accord with the meaning customarily assigned to the concept of law in science and in methodological inquiry. Thus, for example, we would not say that Bode's general formula for the distance of the planets from the sun was a law relative to the astronomical evidence available in the 1770s, when Bode propounded it, and that it ceased to be a law after the discovery of Neptune and the determination of its distance from the sun; rather, we would say that the limited original evidence had given a high probability to the assumption that the formula was a law, whereas more recent additional information reduced that probability so much as to make it practically certain that Bode's formula is not generally true, and hence not a law (Hempel and Oppenheim in Hempel 1965:265).

Many archaeologists, as we have noted, worry about the truth condition for laws because, they say, it is hard to find even well-confirmed statements in archaeology; truth is just too strong a requirement. Fortunately, their worries are misplaced here. The above example shows that astronomy faces the same problem, as, indeed, do all sciences. This need not particularly concern us, for we can speak of lawlike statements, instead of laws, where lawlike statements are those having all the characteristics of general laws, with the possible exception of truth. We *can* do this, but we generally don't. Instead we usually speak of laws, always keeping in mind the epistemological worries about their truth.

Philosophers have been inclined to impose two conditions on lawlike sentences. First, that they should be universal, i.e., they should have the logical structure of "All A's are B's." And second, they should not refer essentially to particular objects. That is to say, in Hempel's words, a statement of a lawlike sentence's meaning should be general in the sense that it should not require

> reference to any one particular object or spatio-temporal location. Thus, the terms "soft," "green," "warmer than," "as long as," "liquid," "electrically charged," "female," "father of," are purely qualitative predicates [and, thus, can appear in lawlike sentences], while "taller than the Eiffel Tower," "medieval," "lunar," "arctic," "Ming" are not (Hempel 1965:268–269).

But Nelson Goodman has shown that this characterization of law-likeness in terms of universality and generality will not do. "That a given piece of copper conducts electricity," he says:

> increases the credibility of statements asserting that other pieces of copper conduct electricity, and thus confirms the hypothesis that all copper conducts electricity. But the fact that a given man now in this room is a third son does not increase the credibility of statements asserting that other men now in this room are third sons, and so does not confirm the hypothesis that all men now in this room are third sons. Yet in both cases our hypothesis is a generalization of the evidence statement (1965:73).

The point is, of course, that in the copper case the generalization is lawlike, whereas in the third son case it is merely accidental. The problem is saying what the difference is between the two, and accomplishing that is a much more difficult matter than has been supposed.

Nelson Goodman (Ibid.: 73) has illustrated this difficulty in a particularly compelling way. He has coined a new term, "grue," which, he says, "applies to all things examined before [a certain time] *t* just in case they are green but to other things just in case they are blue." Then all emeralds examined up to time t are both green and grue, which should lead us to predict that emeralds subsequently examined will be green and grue. But of course they won't, for if an emerald subsequently examined is grue it is blue and not green.

Although there have been many attempts in the literature to deal with this problem, no solution is generally agreed upon. We simply do not seem to have a way of distinguishing, other than by relying on intuition (a notoriously unreliable procedure), those generalizations that are lawlike and thus capable of being confirmed by their instances from those that are not. This problem is absolutely fundamental in confirmation theory. What this means, of course, is that all the talk in the literature on explanation about requiring laws for the major premise in the explanatory argument simply ignores (or moves to another place) the difficulty of saying what a law is.

It should not be supposed that these puzzles about laws have only theoretical significance. Especially in the applied sciences, their force is deeply felt. Geologists, for example, often do not know which of the patterns they have uncovered are lawlike, as opposed to merely accidental, and when this is the case they don't know on which of the patterns to base their predictions. When what is at stake is an accurate assessment of oil and gas reserves, the problem can take on crisis dimensions (Century 1977; 1978).

In archaeology the problem is particularly difficult, for it is extremely hard to tell whether hypotheses put forward to explain given phenomena are even likely to be lawlike. For example, a cluster of hypotheses that have been challenged on several fronts have to do with the establishment of residential patterns from the distribution of ceramics within sites. We can postulate that if the initial hypotheses had been convincingly supported by other cases, they would have been strong candidates for lawlike statements. Problems that have arisen in connection with these hypotheses include incomplete and partial knowledge, the selection of particular ethnographic analogies as the basis for argument, and consideration and elimination of competing hypotheses (*see* Allen and Richardson 1971). Another example of problems associated with a suggested hypothesis believed to be a candidate for lawlike status is seen in Alan Osborn's arguments about "the initial intensive incorporation of marine food resources into hunter-gatherer subsistence strategies regardless of time or space" (1977: 195). This example is of special relevance to the point we are making, as Osborn relies heavily on philosophical positions as well as on borrowed ecological concepts in building and advancing his arguments. This case has been directly challenged (Schiffer 1981), and a number of alternative hypotheses (cf. Richardson 1981; Mosely 1975; and Raymond 1981) have not been eliminated. As a result, the lawlike status of the hypothesis presented by Osborn is in doubt.

The apparent absence of scientific laws in archaeology has led some people to class archaeology with history, as distinct from science. Dealing as it essentially does with prehistory, however, archaeology suffers a great deal from lack of data in a way that history does not. Frequently, this difficulty of finding suitable data against which to test a hypothesis is what makes hypotheses look like mere assumptions or guesses. But the fact that hypotheses in archaeology are so open to criticism does not show that it is wrong to advance hypotheses. It merely shows that the hypothesis in question was not constructed with careful consideration of all the relevant data, that there are too many alternatives no less well substantiated, or that there is not enough information to construct tests that would decide between two conflicting hypotheses.

It has been claimed that in the social sciences, as opposed to the physical sciences, we do not have laws and yet we can have adequate explanations. If we do have explanations in science that do not make essential reference to general laws, as Scriven (1959a: 443–475), for example argues, then such explanations would constitute telling counterexamples to, for example, Hempel's models of explanation, in all of which laws play an essential role. But if we do not have laws in social science, and that has yet to be determined, this could be for one of two quite different reasons. It may be that laws (for some reason such as the intrinsic complexity or the essential unpredictability of human behavior) are not possible in the social sciences, or it may just be that we don't know enough yet and therefore have not discovered the laws that pertain to social behavior.

Still, it may be that, as is often claimed for history and the social sciences, there simply are no universal laws in archaeology that are neither too general nor too specific to be of any interest. It follows only that the D-N model of explanation is probably inappropriate for archaeology. But as we have seen, this does not mean that the work of philosophers of science has no relevance for archaeology. Hempel's (1965:331–496 and 1962) statistical models might still be possible candidates for capturing the nature of archaeological explanation, and Salmon's approach (1971:29–87; *see also* Salmon and Salmon 1979) seems even more promising.

Confirmation

A second problem concerning lawlikeness involves characterizing "confirmation." What do we mean when we say that some evidence *e* confirms some hypothesis *h*? A difficulty in answering this question is that there are two kinds of confirmation concepts that function rather differently (Salmon 1975b). So if we say that a certain hypothesis, or theory, has been confirmed by experimental evidence, we may have either of at least two apparently different meanings in mind. On the one hand, we may mean that the hypothesis is nearly certain in the light of its supporting evidence, i.e., that the degree of confirmation of the hypothesis is high. We will call this, following Rudolf Carnap (1962), the firmness concept of confirmation. On the other hand, we may have wished to say that some particular evidence renders the hypothesis more acceptable than it was in the absence of this evidence, i.e., that its degree of confirmation is higher on the basis of the new evidence than it was on the basis of the previous evidence alone, or that the new evidence is positively relevant to its probability. We will refer to this as the increase in firmness (again following Carnap), or the relevance concept of confirmation.

Clearly these two conceptions of confirmation do not coincide, for a hypothesis might well be confirmed in the second sense without being confirmed in the first sense. If the background evidence gives the hypothesis a low degree of confirmation, new evidence may raise that degree of confirmation without a resulting high degree of confirmation. The reverse can happen too. A hypothesis whose degree of confirmation is already high may not receive increased confirmation by the addition of further evidence, even though the degree of confirmation on the combined evidence may remain high.

Carnap (1962:468–481) pointed out, in discussing Hempel's account of confirmation, that Hempel vacillates between a firmness and an increase of firmness notion. Sometimes Hempel seems to have in mind a firmness notion; but in other places he speaks of favorable or unfavorable data, and of given evidence as strengthening or weakening a hypothesis, and this certainly suggests an increase of firmness view. Even Carnap, in spite of knowing that confusion has occurred in the literature through failure to keep the two concepts of confirmation distinct, was not always consistent in his own use of the verb "confirm" (Ibid.). Further, when conditions which have seemed to many philosophers to be obvious requirements on any adequate confirmation theory are imposed upon the notions of confirmation which we have discussed, serious problems arise.

To see how one difficulty arises for the increase in firmness concept of confirmation we will consider the Special Consequence Condition which Hempel (1965:30–31) has argued any acceptable confirmation theory must meet. That condition amounts to, roughly, the requirement that if evidence confirms a hypothesis h, then that evidence confirms every logical consequence of h (Ibid.: 31). For example, since q is a logical consequence of p and q (the conjunction of p and q), this condition requires that if evidence confirms p and q it must confirm q. But, as Carnap has shown, this condition is not satisfied in the increase of firmness concept of confirmation. The multiplication rule for probabilities dictates that, where p and q are independent, the probability of p and q = (probability of p × probability of q). Thus, we may have evidence that raises the degree of confirmation of p and q by raising the degree of confirmation of p while leaving that of q untouched, or indeed even lowering it.

There are other difficulties as well with the increase in firmness concept of confirmation. In this sense of confirmation, we can get a number of apparently counterintuitive results: (1) each of two evidence statements may confirm a hypothesis while their conjunction or disjunction disconfirms it; (2) a piece of evidence may confirm each of two hypotheses while it disconfirms their conjunction or disjunction; and (3) each of two evidence statements may disconfirm a hypothesis while their conjunction or disjunction confirms it (cf. Carnap 1962 and Salmon 1975b). Thus, the increase in firmness view of confirmation, along with what many have thought was an uncontroversial requirement, yields results that are both counterintuitive and straightforwardly unacceptable.

Difficulties also arise when some obvious requirements are imposed upon the firmness concept of confirmation, the view of confirmation in which a hypothesis h is said to be confirmed by evidence e when h is highly probable given the support of e. Consider here another of Hempel's adequacy conditions, the Conjunction Condition, which amounts to the requirement that "If an observation sentence confirms each of two hypotheses, then it also confirms their conjunction" (Hempel 1965:34). In the firmness view, we can say that if h has a high probability, say r, on e then h is virtually certain and hence confirmed by e. But no matter what particular value is chosen for r, the Conjunction Condition fails, for the probability of each of p and q on e might be higher than r, and yet, because of the multiplication rule for probabilities, the probability of p and q (the conjunction of p and q) on e might be lower than r.

But the really devastating result for the firmness concept of confirmation is that it, together with the Conjunction Condition, leads to contradiction, namely, to the lottery paradox. This paradox, it will be remembered from Chapter 5, is generated in the following way: since the probability of any particular ticket in a million-ticket fair lottery being drawn is only one in a million, in the firmness concept of confirmation we would have to say of each ticket that it will not win. But, by the Conjunction Condition, we must then also say, because each of the tickets will not win that no ticket will win. And since, by hypothesis, the lottery is a fair one and some ticket will win, we get the paradoxical result both that some ticket will win and that no ticket will win.[1] Even though, as we will discuss in greater detail later, high probability is not good enough for acceptance, it does seem, as Kyburg says, as if "It is as rational to accept the hypothesis that ticket i will not win as it is to accept any statistical hypothesis that [we] can think of" (Kyburg 1963:463). Nevertheless, accepting that hypothesis brings us right up against the lottery paradox, and that this is so constitutes a virtually conclusive objection to the firmness concept of confirmation.

As is the case with the increase in firmness concept of confirmation, more could be said about the firmness concept. But there is no need to go into greater detail here. The important point has been made: while the inadequacies of the firmness concept of confirmation are more numerous than that of the increase in firmness concept, neither is unproblematic.[2] If we remain very clear about the distinction between the two concepts of confirmation we can avoid some of these problems with conditions of adequacy. But the results are highly counterintuitive, and we have to give up a good deal that looked obviously true. For the firmness concept, we need to give up at least one of the usual adequacy conditions, and generally we think it easiest to relinquish the Conjunction Condition. But, we must also remember that the Consequence Condition, in the firmness notion of confirmation, presupposes a Consistency Condition—otherwise we might find ourselves in the position of having confirming evidence for every statement whatsoever through having confirming evidence for contradictory statements. From a contradiction, anything follows. In addition, the lottery paradox gives us some reason to think that any account of confirmation, in which a hypothesis will be said to be confirmed if its probability is equal to or above a certain value—assuming that value is less than 1—is simply incoherent.

The issues at stake in the debate over confirmation theory are admittedly complex, but certain things appear clear. First, one source of the difficulties that are generated by imposing adequacy conditions on the two standard concepts of confirmation is the logic that underlies those adequacy conditions. Thus, adequacy conditions which initially sound quite plausible are called into question when the untoward results their logic allows are seen. The Special Consequence Condition, which is the requirement that any evidence which confirms a hypothesis h confirms every hypothesis that follows from h, is, for example, eminently reasonable at first glance. Since, by definition of "consequence" in deductive logic, if the hypothesis h is true then all of its consequences must be true, it does seem as if evidence which confirms h should confirm all of the consequences of h. But, as we saw earlier, this is not so. The following example should make the point more clearly. Let us call the hypothesis "Kivas in the San Juan region of the American Southwest are round and the Northwest Indians wore plastic raincoats" h. On the increase in firmness view of confirmation, the observation of a round kiva in the San Juan region of the American Southwest increases the probability of "All kivas in the San Juan region are round," and, therefore, confirms h. But both "The

Northwest Indians wore plastic raincoats" and "The Northwest Indians wore plastic raincoats or I'm a monkey's uncle" are consequences of *h*, and we surely would not want to say that the observation of a round kiva anywhere confirms either.[3]

It looks in the above (as we have seen before in the positivists' attempts to formulate a formal criterion of cognitive significance and in Hempel's models of explanation) as if perhaps our deductive logic fails to capture what we mean when we say, for example, that a certain statement *follows from* another. How is it that "Northwest Indians wore plastic raincoats or I'm a monkey's uncle" follows from "Kivas in the San Juan region are round and Northwest Indians wore plastic raincoats"? If we presuppose deductive logic, then we can answer that it is necessarily true, by virtue of the meanings attributed to "or" and "and" in that system, that one follows from the other.[4] But if one is tempted to doubt the extent to which the connectives of formal, deductive logic suffice as substitutes for certain ordinary language expressions, then this question begging answer will not do. We are not, of course, calling into question the general usefulness of deductive logic. We are, however, suggesting that here again the assumption of formalizability or of capturing ordinary inference in terms of deductive logic may be inappropriate.

A second source of the difficulties we have discussed in connection with conditions of adequacy is the assumption that the mathematical probability calculus is applicable to confirmation relations. It seems reasonable to say, as Hempel, for example, does, that if evidence "confirms each of two hypotheses, then it also confirms their conjunction" (Hempel 1965:34). As we have seen, however, because of the multiplication rule for probabilities, the Conjunction Condition fails in the increase in firmness concept of confirmation, and the same rule led to the failure of Hempel's Special Consequence Condition in the firmness view. Some philosophers react to these conflicts between our intuitions on confirmation and the mathematical probability calculus by simply dismissing our intuitions on the subject as obviously misguided. But, as we will argue in the next section, the relation that probability should bear to confirmation is not at all clear. We look to confirmation of hypotheses as a guide to acceptance of beliefs and, surely, it is only rational to accept, among competing hypotheses, those with the highest probability. But whether the concept of probability that is appropriate here is the standard mathematical one is a genuine question.

There is still another concept of confirmation—one which we have not yet considered. Sometimes, when we speak of confirmation, we have in mind evidence confirming a given hypothesis *as against* certain competing hypotheses. For example, evidence may select a particular hypothesis as against its contrary (Goodman 1973). (The contrary of a hypothesis *h* has as its antecedent the same predicate as does *h*, but as its consequent it has the *negation* of the predicate in *h*'s consequent. Thus, the contrary of the hypothesis "All kivas are round" is "All kivas are not round.") So, in this approach to confirmation, the observation of a round kiva confirms the hypothesis *h*, "All kivas are round," as against its contrary, *c*, "All kivas are not round."[5] But the same observation does not select between *j*, "Everything which is not round is not a kiva," and *d*, "Everything which is not round is a kiva." And this result is interesting—indeed, counterintuitive, because *j* is logically equivalent to *h* and *d* is the contrary of *j*. Clearly, logically equivalent hypotheses do not, in general, have logically equivalent contraries.

Thus, one of the adequacy conditions often imposed on theories of confirmation, the Equivalence Condition, fails in the selective concept of confirmation. (Hempel states this condition very simply as: "Whatever confirms [disconfirms] one of two [logically] equivalent sentences, also confirms [disconfirms] the other" [Hempel 1965:13].) Now, while the Equivalence Condition seems an obvious enough requirement, some philosophers have argued for a concept of "hypothesis" in which the condition would be inappropriate (cf. Goodman 1966:331). In this view, hypotheses are thought of as specific inscriptions, or utterances. A sentence such as "All kivas are round" would be, then, the hypothesis and not merely, as we usually want to say, one formulation of it. And if this approach is adopted, then logically equivalent sentences need not be treated as identical and need not bear all the same logical relations to other hypotheses. So, for example, "All not round things are not kivas," though equivalent to "All kivas are round," is not the same hypothesis. Instead, since it is an entirely different string of words, it is another hypothesis altogether. And if equivalent hypotheses are nonetheless different, then we need not say that whatever confirms a given hypothesis must confirm all its equivalents. In this approach it also seems natural to say that, for the purposes of selective confirmation, a necessary condition for treating hypotheses as in competition is that their antecedents be the same, and this is intuitively appealing. To the uninitiated, the adoption of a theory of confirmation in which a requirement as apparently essential as the Equivalence Condition is inapplicable may appear slightly preposterous. But if doing so yields a more plausible account of selective confirmation than do the alternatives, then perhaps we have an argument against imposing the Equivalence Condition in the usual way (Hanen 1967). In Goodman's theory of confirmation, it is possible to obtain the effect of the Equivalence Condition in other ways, so that we can save our intuitions both about contraries and about equivalence. The actual working out of the theory, which is complex, can be found in Goodman's *Fact, Fiction and Forecast*, Chapter 4.

We will have more to say about selective confirmation in the last section of this chapter, but we want to note here that on the face of it the selective concept of confirmation looks promising. Further, selective confirmation need not be opposed to the other two concepts of confirmation we have discussed, for it involves comparative degrees of confirmation while the other two focus only on the confirmation of one hypothesis at a time. Indeed, we think that some combination of the increase of firmness (or relevance) and selective views may ultimately prove the most useful, although much more philosophical investigation must be done before a conclusion can be reached on this issue. Thus, a clear answer to the question "What do we mean when we say that 'some evidence *e* confirms some hypothesis *h*'?" is not yet forthcoming. Even so, familiarity with the considerations at stake in such a question could be useful to archaeologists concerned about the foundations of scientific inference.

In his article "Science in Archaeology: The Saints Go Marching In" (Dumond 1977), Don Dumond examined some of the claims New Archaeologists have made about the benefits of applying the scientific method to archaeology. Dumond has approached this task through an analysis of three studies which "attempt to explain patterns of artifact variability in terms of social organization," and he makes some useful observations in this work, such as that

> [E]ven if all hypotheses regarding the interconnections between ceramic attributes and post-marital residence that were ever put forward by Deetz, Longacre, and Hill had been "confirmed"—i.e., not found untrue and found to be at least as likely as all imaginable alternatives—it would still not be possible to accept their work as demonstrating any such relationship. It would only show that such a relationship remained a viable possibility (Dumond 1977:347).

Although Dumond refers to these three cases as presenting hypotheses, the word "proposition" is sometimes used in this context (cf. Hill 1970). Dumond's characterization of confirmation includes a comparative element, and is, thus, somewhat similar to what we have called "selective" confirmation: confirmed hypotheses must be "at least as likely" as the alternatives, and this is not the only point at which Dumond refers to the necessity in confirmation of comparing alternatives. In evaluating the three studies he analyzes in his article, Dumond says:

> All employ a framework that conceives of systemic relationships of cause and effect. All conceive of archaeological communities as containing potential evidence of complex relationships of variables relating to the adaptation of a people to their social and natural environment. All three studies followed a procedure of introducing hypotheses and generating tests of them against data. Only Deetz, however, before accepting his initial hypothesis considered alternative propositions that might explain the patterning he found (Ibid.:345).

While we think that the emphasis Dumond places on the consideration of alternative hypotheses is appropriate, there are at least three weaknesses in his account of confirmation. First, in Dumond's account a confirmed hypothesis need only be as likely as any other. But surely the real benefit of comparative confirmation is gained when it is used to identify the most probable hypothesis, i.e., when it selects the best hypothesis from the alternatives. Thus, a selective account of confirmation is preferable because it yields everything the comparative one does and more.

Second, a hypothesis is confirmed in Dumond's view when it is found to be at least as probable as "all imaginable" alternatives, and this seems an unnecessarily stringent condition to impose. Even if all imaginable alternative hypotheses could be tested—

which, of course, would be impossible since they are infinite in number—such extensive drudgery is not called for. As Salmon says, there are "infinitely many possible hypotheses to handle any finite body of data, but it does not follow that there is any super-abundance of *plausible* ones" (Salmon 1966b:129). It is usually reasonable to concern ourselves with only the likely alternatives (although we do not mean to suggest that it is always easy to pick out the plausible ones from the rest). It should be noted that although this point is not part of his account of confirmation, Dumond nevertheless is aware of it when he writes: "Formal rules for the generation of researchable hypotheses and their implications are few; propositions, hypotheses, implications must articulate [sic] clearly and above all be reasonable in the view of reasonable people" (Dumond 1977:348).

Third, Dumond appears somewhat confused concerning the relationship between confirmation and acceptance, in that he seems to be suggesting we are only entitled to accept those empirical hypotheses that we know are true. He says, in a passage already quoted, that even if archaeologists report that a hypothesis has survived testing and been shown to be at least as likely as all the alternatives, it is still not open to us "to accept their work as demonstrating" the relationship expressed by the hypothesis (Ibid.: 347), and he says this, presumably, because all empirical hypotheses are vulnerable to subsequent falsification—a completely uncontentious proposition if there ever was one. It doesn't follow from this fact about empirical hypotheses, however, that we are never justified in accepting such a hypothesis. It is obvious that hypotheses are regularly accepted in science; they are added to the body of knowledge in science, and decisions are made and further investigations undertaken on the assumption that such hypotheses are true. And this practice in no way conflicts with the open-mindedness for which scientists are renowned, for acceptance in science is always tentative. We adopt the hypothesis that has best survived testing, always ready to abandon it if additional research shows that another is more probable.

In summary, the basic point to be made in this section is that even though philosophy cannot as yet offer solutions to the deep problems that confront confirmation theory, archaeologists interested in scientific method would do well to familiarize themselves with the work in confirmation theory that is in progress. In this way they may at least avoid certain errors that have come to light through philosophical research on confirmation.

The probability calculus and confirmation

The mathematical probability calculus can be set up as a formal system in which the only undefined term is that which stands for probability. All other terms in the calculus have well-established meanings from other branches of mathematics or logic. But why should archaeologists, or any other scientists for that matter, who are interested in the confirmation of their hypotheses concern themselves with a formal system, and in particular the formal system that constitutes the mathematical theory of probability (cf. Doran and Hodson 1975)? Salmon argues that

> the mathematical calculus has been developed with great care and precision over a long period of time and with due regard for a vast range of practical and theoretical problems. It would be rash indeed to conclude on the basis of casual reflection that the mathematical theory is likely to be wrong or irrelevant in relation to potential applications (Salmon 1966b:64).

The meaning of "probability," Salmon thus insists, must be in accord with the mathematical calculus of probability. Were this not the case, he claims, such terms as "probable" would bear no relation to our behavior in betting and other situations where we rely upon that calculus.

Plausible as this sounds, it should be noted that whether or not we should impose the axioms of the probability calculus on a theory of confirmation is at least arguable. First, as we have already noted in this chapter, the rules of the probability calculus give rise to some highly counterintuitive results. The example we used illustrated how the multiplication rule for probabilities led to the failure of rather obvious sounding adequacy conditions in both the firmness and increase of firmness (or relevance) views of confirmation. It is not only the difficulties which the probability calculus creates for satisfaction of adequacy conditions which are relevant here, however. Others are those associated with the calculus itself. Of primary concern is the fact that all talk of probability is problematic, since no adequate interpretation of probability has been given. That is to say, no difficulty-free account of what it means to say that a particular hypothesis is probable in light of a given piece of evidence has yet been put forward. Each of the three leading interpretations of the formal system of probability—the logical, frequency, and subjective interpretations—fails to satisfy some seemingly obvious requirement.

The philosophical literature abounds with detailed discussions of each interpretation's inadequacies (cf. Salmon 1966b), and there is no need for us to go into those here. Our basic point can be made independent of the specific problems to which the interpretations of the probability calculus give rise. If we do not have an acceptable account of what it means to say that hypothesis h is probable on evidence e, then it hardly seems reasonable to insist that the concept of probability involved in confirmation accord with the mathematical probability calculus. And if the imposition of the calculus upon confirmation functions yields counterintuitive results—as, for example, the multiplication rule for probabilities does—then we have reason to believe not only that the mathematical calculus need not be the governing concept of probability in confirmation theory, but perhaps even that it should not be.

It should be noted here that it is not only in confirmation theory that the appropriateness of the mathematical probability calculus is being called into question. Recent work in the philosophy of law is aimed at showing that the concept of probability used by advocates, judges, and juries is not that of the mathematical calculus. Jonathan Cohen (1977), for example, has taken this position, and some of his arguments involve illustrating how adoption of the multiplication rule for probabilities for legal reasoning leads to difficulties. One particularly forceful point he makes concerns the results of imposing the rule on deliberations in civil cases where the plaintiff has to establish his or her case "on the balance of probability." The most natural way to construe this rule is as requiring that the probability of the plaintiff's case be greater than the probability of the defendant's—thus, greater than .5. But if there are two independent elements to be established in the case, say the circumstances of an automobile crash and the terms of the driver's insurance policy, then, in order to achieve an overall probability greater than .5, at least one of the elements will have to be determined by a probability substantially greater than .5. For this, .71 for each will just do, and various other combinations, such as .6 and .9, will suffice. But this would be to impose a stronger requirement than proof on a balance of probabilities, which does not accord with actual legal practice (Ibid.:58–59).

From this and other examples, it would appear that the multiplication rule for probabilities is as problematic in the law as it is in the confirmation of scientific hypotheses and, thus, that the concept of probability used in common law jurisdictions is not the mathematical one. This is relevant to the question of the application of probability to archaeology in two ways. First, the fact that there are problems of application in both natural science and law suggests that there may be such problems in other areas as well, and that we should not simply assume applicability in archaeology. Second, legal inference is in some respects quite similar to archaeology, so perhaps analogies might be drawn between the ways in which probabilities are calculated in the two areas.

A warning is perhaps in order here: we are not suggesting that, as a mathematical system, the probability calculus is somehow defective. Instead, we are calling attention to the need, not for a better calculus, but for a concept of probability that encompasses, but is broader than, the mathematical one. There may be stages in the confirmation of scientific hypotheses, or in legal proceedings, where a straightforward application of the mathematical calculus is appropriate. (Indeed, especially in the case of science, there are sure to be.) But when the question before a scientist, or a judge or jury, is whether to *accept* a particular hypothesis, it appears that the mathematical concept of probability is not sufficient. As we have said earlier, we rely on confirmation of hypotheses as a guide to acceptance and belief, and in some sense of "probability" it must be true that it is only rational to accept, from among competing hypotheses, those with the highest probability. The unanswered question is: What is that sense of "probability"?

Much of what has already been said about problems with the various accounts of probability applies to statistics as well. But, as we have already indicated with respect to problems of imposing the mathematical probability calculus on confirmation of scientific hypotheses, it is not statistics *per se* that are philosophically worrying. It is, instead, the manner in which statistical results often are applied that is suspect. Without doubt, statistical tests can be of assistance in sorting accidental from lawlike regularities, but they cannot be relied upon alone to justify acceptance of hypotheses.

Of greater interest perhaps than the obvious fact that statistical tests have, as Giere puts it, "a certain probability for leading us astray" (Ibid.:65) because conclusions reached on the basis of them may be false, is the fact that those tests can lead us astray in two importantly different ways. We can be led to mistakenly reject or accept a hypothesis. In statistics the former is called an error of Type I and the latter of Type II, and the difference between the two can be characterized as the difference between, in the first case, failing to say what is true and, in the second, saying what is false.

Different statistical tests will show different frequencies of these errors with any given process, and the same statistical tests will show different frequencies of these errors with different processes. So which statistical test to use with any particular set of data will depend upon the frequency and kind of error deemed appropriate for the purposes at hand. And what will be deemed appropriate for those purposes will depend, presumably, on what is at stake. In the law, for example, it has been traditionally held that it is better to let ninety-nine guilty persons go free than to punish one innocent individual. In other words, recognizing that mistaken decisions are inevitable, the law has sought to minimize errors of Type II by trying to avoid mistaken acceptances of guilt. It is also easy to imagine circumstances in which we would want to minimize errors of Type I, instead of Type II. For example, if we are trying to uncover carriers of a deadly, contagious disease for purposes of quarantine we might rather mistakenly quarantine a large number of disease-free persons (i.e., mistakenly accept false hypotheses) than mistakenly leave unquarantined a large number of diseased persons (i.e., mistakenly reject true hypotheses). In archaeology, too, the choice of which type of error to minimize might arise. Suppose, for example, that we have evidence of a site very rich in archaeological data believed to be particularly relevant to solving an important problem. We would, presumably, prefer to dig more, rather than less extensively to ensure that significant data would not go unfound. In such a case, archaeologists could be said to be trying to minimize the rejection of true hypotheses (Type I error).

Again, archaeologists may be seen as attempting to minimize Type I error (the rejection of true hypotheses) when they make decisions, in part based on pragmatic considerations, to broaden a data base that is to be used for statistical testing of a hypothesis. Pragmatic considerations can also lead to what amounts to the rejection of true hypotheses. For example, the statement "All archaeological sites can contribute to scientific understanding of the past" is something of a truism, but one that nonetheless cannot serve as a charter for archaeological behavior. It is impossible for archaeologists to treat all sites equally, and many sites are lost to archaeological research because of inadequate resources in the discipline, on the one hand, and factors beyond the control of the discipline, on the other.

Confirmation, explanation, and acceptance

We are inclined to think that the most interesting and useful notion of confirmation will combine selectivity and relevance, or, as we have called the latter, increase in firmness. It is necessary to think of selectivity in reasonably broad terms. It is not just a matter of selecting a hypothesis as against its contrary, for the notion of selective confirmation is capable of generalization: where we have three or more hypotheses that are mutually exclusive and exhaustive of the relevant possibilities, confirmation of one of the alternatives provides selective confirmation of it as against the competing alternatives, however many there are. The problem is to get a suitable set of hypotheses from which to make the choice, and this depends in part on plausibility considerations.

One might think of confirmation rules in general not so much as acceptance rules but as rules for selecting hypotheses from among alternatives by eliminating the alternatives. The selected or best hypothesis, then, is the one which survives elimination in a given situation. It is the one that is best confirmed in the circumstances, not because we have more positive instances of it than of alternatives (although that may be the case), but because we have instances that select it as against the alternatives, or because we have rules for elimination of competing hypotheses, leaving the one in question uneliminated. These rules would be concerned with more than just how many positive and negative instances of each have been observed. Other criteria would include fit with other things we believe, possible strengths and weaknesses revealed by analogies with past successes and failures, apparent potency (likelihood of leading to further useful developments), and economy (accomplishment of the most with the least). Many of these considerations bear on practical as well as theoretical reasoning.

According to Salmon, the standard objection to the sort of confirmation we are discussing (which he calls "induction by elimination") is that "it is impotent in the face of an unlimited supply of possible hypotheses, for we never arrive at a unique hypothesis as a result." He notes that while "this objection is valid against any form of induction by elimination that proceeds by trying to eliminate from the class of all possible hypotheses, it is not pertinent to [all eliminative inference]" (Salmon 1966b:129). There are, he emphasizes, "infinitely many possible hypotheses to handle any finite body of data, but it does not follow that there is any superabundance of *plausible* ones. Indeed, in practice it is often extremely difficult to think up even a couple of sensible hypotheses to explain a given problematic datum" (Ibid.:129). Still, there obviously must be restrictions on which hypotheses can be taken seriously as possibly providing alternatives in the first place. A rough approximation might be to say that those hypotheses can be taken seriously that could serve as *explanations* of the evidence.

The link between confirmation and explanation has often been noticed, and we have remarked upon it earlier in this chapter as well as in Chapter 5, but it has seldom been made explicit in a systematic way. One link is through the concept of law, which is commonly believed to be required for explanation and is also widely supposed to have something to do with confirmation. Braithwaite addresses this connection in arguing that a hypothesis, for which the only evidence is evidence of instances of it, is not to be regarded as a law:

> A hypothesis to be regarded as a natural law must be a general proposition which can be thought to *explain* its instances; if the reason for believing the general proposition is solely direct knowledge of the truth of its instances, it will be felt to be a poor sort of explanation of these instances. If, however, there is evidence for it which is independent of its instances, such as the indirect evidence provided by instances of a same-level general proposition subsumed along with it under the same higher-level hypothesis, then the general proposition will *explain* its instances in the sense that it will provide grounds for believing in their truth independently of any direct knowledge of such truth (Braithwaite 1970:62).

Now, if lawlike hypotheses are well-confirmed hypotheses, as we have earlier suggested they are, then Braithwaite can be taken as claiming that confirmation and explanation are related in that evidence confirms those hypotheses which could be used to explain the evidence. Further, as Braithwaite says and we have also noted earlier, in order that it have explanatory power, a hypothesis must be more than just a summary of its observed instances. It appears that confirmable hypotheses must also go beyond the data on which they are based. Confirmable hypotheses have been characterized as those which are capable of receiving support from new instances, and this is impossible with mere summaries of observations; there cannot be additional instances of such statements, for they are already exhausted.

It may be useful here to reflect on where this discussion of the relationship between confirmation and explanation is taking us. As we have argued earlier, the point of science—indeed, of any inquiry—is the acquisition of additional information about the world, or, in other words, the expansion of our body of knowledge. And the issue in confirmation theory that we have been focusing on is, "When are we justified in adding a hypothesis to our stock of knowledge?" It won't do, of course, to answer "When the hypothesis is true," because we can never be sure that what we believe to be true now won't be shown false in the future. Instead, what we are looking for are criteria according to which we should rationally accept or reject hypotheses, at least provisionally. And there is no suggestion here that acceptance should ever be regarded as more than provisional—as providing a justification for admitting the hypothesis in question (tentatively) to our body of knowledge.

If we take the expansion of our stock of information about the world as the goal of doing science, then that activity must involve inductive inference, for, as we have already suggested, in an important sense deductive reasoning can never take us beyond what is already known. Ampliative inference is inductive, and the question becomes "What sort of inductive inference gives us knowledge?" Our consideration of this matter suggests that such inference (1) should be eliminative and (2) should provide us with explanations. Interestingly, when this question has been approached, not through the philosophy of science but in a general epistemological context, the same conclusions have been reached. Gilbert Harman, who has tried to develop an account of how theories of knowledge and of reasoning must be adapted to one another, has claimed that "Knowledge depends on whether reasoning justifies acceptance of one's conclusion" (1970:83). The sort of inductive inference that, according to Harman, does give us knowledge is inference to the best of competing explanations. Harman writes:

> If . . . we think of [permissible inductive inference] as an inference to the best explanation, we can explain when a person is and when he is not warranted in making the inference from "All observed A's are B's" to "All A's are B's." The answer is that one is warranted in making this inference whenever the hypothesis that all A's are B's is (in the light of all the evidence) a better, simpler, more plausible (and so forth) hypothesis than is the hypothesis, say, that someone is biasing the observed sample in order to make us think that all A's are B's. On the other hand, as soon as the total evidence makes some other, competing hypothesis plausible, one may not infer from the past correlation in the observed sample to a complete correlation in the total population (1965:90–91).

The convergence of views of knowledge in epistemology and the philosophy of science is encouraging, but the suggestion that what we are after is inference to the best explanation is, nonetheless, difficult to assess in the absence of a more adequate account of explanation than we at present possess.

The most intuitively satisfactory cases of explanation are clearly the causal ones. Since the concept of cause involves at least correlation, if not law, it does seem as if causal explanations are deductive-nomological (although perhaps only implicitly so). This is probably why so much attention has been focused on D-N explanation and relatively so little on statistical explanation, but if we take a closer look at the three kinds of difficult cases for statistical explanation discussed in Chapter 5, some interesting insights emerge.

(1) The paresis case. Here we suppose that something causes paresis in those individuals who get it, even though we don't know what that cause is. That is to say, we suppose that if we were to divide the class of latent syphilitics into those who contract paresis and those who don't, eventually we should be able to find some factor or factors present in one class and absent from the other which explains why one group gets paresis. And if we could find some such factor or factors, then we would suppose we had identified the cause of paresis, and thus explained its occurrence. The

statistical connections that are explanatory seem to be those we have reason to think are causal.

(2) Chance cases. What should we say of occurrences such as getting a long run of heads when tossing a fair coin? Here, all we seem to be able to (or need to) say is that what happened happened by chance (Jeffrey 1971:24). Its happening was improbable, but nonetheless one of the possible outcomes of a certain process, and that's the end of the matter. On the other hand, one might object that to say something happened by chance is just to say we don't know why it happened, or we don't know what caused it. But there must be a cause, and if we knew enough we could find out what it is. Thus, in the case of tossing a coin, if we knew about all the factors operating on the coin—the force with which it was tossed, spin, air currents, and so on—then we could predict with certainty how it would land. We would then have a complete causal explanation as well, and such explanations are possible at least in principle in such cases.

(3) Quantum cases. One might even want to take this sort of view with respect to cases in quantum physics. But here the position looks much less plausible. If we think of the radioactive decay of a particular atom in a metallic substance, there is explanatory relevance in pointing out that the atom that decayed was a uranium 238 atom even though the probability of any particular uranium atom's decaying in a short space of time is very small. And in this case we seem to have no reason to think we will someday be able to fill out the explanation further. If this is so, then this sort of case does seem to be different from the paresis case, whatever we want to say about coin tossing and other "chance" cases.

One difficulty here may be that we are asking for too much. Where the best we can provide is a statistical explanation, it may be that we shouldn't speak of explaining individual events. There is no reason, we may say, why this atom rather than that one decayed. And in all cases where the probability of the occurrence was less than 1 there may always be the possibility that we do not have sufficient information to predict the event in advance. Explanations, in these cases, that confer high probability on the explanandum-event look convincing; but they are no better *as explanations* than accounts that confer only low probability on the explanandum-event. Perhaps in the statistical cases we should be looking only for explanations of classes of events.

In the social sciences we often can't predict individual events—for example, the delinquency of any particular teenager, and our explanations amount only to assigning probability values to *classes* of events. The same may often be true in archaeology. For example, let us say that most seventeenth-century Zuni families practiced matrilocal residence; we cannot with complete confidence predict that any particular Zuni household from this period will exhibit matrilocality. About ninety percent of Zuni households were matrilocal early in this century, but that leaves about ten percent that were not (Kroeber 1916). So the best we can do is to predict that of the seventeenth-century Zuni residences yet to be discovered ninety percent will be matrilocal, assuming that the early twentieth-century analogy is valid. In answer to the question "Why is any particular seventeenth-century Zuni household matrilocal?," the best explanation we may be able to offer is that ninety percent of such residences are of this kind. There may equally be no very satisfying explanation of why any particular Zuni household is not matrilocal.

The traditional response to this difficulty would be to suggest that we simply have incomplete knowledge in these cases: that if we knew more we could fill out the account into a genuine causal one, or at least a high-probability statistical account. But that the latter tells us any more than a low-probability statistical account is becoming less and less plausible. If that is so, then we may have two kinds of case, the genuinely causal D-N ones and the statistical ones, which again divide into two: those that are cases of incomplete knowledge but for which we expect to be able to fill out genuine causal accounts, and those for which it makes sense to try to explain groups of events rather than individual ones. Further, one could argue that, at least for purposes of explanation, our interest in individual events is always an interest in them as they exemplify particular classes of events; and it may even be that statistical explanations are, at bottom, causal.

Another and perhaps more disturbing possibility is that our difficulties with statistical explanation are evidence for the view that, though we do have particular explanations, we do not (and perhaps cannot) have a *theory* of explanation. If we are forced by the difficult cases into abandoning the idea that the particular event in question can be explained, then it is hard to see how we can claim to have a general account of explanation. When we remember that all or nearly all cases that science deals with are in some sense statistical, or at least nondeductive, either from a theoretical or a practical point of view, the difficulties look enormous.

With respect to confirmation we are in no better position, and this could hardly be otherwise since explanation and confirmation are such closely related concepts. We have argued that confirmation can perhaps best be viewed as a process which yields the generalizations we need for explanations; a confirmed hypothesis, then, is one that could be used to explain the data which support it. Thus, we have suggested, confirmation can be seen as inference to the best explanation, and selective confirmation as inference to the best of competing explanations.

Inference to the best explanation is, in the nicest cases, inference to causes; but causes are not always available. Sometimes, it would appear, explanations are essentially statistical, as, perhaps, are the cases from quantum physics that we have considered. Sometimes they fail to be strictly causal for other reasons—for example, for lack of information. It may be that with additional research we will be able to give a causal explanation of why a particular syphilitic contracted paresis, but for now, a statistical account is the best that can be offered. In any case, however, we are still looking for laws, or at least for generalizations that may be lawlike, and lawlike hypotheses are usually thought to be those that are confirmable— capable of receiving support from their instances.

We are, therefore, interested in confirming lawlike generalizations rather than accidental ones. In the statistical case, lawlike hypotheses will usually, but not always, have high probability. That high probability is neither necessary nor sufficient (and certainly not both necessary and sufficient) for either explanation or acceptance is strongly suggested by the lottery paradox and by explanation examples we have considered in the last chapter. These cases also point to the importance of the increase in the firmness (strengthening) or relevance notion of confirmation as opposed to the firmness notion. For if high probability is neither necessary nor sufficient for explanation or acceptance, it is difficult to see how it could be either (and virtually impossible to see how it could be both) for the closely related concept of confirmation.

Both confirmation and explanation thus seem to require a notion of relevance that has not yet been fully explicated. In the context of confirmation we look for evidence that strengthens the hypothesis—that makes its degree of credibility greater than it was on background evidence. In the context of explanation, we look for an explanans which is such that the explanandum is more probable given the explanans than without it (or given any competing explanans), though of course the explanandum may still not be highly probable. This would be to explicate both confirmation and explanation in terms of the increase of firmness notion of confirmation.

Further, although the conditions of adequacy usually proposed for confirmation and acceptance are extremely plausible, as we have seen, there are good reasons for rejecting some of them. We should not expect to resolve conflicts of intuition about adequacy conditions a priori; rather, we need to develop adequate theories of confirmation and acceptance and to see which of such conditions are met. If one could construct an acceptable system without having to impose the conditions, this might be one way out of our difficulties (cf. Bar-Hillel 1968; Carnap 1963; Jeffrey 1956:237–246; but *see also* Kyburg 1968 and I. Levi 1967 for contrary views).

It has often been claimed that we do not require acceptance rules, and, thus, that many of the problems we have been discussing can be avoided. Sometimes the underlying claim is that we cannot have them anyway, as probability conclusions cannot be detached (*see* Chapter 5). For science as well as for general epistemology, we do in fact seem to accept or reject hypotheses, at least provisionally; therefore, theories of confirmation and acceptance are needed.

The most promising candidate for a rule that does what is wanted of acceptance rules, we have suggested, is an elimination rule—one that allows us to choose the best from among competing hypotheses by ruling out the others. We are not the first to suggest this course in archaeology (cf. Hole 1973). Within philosophy it is of interest to note that there is an analogy between thinking of confirmation as eliminative and Popper's views on corroboration of hypotheses. In Popper's view (as we have discussed in greater detail in Chapter 3), scientists "jump to conclusions" and then subject those conclusions—hypotheses—to tests. The hypotheses that survive testing, i.e., those that are not falsified by empirical observation, are corroborated, or, as we would say, confirmed (cf. Popper 1962: 33–59). In both approaches the emphasis is on showing that a hypothesis has failed to meet some standard, as opposed to showing that it has achieved, for example, a high probability of truth. There are significant differences between the two views, however. In the selective confirmation approach, hypotheses are eliminated from an initial set of plausible alternatives, leaving one that, in contrast to its competitors, is confirmed. But in Popper's view it is an individual hypothesis that survives or fails testing. Thus, though all of its competitors might have been even more thoroughly refuted by the results of observation, a hypothesis which fails is eliminated by Popper's method.

There is a yet more dramatic difference between the two. In the view of selective confirmation that we have been discussing, restrictions are imposed ensuring that only plausible hypotheses, from among the frequently numerous possible alternatives, are considered. But in Popper's view, the less plausible the hypothesis the better, for, as he argues, we learn more if a very improbable hypothesis survives testing than if a probable one does. This is true enough. By definition, there is less likelihood that a wild conjecture will survive testing than that some of the most plausible candidates will, however, and this fact is significant. If a patient is suffering nausea and acute abdominal pain, an inflamed appendix is among the more plausible diagnoses. Medical science might stand to learn more if it "jumped" to the conclusion that the patient had a brain tumor and tested that, as opposed to the appendicitis hypothesis. The overwhelming odds are, however, that if investigators proceeded in such a way the only result would be a person dead of a burst appendix. Since in science our investigations are often concerned with pressing human problems, it would seem absurd, as well as irresponsible, to conduct all scientific research along Popperian lines.

If we think of inductive inference as eliminative, then sometimes a given hypothesis will be eliminated where some more general hypothesis from which it follows escapes elimination. This may be a case where a hypothesis eliminated by our confirmation rules is nonetheless acceptable in our body of knowledge because it follows from another hypothesis that is itself acceptable. This, however, is clearly too simple, for in the sort of case under consideration there are at least two possibilities. One is that the more general hypothesis requires reexamination which might lead to its rejection; we might, for example, find that what had been regarded as evidence for it should not be so regarded, or that it was not as strong as we had supposed, or some such thing. We may, that is, find that there are reasons for rejecting some hypotheses which had been previously accepted through not being eliminated by our rules.

The other possibility, of course, is that the hypothesis' having escaped elimination was entirely appropriate, and that it should be incorporated into our body of accepted beliefs. In that case, the elimination of its logical consequences is overruled, for rationality seems to demand that where p is accepted and q follows from p, q should also be accepted. This is one way of obtaining the effect of certain standard conditions of adequacy—in this case the special consequence condition—without imposing them a priori in the very development of our theory of confirmation. Not all confirmed hypotheses are acceptable, nor are all acceptable ones confirmed. For example, hypotheses all of whose instances have been examined may be acceptable but are not capable of being confirmed, at least in the increase of firmness sense.[6] Questions about confirmation are different from questions about acceptance, then, and it is always at least an open question whether a hypothesis with a high degree of confirmation should be accepted or one with a low degree of confirmation rejected.

A hypothesis with low probability may or may not be considered confirmed—it depends partly on whether its probability is higher than it would be without the evidence in question. A hypothesis with low probability may or may not be acceptable, and this depends partly on whether it provides the best explanation of the evidence. Accepted hypotheses can later be rejected on the basis of new information, and previously rejected ones can be accepted either because new information overrules the previous rejection or because the rejected hypothesis follows from an otherwise acceptable one. Acceptable hypotheses are sometimes false and may yet provide the best explanation of the evidence. If you think you see a rhinoceros charging down the street, the best explanation may be that you are hallucinating even though the correct explanation is that there is a rhinoceros there, recently escaped from the local zoo.

There are, of course, a number of deep problems in confirmation theory about which we have had very little to say. In the first place, we have said almost nothing about how we are to explicate "cause" and "relevance." It would be very nice if this could be done in terms of statistics, but we are not at all confident that it can be. And, perhaps more importantly, we remain very much in need of an account of how to choose the appropriate class of hypotheses to which to apply our inductive rules in the first place since, for one thing, confirmation in the increase of firmness sense may not be sufficient for acceptance if we have not considered all of the relevant alternatives.

The importance to archaeologists of what we have been saying in Chapters 5 and 6 is that the concepts underlying the philosophy of science—"explanation," "confirmation," "probability," "acceptance," etc.—are problematic. But if we are to develop a proper understanding of the theoretical foundations of archaeology, it is important to grasp these difficulties and to come to appreciate their significance for particular archaeological investigations. Further, though a great deal of archaeology can be done without questioning the theoretical underpinnings of the discipline, many practicing archaeologists, like practicing chemists, can work with the common-sense versions of concepts such as "confirmation" and not run into much trouble. Still, practice and theory, especially in a developing discipline, are not easily divorced from one another. For example, in contemporary archaeology what is to count as an archaeological law is a live, and not merely an academic issue. Whether the usual requirements in the philosophy of science, atemporality and aspatiality, are too stringent, or even inappropriate for laws in archaeology is a question that has practical bite. Thus, it seems reasonable to think that for archaeologists—more than for, say, chemists—a thorough understanding of the philosophical issues at stake in their discipline is desirable.

Analogy, model, and metaphor

We have suggested in the above that the most fruitful approach to finding confirmed generalizations may be that of selective confirmation and also that essential to this method's success is the appropriate choice of initial hypotheses from which all but the surviving, i.e., selected one are eliminated. There are, of course, numerous sources of hypotheses in archaeology, but one of the most important is the results of ethnological and ethnographical studies. In fact, many of the generalizations about prehistoric cultures put forward are drawn by analogy with living primitive culture. In this respect, archaeology has much in common with the better established sciences, in which the use of analogy is common. Notwithstanding the frequency with which analogy appears in science, however, the role of analogical reasoning is somewhat controversial and a number of issues are debated in the growing body of literature on the subject (cf. Black 1962; Hesse 1966; Achinstein 1968; Leatherdale 1974; and MacCormac 1976).

Analogies often appear in science in the form of models or metaphors;[7] billiard-ball movement, for example, is frequently used to model the kinetic theory of gases, and some of the terms central to physics have analogical aspects. Illustrative of the latter is the word "force"; as Earl MacCormac says: "Newton probably took the 'force' of [his second law of motions, $f = ma$] to be a concept given intuitively by analogy to man's muscular 'force'" (1976:34). It has been suggested that the analogical aspects of both model and metaphor are crucial to their effectiveness. This view is held by Leatherdale, for example, who has written:

> It is implicit in much of the writing on metaphor and model that analogy is a more fundamental and simple concept than metaphor or model. One talks . . . of metaphor "expressing an analogy" or being "grounded on analogy." Similarly, a model works "by analogy" or exhibits an analogy with what it is a model of, or for (1974:1–2).

We will discuss both metaphors and models in the course of our consideration of analogical reasoning in science.

While, as we have already noted, analogies are common in science, it has been claimed that they are inappropriate to that context. One objection is that analogy, as it is used in metaphor, for example, is insufficiently precise for scientific purposes. As MacCormac says: "[Metaphors] were thought to be the primary possession of poets who wished to dream about the world rather than to explain it in a theory" (1976:36). But to restrict analogy to the realm of art clearly seems incorrect, for analogical reasoning is pervasive, and not just in art and science. For example, it is fundamental to the workings of the law. As Edward Levi writes:

> The basic pattern of legal reasoning is reasoning by example. It is reasoning from case to case. It is a three-step process described by the doctrine of precedent in which a proposition descriptive of the

first case is made into a rule of law and then applied to a next similar situation. The steps are these: similarity is seen between cases; next the rule of law inherent in the first case is announced; then the rule of law is made applicable to the second case (1949:1–2 footnote omitted).

That such reasoning is analogical emerges distinctly when Levi's description is compared to the following characterization of analogical argument in general:

$$A \text{ is } P, Q, R, S$$
$$B \text{ is } P, Q, R$$
[Therefore] $\quad B \text{ is } S$

Applying this model to the legal context, we could say that A is the already decided case which has been identified as most similar to the case before the court, B, sharing with it features P, Q and R; S is the rule of law in A which is applied to B. Courts, then, can be seen as applying to the case at bar the rule of law found in earlier cases to which cases before them are most analogous. Thus, there are many areas where analogy is relied upon, which suggests that it would hardly be reasonable to claim that models and metaphors can be ruled out of science solely on the grounds that the analogies on which they are based are unsuitable to "real-life" matters (*see also* Thagard 1978:89). After all, people have been hung on the basis of them. Further, if metaphor involves the use of "a known linguistic expression to stand for [an] unknown," as MacCormac (1976:34) puts it, then it would be surprising if there were not at least some occasions on which metaphor, and hence analogy, was essential in science.

In fact, it has been argued that metaphor is basic to science. MacCormac is one who has taken this view, claiming that "the very nature of science . . . is such that scientists need the metaphor as a bridge between old and new theories" (Ibid.). He begins the case for this conclusion with the observation that to the extent that a theory determines the meaning of the terms within it "then when a theory changes the meaning of terms within it will change also" (1976:27). MacCormac claims that theories fully determine the meaning of the terms used within them, but one need not go all the way with him on this point. All that is needed in the context is a blurring of the distinction between theory and observation. The problem, then, is how to understand the terms of a new theory; MacCormac's answer is, through the use of metaphor. He writes:

> Only by some such device as a metaphor is it possible to develop new meanings in new theories that are intelligible. Metaphors have a hypothetical nature; they suggest possible new meanings to us. In attempting to describe the unknown, the scientist must use terms that are known to us. Sometimes this is done mathematically, but often ordinary words are stretched in their meanings to accommodate new hypothetical understandings (1976:36).

And he concludes that

> Without the possibility of proposing new hypothetical meanings, science could not create new theories. And without the possibility of formulating such new hypotheses by extending the meaning of terms that we already comprehend [through metaphor], science could not produce new theories that are intelligible (1976:37).

Even if one accepts that analogy, however, through the use of metaphor or model, has a proper, even a necessary, role to play in science, there is still the question of precisely what role that is. Hempel (1965:331–496) is among the philosophers who have addressed this question, and it is his view that all references to analogies that appear in scientific explanations are inessential. According to Hempel, any reference to an analogy that appears in a scientific explanation can be eliminated without damage to the explanation. Hempel does allow, however, that analogies can be useful to scientific investigation in three important ways. First, they can make for "intellectual economy." If a similar property has been identified in both A and B and the consequences of that property for A are known, then by analogy the consequences of that property for B can be deduced. Second, they may "facilitate one's grasp of a set of explanatory laws or theoretical principles for a new domain of inquiry by exhibiting a parallel with explanatory principles for a more familiar domain," and "in this manner, they can contribute to the pragmatic effectiveness of an explanation" (1965:441). Third, analogies may prove useful in the generation or "discovery" of hypotheses. As Hempel says: "they may provide effective heuristic guidance in the search for new explanatory principles. Thus, while an analogical model itself explains nothing, it may suggest extensions of the analogy on which it was originally based" (Ibid.).

Hempel's reasoning in support of the claim that analogies need never appear in scientific explanations depends for its success upon there being an absolute distinction between discovery and justification. He argues that while a plausible hypothesis may be suggested to a scientist by an analogy, the explanatory force of that hypothesis is completely independent of the analogy which gave rise to it. So, an analogy between an ecosystem and a culture may suggest to the archaeologist a hypothesis about cultural change, but whether that hypothesis explains the cultural change has nothing to do with parallels that may exist between ecosystems and cultures. Hempel writes:

> In order to appraise the explanatory significance of . . . analo-
> gies . . . let us suppose that some "new" field of inquiry is being
> explored, and that we try to explain the phenomena encountered in
> it by analogical reference to some "old," previously explored do-
> main of inquiry. This calls for the establishment of a . . . [regularity]
> between . . . the old field and . . . the new . . . But once this has
> been done, those laws can be used directly for the explanation of the
> "new" phenomena . . . For the systematic purposes of scientific
> explanation, reliance on analogies is thus inessential and can always
> be dispensed with (1965:438–439).

Thus, Hempel's view on the role of analogy in science is two-pronged: for the purposes of discovery it can be very useful, but for those of justification it is irrelevant.

As we have seen to be the case in a number of other areas, Hempel's views on this subject have not been universally shared. In fact, a major debate in the history of science centers on whether models, which seem to "work" only because they involve analogy, are "*mere* aids to theory-construction" (Hesse 1966:3). In the classic instance of this controversy, the position of philosopher and physicist Pierre Duhem was attacked by physicist N.R. Campbell (cf. Duhem 1954; Campbell 1920). Their views on the subject are somewhat anachronistic—both wrote during the early part of this century and were concerned solely with mechanical models; nevertheless, we can see in their work the foundation of later approaches, such as those of Hempel and MacCormac, which we have already considered. Hempel, for example, shares Duhem's notion which we discuss below that while models may be of use in the discovery of hypotheses, or theories, they are irrelevant to questions of justification. And MacCormac's view that the language of science is, at least in many cases, inescapably analogical is similar to Campbell's claim, also discussed below, that models are necessary to the intelligibility and extension of scientific theories.

According to Duhem:

> models drawn from familiar mechanical gadgets may be useful psychological aids in suggesting theories . . . But this admission implies nothing about the truth or significance of the models, for many things may be psychological aids to discovery, including astrological beliefs, dreams, or even tea leaves, without implying that they are of any permanent significance in relation to scientific theory (Hesse 1966:2–3).

In contrast, Campbell (1920 and 1953) insisted that models are more than just helpful in the formulation of hypotheses; they are absolutely essential to the intelligibility of theoretical explanations and to the extension of scientific theories to explain and predict new phenomena. The debate continues, with contemporary philosophers and physicists split on the role of analogy in both model and metaphor. For example, Mary Hesse, who discusses the Duhem-Campbell controversy at length in her book *Models and Analogies in Science*, writes:

> many physicists would now hold in essentials with Duhem and would claim that Campbell's position has been decisively refuted by the absence of intelligible models in quantum physics; indeed, many would claim that something like Duhem's position must necessarily be the accepted philosophy underlying modern physical theory (1966:5).

But Hesse herself remains unconvinced. It is her belief that "an element of truth remains in Campbell's insistence that without models theories cannot fulfill all the functions traditionally required of them, and in particular that they cannot be genuinely predictive" (Ibid.).

In our view, analogy is an integral part of scientific explanation, and not just in the context of discovery where it seems beyond question that analogical reasoning assists in the generation of new hypotheses. Arguments, such as those of MacCormac, that the language of science is inherently analogical are persuasive. As Max Black (1962:25–47) has argued, it seems as if analogy, at least in the form of metaphor, is not just a poetic way of saying what one could say as well in literal terms. If this is so, metaphors cannot be translated without loss of content to literal language. Analogy, then, appears to be just as necessary to justification as it is useful to discovery. As we have indicated elsewhere, we do not want to say that discovery and justification are completely divorced from one another. Consequently, to admit a legitimate role for analogy in discovery is already to concede it a place in justification. Of course, Hempel would respond that notwithstanding all other consider-ations, in the final analysis "What remains as the principal require-ment for scientific explanation is . . . the inferential subsumption of the explanandum under comprehensive general principles, irre-spective of the analogies these may display to previously estab-lished laws" (1965:445). But if the propositions which constitute an explanation are incomprehensible without reference being made to analogies with "previously established laws," then surely Hempel's view that analogies are always eliminable from explana-tions is incorrect. To adopt his view is to take the rejection of pragmatic elements in explanation too far.

Still, if analogies cannot be eliminated from scientific explana-tions, there nevertheless have been a number of concerns about the use of analogy in science. Archaeologists in particular have repeatedly noted any number of shortcomings of analogical rea-soning as applied within the discipline, and Orme (1973) has noted the historical distrust of its use. Most of the critical discussions have centered on ethnographic analogy, but much the same range of complaints can equally well be applied to analogies drawn from other sources. It has been argued, for example, that much of the archaeological record is concerned with adaptations for which there are no reasonable ethnographic analogies. Ethnographic analogy has been seen as unnecessarily limiting the range of ar-chaeological inferences; and ethnographic or, for that matter, other analogies may themselves be incomplete or lacking in sufficient detail to provide the basis for a useful inference for archaeological purposes. Selection of the most appropriate ethnographic analogy from among those potentially available is a problem with both logical and substantive aspects. Some archaeologists are concerned about how much time, under what conditions, can be bridged by direct historical analogy. Others have attempted to specify conditions or rules for selecting ethnographic analogies, such as requiring similar environmental conditions or subsistence strategies in order to avoid some of the more obvious pitfalls of bad analogy.

Many archaeologists who espoused positivism held the view that analogy belongs solely to the context of discovery, having no role in justification—not unexpectedly mirroring Hempel's views. Binford (1967:10) argued that the "final judgement" of an hypothesis "must rest with testing through subsidiary hypotheses drawn deductively." We have already argued in several contexts that the firm lines of demarcation established by the positivists between the processes of justification and discovery should be seen as somewhat overlapping, although it is useful to keep them separate for purposes of clarity.

Many points have been made about the pitfalls of using analogy in archaeology, and it seems important to discuss a few that have particular relevance to the problem of confirmation. The incorporation of analogies into hypothesis formulation is, of course, not a disputed issue, since it is widely accepted that inspiration for hypotheses to be tested can originate in even unlikely places. One of the topics relevant to confirmation concerns the principle of uniformity . . . (or uniformitarianism as it is sometimes called), which is a particular case of analogical reasoning. (This is contrary to Gould's position, where he says analogy should not be mistaken "for cultural uniformitarianism" or used "as a substitute for . . . science" [in Gould and Watson 1982:376].) It can be argued that all inductive reasoning—i.e., reasoning that goes beyond observations or facts accepted as true—involves the assumption of continuity in at least some respects from one point in time to another; and, at a very basic level, this is the principle of uniformity. If this argument is admitted, then, since some scientific hypotheses are testable and therefore subject to some aspects of confirmation, it follows that some hypotheses suggested in the light of this principle are, at least in theory, testable. Quite clearly, hypotheses generated on the basis of the principle of uniformity in conjunction with other implicit and explicit premises may be wrong, just as any empirical proposition, no matter how discovered, may be false.

This merely emphasizes again the importance of critically evaluating hypotheses before tentatively and provisionally accepting them into our body of knowledge.

The principle of uniformity at this most general level seems fundamental to all science as well as to everyday life. A chemist would soon abandon his lab if particular chemicals behaved differently today from yesterday without there being some reasonable explanation. The seekers of laws in archaeology and elsewhere must subscribe at some level to the principle of uniformity, as Richard Watson has correctly pointed out:

> It should be noted that archaeologists who do propose laws of cultural evolution or new laws of human behavior assume—as do other scientists—that the past was like the present, and that although combinations and rates may have been different then from what they are now, the basic behavioral characteristics of men and material were not different then from what they are now. This may in fact be wrong. But if human nature and the environment were radically different in the past from what they are now, we assume that there has been lawlike change from the past to the present that can be derived and understood from its physical remains. This general uniformitarianism is the primary procedural or methodological assumption of archaeology, as it is of all other sciences (1976:62).

The spectrum of views in archaeology about the principle of uniformity runs the gamut from full acceptance by the seekers of laws, through a qualified acceptance of the principle recognizing that "archaeologists cannot use the principle of generic uniformity at all as extensively as geologists do to move from present to past and past to present" (Patty Jo Watson in Gould and Watson 1982: 360), to an even more qualified acceptance for the moment as an encouragement to archaeologists to accept and apply the general principles of empirical sciences (Gould in Gould and Watson 1982:365–371). As is the case with laws, the problems are less with the principle of uniformity per se than with identifying particular principles to apply, or more significantly, to test.

The fact that analogical reasoning is an inductive or ampliative form of inference means, of course, that the truth of the conclusion cannot be guaranteed. Unlike deductive inference, the arguments are not either valid or invalid, but are of varying degrees of strength. When Gould, for example, rejects analogical reasoning as "wishful thinking," he seems to be responding to some of the same motivations that led to an earlier espousal of positivism— although there are some important distinctions between the two positions, as Wylie (1982a:388–389) notes. Nevertheless, Gould seeks to move archaeological discourse to another plane that somehow escapes the uncertainty of inductive inference, and this was a most significant motivation in the archaeological adoption of positivism. Both Watson (in Gould and Watson 1982a) and Wylie point out, we believe correctly, that Gould's own program is actually one of controlled analogical inference, albeit by other names. Gould seems to have a rather narrow concept of analogy, but given the broader view that we have been espousing, Gould's line of research can be seen as analogical. As a sidelight, and as Wylie makes clear, Gould's testing program involving "anomalies" is directed more at falsification, "the objective of which is not so much to shore up and 'confirm' an hypothesis as to test for its weaknesses or inaccuracies and the limits of its applicability" (Wylie 1982a:383), a position reminiscent of Popper's.

If, as we have argued, analogical reasoning is fundamental to science and must play a role in at least many if not all episodes of hypothesis testing, confirmation, and knowledge acquisition, the question arises as to how to implement the best safeguards against the worries that accompany its use. The point is to avoid or minimize methods that provide only speculative results on the one hand, or the selection of low-level (i.e., descriptive) methods, on the other (cf. Wylie 1985:81).

Ascher, speaking of the "new" analogy, is among those who have attempted to impose "boundary conditions for the choice of suitable analogs" (1973:405), citing the remarks of several commentators to support his claim that archaeologists are increasingly aware of the need for appropriate boundary conditions:

> According to Clark the archaeologist should " . . . restrict the field of analogy to societies at a common level of subsistence," and should " . . . attach greater significance to analogies drawn from societies existing under ecological conditions which approximate those reconstructed for the prehistoric culture under investigation than those adapted to markedly different environments." Willey would select cultures on " . . . the same general level of technologi-

cal development, perhaps existing under similar environmental situations." V. Gordon Childe advised that an analog " . . . drawn from the same region or ecological province is likely to give the most reliable hints . . . " In summary, then, the canon is: seek analogies in cultures which manipulate similar environments in similar ways (Ascher 1973: 405–406, references omitted).

This leads us again to Salmon's arguments regarding relevance:

> The strength of an analogy depends principally upon the similarities between the two types of objects being compared. Any two kinds of objects are alike in many respects and unlike in many others. The crucial question for analogical arguments is this: are the objects that are being compared similar in ways that are *relevant* to the argument? To whatever extent there are *relevant similarities*, the analogy is *strengthened*. To whatever extent there are *relevant dissimilarities*, the analogy is *weakened* (1973:98).

Salmon further notes that "to evaluate the strength of an analogy it is necessary to determine the relevance of the respects in which the objects of different kinds are similar" (1973:100). This, of course, is the difficult part, for we have no readymade criteria for determining relevance, and interesting growth in science occurs sometimes on the basis of newly discovered analogies that might, a priori, have been regarded as not relevant.

One point to consider in using analogies is their degree of specificity. In his discussion of the role that analogies play in explanation, Hempel discusses the relative uselessness of very general analogies. As an example, he cites the following:

> An osmotic growth has an evolutionary existence; it is nourished by osmosis and intussusception; it exercises a selective choice on the substances offered to it; it changes the chemical constitution of its nutriment before assimilating it. Like a living thing it ejects into its environment the waste products of its function. Moreover, it grows and develops structures like those of living organisms, and it is sensitive to many exterior changes, which influence its form and development. But these very phenomena—nutrition, assimilation, sensibility, growth, and organization—are generally asserted to be the sole characteristics of life (1965:439).

Hempel is critical of this analogy because only "regularities of a vague qualitative kind" are illustrated there. These regularities include the facts that:

> organisms grow and decay, and so do their osmotic counterparts; there is an exchange of materials between each of the models and its environment; there is some measure of repair of injuries in organisms and in their physico-chemical models, and so on. Because of their lack of specificity, generalizations of this kind do not have much explanatory force. In this respect, the analogies here exhibited are vastly inferior to those between water waves and electromagnetic waves, for example, which rest on . . . [similarities between] two extensive theories formulated in mathematical terms (1965:440).

Clearly, the problem here is not only the absence of quantitative or extensive theoretical basis for the analogies under discussion as Hempel suggests. In addition, it is the vagueness of the comparisons, and the fact that what is predicated of osmotic growths is so general as to be attributable to a vast range of things. This lack of selectivity makes the analogy one of questionable usefulness. Further, presumably precisely what is at issue is whether an osmotic growth *is* a living organism, and this can hardly be settled by appealing to a positive analogy while disregarding negative ones.

A further problem is that such broad analogies are often used to provide explanations which seem to cover all possible cases. It has often been claimed in philosophy that that which explains everything—including contradictory results—explains nothing. For example, where the evidence cited to explain a child's actual delinquency is the same as that which would have been offered had he never behaved in a seriously antisocial way, the account based on that evidence cannot be explanatory. Of course nothing would be better than a theory that explains both all that happens and, when new facts are uncovered, those too. But that would not involve the explanation of mutually exclusive states of affairs. To explain is, very roughly, to show why something could occur, did occur, or could have occurred, *as opposed to* something else. But if a putative explanation can account for any possibility at all then there is no such contrast, and the suggestion is, then, that the "explanation" in question is no explanation at all. Our concern here is with "explanations" that account for too many, including self-contradictory, states of affairs. In an earlier chapter we considered briefly a related difficulty which arises when several hypotheses are put forward as equally good accounts of a single phenomenon. The underlying problem in both cases is the same: in neither has the specific set of processes which resulted in the specific event which we seek to understand been identified.

Reliance on very broad laws (or analogies) is roughly the basis, for example, of Popper's view that psychoanalytic theory, Adler's "individual psychology," and Marxism are not explanatory. As Popper wrote, years after he first began to doubt the scientific value of such theories:

> [W]hat worried me [about them] was neither the problem of [their] truth, at that stage at least, nor the problem of exactness or mea-surability. . . . It was precisely this fact—that they always fitted, that they were always confirmed—which in the eyes of their ad-mirers constituted the strongest argument in favour of these theo-ries. It began to dawn on me that this apparent strength was in fact their weakness (1962:34–35).

Popper's complaint is, in essence, that each of the theories in question "explains" too much and, therefore, nothing at all.

In archaeology, too, there is the danger of working with analo-gies or concepts so broad that they explain nothing, as Ascher has noted. A category of analogies that "includes analogies to proper-ties common to all men such as the need for capturing energy and the possession of a language [is], for purposes of interpretation, meaningless. One does not need to undertake archaeological in-vestigation to know that the individuals in a particular culture engaged in these activities" (1973:404). Because, as Ascher says, "The question which the archaeologist seeks to answer is what were the *particular* patterns of a prehistoric people in carrying out these and similar activities" (Ibid.), analogies involving "proper-ties common to all men" have little, or no, empirical value. This is surely correct.

There is, of course, much more to be said about the usefulness of analogical inference in archaeology.[8] Perhaps the most important point to note is that analogical inference is no more intrinsically suspect or subject to error than is any other form of ampliative inference, nor is there any reason to treat it as a simplistic transla-tion of the unfamiliar into something familiar. As much of the foregoing discussion has indicated, care must be taken in the use of analogy, but this again does not distinguish this form of nondeductive inference from any other. In particular, background knowledge must be incorporated and systematic testing of analogies undertaken. Relevance is perhaps the chief desideratum, and in order to determine relevance, it is necessary to explore the negative and neutral aspects of analogies as well as the positive ones. Clearly causal, functional, and structural analogies are more likely to be theoretically relevant than are "mere" similarities, so we need to weigh similarities against differences not only in number but also in type in order to determine the strength of an analogy. What seem to provide strong bases for archaeological inference, then, are arguments by relational analogy that involve considerations of relevance. As Wylie (1985:105) says: "The real value of relational forms of analogical inference is . . . that *where* they involve a discriminating, selective transposition of information from source[s] to subject, they can be a profoundly creative, expansive form of interpretive argument." This is essentially because such arguments can be built up by using analogies of a small number of features from several different sources, thereby building a model that may be importantly new, substantially different from any known contemporary analog, and yet plausible enough to be worth serious consideration as a reasonable account of the phenomena. The next step, of course, is to apply whatever tests are possible in order to confirm or disconfirm the account.

Confirmation and the argument about realism

Among the concerns that critics of analogical inference in archaeology have voiced is that such inference is particularly prone to error and thus that the models it helps us to build may have no tie with reality. This general concern—that of the relation between a model or theory and the world it purports to describe—returns us to the controversy between scientific realism and its alternatives that occupies so much of contemporary philosophical discussion.

Scientific realism is often alleged to underlie our normal understanding of scientific theory and practice. When we accept a scientific theory—it is agreed—we assert its truth and believe that the entities it postulates actually exist. According to van Fraassen, this argument is normally based on some version of the rule of inference to the best explanation. One question that must be addressed is whether inference to the best explanation is indeed so closely tied to scientific realism. Given that we have earlier argued for the usefulness of the rule of inference to the best explanation in archaeology and against scientific realism (or at least against the proposition that only realism can make sense of scientific activity), the question is whether this separation of the two can be maintained.

Van Fraassen characterizes a simple version of the rule of inference to the best explanation thus: "Let us suppose that we have evidence E, and are considering several hypotheses, say H and H'. The rule then says that we should infer H rather than H' exactly if H is a better explanation of E than H' is" (1980:19). (We omit here the qualifications needed to avoid inconsistency. Roughly, they require that we seek the best overall explanation of all the available evidence.) Van Fraassen's suggestion, not surprisingly, is that we need not believe that the theory which best explains the evidence is true, but only that it is empirically adequate—that the observable phenomena are as the theory describes. For him, to accept a scientific theory is to believe that it is empirically adequate. Furthermore, the rule of inference to the best explanation dictates a choice only from within a set of rival hypotheses. Once we are committed to accept one of a range of hypotheses, the rule can be applied to tell us which from among the hypotheses in that range we ought to choose. The point is that for the realist and the anti-realist the range of hypotheses from which to choose will be different: the anti-realist will want to choose among hypotheses that claim not truth, but empirical adequacy for particular theories. Van Fraassen claims, then, that it "should at least be clear that there is no open-and-shut argument from common sense to the unobservable. Merely following the ordinary patterns of inference in science does not obviously and automatically make realists of us all" (Ibid.:23).

There are a number of other arguments for scientific realism which van Fraassen considers and rejects. The only one we wish to mention here is an argument given by Hilary Putnam (1975). (Putnam has since abandoned realism.) It is that

> realism is the only philosophy that doesn't make the success of science a miracle. That terms in mature scientific theories typically refer. . . , that the theories accepted in a mature science are typically

approximately true, that the same term can refer to the same thing even when it occurs in different theories—these statements are viewed by the scientific realist not as necessary truths but as part of the only scientific explanation of the success of science, and hence as part of any adequate scientific description of science and its relations to its objects (Putnam 1975:73).

Van Fraassen refers to this as the Ultimate Agreement because it tries to answer the question why we have successful scientific theories at all. Van Fraassen's response is that science is itself a biological phenomenon—an activity that facilitates the interaction between human beings and their environment. Thus the success of current scientific theories is no miracle. In Darwinist terms, scientific theories are in competition with one another and only the successful ones survive. That a species exists is evidence that it has been able to compete successfully with its natural enemies; similarly for a scientific theory.

There are problems with van Fraassen's views, of course, and we do not wish to claim that his version of anti-realism is correct. One problem with it, for example, is that it depends upon a notion of observation as theory-independent, which we have already had occasion to question. Nevertheless, this is not a respect in which it is worse off than most versions of realism, so it provides a reason for us to seek a more adequate formulation of anti-realism rather than a reason to embrace realism. We hope to have achieved two ends in this section (and in Chapter 5): first, to have provided reasons for believing that it is not necessary to accept scientific realism, and second, to have provided reasons for believing that scientific realism no more provides the basis for successful theorizing in archaeology than did the earlier logical positivism. The debate between realists and anti-realists is interesting in itself and will no doubt continue for some time to come, but its importance consists in drawing lines along which we can continue the attempt to sort out the conceptual foundations of science (and, indeed, of knowledge itself). There is no reason for archaeologists to suppose that they must accept realism as the only framework within which to make sense of science, or even that they need have a firm view about how the controversy over scientific realism is to be settled, in order for them to make systematic or theoretical progress in archaeology. Most of us, no doubt, do have such views, even if they are only implicit in our work and not, perhaps, very fully formed; but there is much that can be understood about the nature of theory in archaeology without having a final answer to the question of the relation between theory and the world. As van Fraassen says:

> we can distinguish between two epistemic attitudes we can take up toward a theory. We can assert it to be true (i.e., to have a model which is a faithful replica, in all detail, of our world), and call for belief; or we can simply assert its empirical adequacy, calling for acceptance as such. In either case we stick our necks out: empirical adequacy goes far beyond what we can know at any given time. (All the results of measurement are not in; they will never all be in; and in any case, we won't measure everything that can be measured.) (1980:68–69).

The difference is that the assertion of empirical adequacy is weaker than the claim of truth, and acceptance involves a lesser commitment than does belief. The preference for one or the other mode may ultimately be as much a matter of epistemic style as of ultimate correctness.

Notes

1. Actually, also required to generate the lottery paradox here is a Consistency Condition, Hempel's version of which is: "Every logically consistent [i.e., not self-contradictory] observation report is logically compatible with . . . all the hypotheses which it confirms" ("Studies in the Logic of Confirmation," p. 33). Since the need for the condition is not peculiar to this problem, however, there is no call to develop the point further.

2. For a more thorough discussion of the problems which arise through imposition of the usual adequacy conditions on the firmness and increase in firmness concepts of confirmation, see Hanen 1971 and 1975.

3. The following should make clear how this result is achieved. Let us call "Kivas in the San Juan region of the American Southwest are round" p and "The northwest Indians wore plastic raincoats" q. (Note that the conjunction of p and q, which is p and q or, symbolically, p & q, is the same sentence as that which we have called h in the text: "Kivas in the San Juan region of the American Southwest are round and the northwest Indians

wore plastic raincoats.") Since, in the increase in firmness view, the observation of a round kiva in the San Juan region increases the probability of p and leaves the probability of q unchanged, that observation confirms $p \& q$ (and therefore h). According to the rules of deductive logic that govern the connective "$\&$," q follows from $p \& q$. By the Special Consequence Condition, then, the observation of a round kiva confirms q. That is to say, in the approach we are considering, the observation of a round kiva increases the probability that the northwest Indians wore plastic raincoats. Worse, we can let r denote any statement whatever—for example, "I'm a monkey's uncle"—and, again by the rules of deductive logic, q or r or symbolically $q \ v \ r$, is a consequence of $p \& q$. Thus, the observation of a round kiva confirms the statement "The northwest Indians wore plastic raincoats or I'm a monkey's uncle."

4. The connectives of formal, deductive logic are the symbols used to generate more complex out of less complex statements. In note 3 above, two such connectives were used—"$\&$" and "v," which are usually translated as "and" and "or" respectively. The question we raise in the text concerns the extent to which these and the other logical elements actually do capture what we ordinarily mean by expressions such as "and" and "or."

5. The concept of confirmation being used here is Hempel's satisfaction criterion of confirmation. In this view, an observation report—for example, "Kiva (a) is round," confirms a hypothesis—for example, "Kivas are round," if the particular object mentioned in the observation report, here kiva (a), has the characteristic it would have if the hypothesis was true, i.e., roundness. (For details of Hempel's satisfaction criterion, see his "Studies in the Logic of Confirmation," p. 37.) In this case, it does: if "Kivas are round" is true, then it follows that kiva (a) should be round. The observation report states that it is. On the satisfaction approach, therefore, the observation that kiva (a) is round confirms the hypothesis that kivas are round.

6. To see how a hypothesis, all of whose instances have been examined, may be acceptable but not capable of being confirmed in the increase of firmness sense of confirmation, consider the following hypothesis, which we will call h: "Women who take drug X while pregnant give birth to children who are seriously deformed." Then, let us assume both that h is acceptable in light of all the available evidence and that drug X has been taken off the market so that no further instances of this hypothesis will ever be available. If h is confirmable in the increase in firmness sense of confirmation, then there must be additional evidence on the basis of which the probability of h is increased. But this is not possible in this case, because there can be no such additional evidence.

7. There are, of course, distinctions to be drawn among the terms

"analogy," "model," and "metaphor," but these are not strictly relevant to our purpose here. For careful treatments of these matters, *see* Black (1962) and Goodman (1968).

8. Besides the works already cited on this topic, we refer the reader most particularly to the recent papers by Alison Wylie (1982a and 1985) and the bibliographies cited there.

7 / Some Critical Archaeological Cases

O ne of the ways in which both philosophers and scientists have explored the relationships between philosophical abstractions and the actual doing of science is by looking at case histories. We turn now, following the more general and philosophically oriented discussions in the earlier chapters, to a consideration of archaeological cases.

Only a few of the cases used here focus specifically on a single publication or a single person's work. We have found it useful to target an issue, a controversy, a hypothesis, or a topic and explore the methodological issues raised. In this approach, the cases are less detailed explorations of a particular piece of research than is often the case in the history of science. Rather, they are frameworks for examining methodological issues that face archaeologists. Clearly it would have been possible to develop the argument of this book around a detailed study of just one or two examples, but it has seemed to us more useful to touch on a number of cases in order to show both something of the range of methodological problems that arise in archaeology and the way in which some of them keep recurring in different contexts. Quite naturally, this interest has made it necessary that we focus primarily on problematic examples rather than on totally noncontroversial research, and this seems the appropriate way to make the connection between the rather abstract interests of philosophy of science and the more concrete situations in which archaeologists may want to apply them.

Choosing the cases was difficult with the entire world of archae-

275

ological reporting offering numerous tantalizing leads. We do not claim that the cases presented here are either representative of all archaeological work or that they exhaust the methodological issues that could be addressed. We do feel, however, that the range of cases offers a partial and perhaps not atypical view of many of the problems faced in the discipline.

Judging from these cases, at least, archaeological methodology is not just a direct mapping of philosophical concepts onto archaeology. Instead, a subset of philosophical issues emerges as being of great concern to archaeologists, while many major philosophical concerns, including some discussed in earlier chapters, appear rarely or not at all in these cases. This may merely mean we have selected our cases in such a way that we are ignoring certain issues. As we have argued throughout previous chapters, however, there seems to be no inherent or compelling reason to expect archaeology to behave in the manner characterized by positivist philosophy of science, and the cases we have chosen present a number of differences from that model.

Formal explanation of the sort described by the D-N model appears to play a minor and somewhat self-conscious role in archaeological reporting, in spite of the emphasis on that model in the first wave of archaeological awareness of philosophical issues. It is not that explanation is not an ongoing archaeological concern, but that the D-N model is not the most appropriate one for archaeology; at least it is not the most appropriate for much of the kinds of work archaeologists do.

We have suggested earlier that inference to the best explanation provides a suitable characterization of much inference in archaeology as well as in other areas. As Paul Thagard has said: "[I]nference to the best explanation consists in accepting a hypothesis on the grounds that it provides a better explanation of the evidence than is provided by alternative hypotheses. We *argue* for a hypothesis or theory by arguing that it is the best explanation of the evidence" (Thagard 1978:77). In this account explanation is unquestionably viewed as crucial to the pursuit of knowledge, but not in the way many people have taken it to be. The appropriate focus is not on models of explanation, but on the choice, among alternative explanation candidates, of that one which best accounts for the data. Frank Hole (1973) has made much the same point, drawing on Platt's (1964) article on "Strong Inference," which in turn was stimulated by Chamberlain's (1897) classic article, "The Methodology of Multiple Working Hypotheses." Indeed, archaeologists often operate within a framework of "inference to the best explanation," although they seldom call it that, as is demonstrated in the following cases.

These cases reflect several recurring themes which we believe are widely found in archaeological research and writings. One has to do with the nature of evidence and acceptance of ideas into the working body of knowledge. Interestingly enough, various non-scientific or contextual/sociological factors play a significant role in what is accepted, and in requirements for providing support for hypotheses. Based on Kuhn's view of science, it is to be expected that such factors as core beliefs or shared views as well as other sociological factors would influence scientific procedures (*see* Chapter 4). Another such factor that proves to be important for scientific progress concerns the desirability of achieving a critical mass of researchers who are working toward the elucidation of a particular problem. These influences are easier to see in controversial cases, and this provides a further reason for including a number of controversial cases here.

With regard to more standard methodological issues, the kinds of problems that recur repeatedly in these cases and, indeed, across the discipline, include the role of testing, the importance of induction and confirmation, the question of what is to count as evidence in particular contexts, the intertwining of inductive and deductive strategies, the extent to which new data dictate new hypotheses or can be integrated into existing hypotheses, the ties between substantive and methodological issues, and the evaluation of conflicting knowledge claims. The crucial importance of the evaluation of evidence and inference reinforces some of the most insightful arguments of the New Archaeology. Ideas of taking hypotheses back to the data, testing, revision, and expansion of the hypotheses rather than limiting one's hypotheses to the plausible are implicitly seen in actual archaeological cases. Critical evaluation of individual hypotheses or conflicting knowledge claims does not always provide clear-cut choices for several reasons, some of which are discussed below.

Archaeologists often face problems of incomplete or ambiguous data, pragmatic problems of access to data, and problems of retrieving existing data. Indeed, data management and data retrieval have become crucial issues in the field, although the particular cases used here show this in less definitive ways than do other examples (but *see* Cordell and Plog 1979; *see also* Renfrew 1983). Decisions must be made about scope and scale of both hypothesis and data. As the hypotheses become more broadly framed and the data base becomes larger, it becomes increasingly difficult to control adequately the variables. Unusual combinations of highly specialized bodies of data or theory may be difficult for colleagues to evaluate.

Hypotheses always rest on a series of underlying assumptions and/or corollaries. Even if a particular hypothesis gains support or is not disconfirmed, the underlying assumptions can often be questioned and alternatives suggested. As there are enormous numbers of variables underlying even simple and straightforward hypotheses, the question of choice among alternatives becomes quite complex.

It can be argued that research strategies and orientation are as important as other methodological factors, because it is this basic orientation (which is at least partly sociologically grounded) that sets the course of inquiry to a significant degree. Other archaeologists will tend to use, emulate, or criticize work from a particular geographical area or work espousing a particular theoretical approach. Although such reasons do not override methodological issues, communality of interests is a significant variable in the business of building archaeological knowledge.

As a final word before considering individual cases, we would like to say that we find in these cases, as well as in our more general familiarity with archaeological research and reporting, that methodological concerns within the discipline range over a wide spectrum. The recurring themes we perceive, some of which have received little attention in earlier writings on archaeological methodology, include the very real questions of the nature of evidence and its context, consideration and choice from among alternatives (and this applies to every aspect of the scientific enterprise from assumptions underlying hypotheses, through hypothesis formation, to data selection, testing, and reassessment), and the importance of certain sociological factors such as the influence of core beliefs and the achieving of a critical mass in inquiry.

The business of science is not to provide us with isolated explanations that satisfy some more or less rigid set of adequacy criteria.

Rather, it is to make as comprehensible as possible the phenomena we observe through identification of the best from the available explanations of them. As we review these cases, it seems to us that many archaeologists understand this fundamental notion.

An example of archaeological explanation

The particular significance of this case for our purpose is, first, that it illustrates the fact that hypotheses can arise from indications in the data; second, a particularly careful and sound methodology of testing was used; and third, the explanation in this case was chosen over an alternative as better accounting for the data—thus illustrating "inference to the best explanation." This case also illustrates certain contextual factors such as the way core beliefs can shape one's preferences in hypothesis formation, and the way that a controversial explanation put forward reluctantly by someone who previously opposed such explanations can facilitate acceptance.

Emil Haury's paper, "Evidence at Point of Pines for a Prehistoric Migration from Northern Arizona," was originally presented at a symposium on migrations at the 1958 American Anthropological Association meetings (Haury 1958). In examining the remains at Point of Pines, Haury discovered archaeological evidence which seemed at odds with the generally accepted account of that site. In attempting to explain these anomalous phenomena, he formulated a hypothesis of migration from the Kayenta region south to Point of Pines.

Haury was not predisposed to explain the Point of Pines remains in terms of migration. Indeed, he had previously been very skeptical of migration hypotheses generally. At Point of Pines, however, he was faced with substantial evidence suggesting that population movement had occurred. The phenomena he observed which prompted his migration hypothesis included: (1) an L-shaped block of about seventy rooms that were larger than, and lacked the architectural features typical of, other rooms of the same period; (2) the pottery associated with this block of rooms, including local vessels, vessels from the Kayenta region to the north, and, most significantly, vessels made of local clays but painted in the northern tradition; (3) a D-shaped kiva that was unique among the Point of Pines kivas but of a form known in northern Arizona (furthermore, the pottery from this unusual kiva matched the pottery in the room block); (4) fragments of charred wooden tools made from trees with growth characteristics unobserved in tree-ring evidence from Point of Pines; (5) botanical evidence in the form of varieties of squash and corn more commonly found in northern Arizona during the period in question; (6) tree-ring dates establishing the period of occupancy of the excavated rooms. Less relevant evidence that did not contribute directly to the migration hypothesis or its testing but that served in a supporting role included, among other things, the fact that the room block was burned after the harvest had been stored (i.e., in the fall) and evidence from a ceremonial cache some twenty-five miles southwest of Point of Pines that contained a pot made of Point of Pines materials but painted in the northern style.

Faced with data which strongly indicated migration, Haury formulated a list of conditions, derived from fairly standard diffusionist principles, that he required be met before acceptance of such a hypothesis could, in his view, be justified. As Haury wrote:

> A migration is the probable, though not the only, explanation in the archaeological record of past people:
> 1. if there suddenly appears in a cultural continuum a constellation of traits readily identifiable as new, and without local prototypes, and
> 2. if the products of the immigrant group not only reflect borrowed elements from the host group, but also, as a lingering effect, preserve unmistakable elements from their own pattern.
> The probability that the phenomena outlined above do indeed represent a migration, rather than some other force that induces culture change, is increased:
> 1. if identification of an area is possible in which this constellation of traits was the normal pattern, and
> 2. if a rough time equivalency between the "at home" and the displaced expressions of the similar complexes can be established (Haury 1958:1).

Haury then proceeded to see whether these criteria were met by the evidence from Point of Pines that he had assembled. He concluded they were, and, consequently, an explanation of the Point of Pines archaeological record in terms of migration was justifiable. Any one of the lines of evidence gathered at Point of Pines might have been explainable without recourse to a migration hypothesis: local development, exchange patterns, convergent innovation, and so on could have been invoked. It was the fact that so many indications suggested a major break in the expectable pattern, and that these were localized spatially and within a limited temporal period, that led to the migration hypothesis.

The approach Haury took towards his work at Point of Pines is thus clear: the data he uncovered suggested a possible explanation to him, he formulated what seemed a reasonable hypothesis with specific test implications along the lines indicated by the evidence, and then he returned to that evidence to test his hypothesis. That the hypothesis and its testing involved the same data base is somewhat worrisome given the usual view that separate sets of data should be involved in these two procedures. As this case illustrates, however, this is not always possible or necessary.

Several features of the process Haury employed are worth expanding upon, because they illustrate the importance of points made in earlier chapters. First, as already noted, Haury did not begin his work at Point of Pines with a migration hypothesis in mind; instead, an explanation of this sort was suggested to him by the data he observed. In Chapter 2 we argued that the supposed distinction between inductivism and deductivism was largely illusory because scientists cannot collect the data necessary to their studies without making reference to *some* hypothesis or other; that—in sum—facts just don't speak for themselves. Haury's work is not an exception to this claim. He began his investigation of the Point of Pines remains with some notion of what he might encounter—data to support a theory of local development, but the evidence he uncovered drew him elsewhere.

More importantly, scientists work back and forth between evidence and hypothesis, often revising the latter in light of the former, as well as sometimes looking for different types of evidence as a result of changes in the hypothesis. We could then say that the Point of Pines evidence pushed Haury to a conclusion he did not anticipate: "As one who has always viewed archaeological evidence for migration sceptically, I find myself now in a situation where suspended judgment is no longer rational" (Haury 1958:6). Or we could just as well say that he had implicitly formulated a hypothesis of local development which was found inadequate to the body of evidence before him, causing him to formulate a replacement hypothesis. Characterized in this way, it is reasonable to view Haury's initial hypothesis as having been falsified by his data. Or, it would be more correct to say that *part* of Haury's hypothesis was falsified, because the rest of the extensive materials from Point of Pines continued to be viewed as a result of local development. Another way to view Haury's formulation and testing of the migration hypothesis is to say that he earlier felt such a hypothesis had very low prior probability; finding it to be strongly supported by the evidence caused a dramatic change in his evaluation of the credibility of the hypothesis.

A second significant feature of Haury's methodology is a more strictly theoretical one. Haury, in essence, said: "If migration was the cause of the archaeological remains at Point of Pines then the conditions I have imposed should be met; those conditions are met; therefore, the migration hypothesis is plausible." And this pattern of reasoning is, of course, paradigmatic of confirmation in science. The logic which most obviously underlies this reasoning, where "H" is the hypothesis "Migration was the cause of the archaeological remains at Point of Pines" and "O" is the observation that "The conditions imposed were met," is:

If H then O
O
‾‾‾‾‾
∴ H

Given the central role reasoning of this form plays in confirmation, it is a matter for concern that, as we have noted especially in Chapter 2, the pattern is deductively invalid. One should be reluctant to say that valid inductive inference—which is straightforwardly essential to confirmation in science—is just invalid deductive inference.

It may be, however, that this apparent difficulty can be resolved if we take a closer look at the logic of confirmation, for it seems that scientists are saying more in the justification of their conclusions than is revealed by the above logical structure. Implicit in the cases they build for acceptance of their hypotheses seems to be, as well, the claim that "If the hypothesis was not true, then probably the phenomena suggested by the evidence would not have occurred." Or, to take Haury's example, "If migration was not the cause of the archaeological remains at Point of Pines, then the conditions imposed more than likely would not have been met." And this argument can be seen as an inductive analogue to the deductively valid inference called *modus tollens*:

If not H then not O
O
___
∴ H

Ronald Giere (1979:88–94) has argued that a probabilistic version of this inference is part of every good test of a theory. That is, when we talk of confirming or testing hypotheses in science, we want to know both that the evidence supports the hypothesis (or that the hypothesis "fits the facts") and that, if the hypothesis were not true, then the facts or evidence would likely have been different. This is related to the idea that we want to be sure the hypothesis we accept or believe to be true should be the one that provides the best explanation of the data, in that, when we must select one from among several hypotheses each of which "fits the facts," a way to do this selecting is to ask whether, if the hypothesis were false, the data would likely (or almost certainly) have been different. If so, we have some confirmation of the hypothesis.

One thing that is especially interesting here is that Haury was not interested only in explanation in his work on Point of Pines, but also in confirmation, and his work illustrates the close relationship that we have suggested inevitably exists between the two. As we noted earlier, it seems that what we are after for explanatory purposes is laws, or at least generalizations that may be lawlike. And one plausible characterization of lawlike hypotheses is that they are those that are confirmable—capable of receiving support from their positive instances. Although not primarily concerned with discovering a universal generalization, Haury went some way towards articulating a law candidate in setting out his conditions for acceptance of a migration hypothesis. Indeed, Haury's account of Point of Pines *is* lawlike in that, if similar conditions to those at that site were found elsewhere, Haury would presumably insist that we should draw the same conclusions regarding an occurrence of migration. Haury's explanation bears some similarity to an "ideographic explanation" in that it is essentially a historical explanation of a unique event that is explained by conditions that are, as well, more or less unique. It is interesting to note that the hypothesis in question has received further attention recently that extends the inference.

In 1983, Alexander Lindsay, Jr., read a paper at the meetings of the Society for American Archaeology in which he reexamined Haury's original hypothesis in the light of more extensive information than was available in 1958. He found that the original hypothesis holds up (is not disconfirmed), and he added other supportive evidence. Presumably using the same criteria, Lindsay then postulated another population movement from the Kayenta region in the San Pedro valley to the Reese and Davis ruins, as well as noting evidence in the Gila valley near Safford at Pueblo Viejo and at sites near Duncan that could represent the displacement of Kayenta people from Point of Pines or separate movements out of the Tusayan region. He considered the population movement hypotheses in all these cases to be preferable to alternative hypotheses of aberrations in Salado development or trade networks between the sites in question and central Arizona. Lindsay therefore undertook further evaluation of the initial hypothesis; he fit new data into the hypothesis and he selected the population migration hypothesis over alternatives in an implicit use of "inference to the best explanation."

Although this migration hypothesis is being reevaluated and extended to other cases after twenty-five years, it remains true that no nomothetic explanation of migration has been advanced. More generalizing hypotheses or perhaps even nomothetic explanations could be developed, however, from repeated and critical tests of specific cases, distillation of the variables involved in ethnographic and historical population movements, and other sources from which theories could be formulated, such as ecological models. Other disciplines (rural sociology, for example) offer a large body of data as well as various levels of hypotheses and theories on this topic which might prove useful in the development of more encompassing hypotheses for archaeologists who must inevitably select what is useful from a bewildering array of options. The hypotheses and theories must then be analyzed for archaeological analogues, and the hypotheses must be continually taken back to archaeological data. Not only population movement itself is of interest, but its role in culture change and the stages of adaptation occurring during movement and subsequent resettlement are of great importance to understanding other cultural processes, as

Schwartz (1970) has shown in his review of ethnographic and other information which he uses to build a basis for archaeological inference about the impact of migration on the "post migration" cultures.

It is of some interest to note that Haury's study of Point of Pines predates by a number of years the emergence of what has been called the New Archaeology. As we noted in Chapter 1, concern about the theoretical foundations of archaeology did not begin with that movement, although interest in such problems became considerably more widespread with its appearance. Another aspect of Haury's work to which we want to draw attention is a sociological one. Haury's extensive personal familiarity with Arizona archaeology surely contributed to his recognition of, for example, the distinctions between local Point of Pines pottery, local pottery painted in Kayenta style, and Kayenta pottery. Archaeologists less familiar with a broad range of archaeological materials could easily have missed these differences, and these are the "indications" that triggered the formulation of the migration hypothesis. As will be discussed again later in this chapter, migration has been looked upon with great disfavor as an explanatory concept by many archaeologists during the last few decades, probably because the concept was so dramatically abused. Haury was very skeptical of the notion because he was sensitive to these earlier abuses. Consequently, he approached the Point of Pines problem in a very conservative manner. In Raymond Thompson's words: the Point of Pines "sequence was interpreted as a purely local development until the accumulation of an overwhelming body of evidence forced a consideration of the possibility of migration" (Thompson 1958a: vii).

The fact that Haury concluded that a migration had occurred at Point of Pines undoubtedly made that proposition more palatable to other skeptics. As a convert to a view he had previously opposed, Haury's considered judgement that a migration hypothesis was inescapable was much more readily accepted than would otherwise have been the case. Haury's previous convictions on the subject of migration were, of course, entirely irrelevant to the soundness of his research technique or his reasoning. The acceptance of propositions into our body of knowledge often turns upon such extrinsic factors, as well as more strictly methodological ones, however. Haury's earlier suspicion of migration explanations facilitated the acceptance of just such an explanation, but his careful methodology was even more significant.

Relevant evidence and the discovery of general laws in archaeology

This case is used to explore questions of scale, applicability of similar hypotheses in different contexts, and, most fundamentally, the question of what counts as evidence. Other issues touched upon include the problem of amalgamating the insights resulting from geographically widely separated research into generalizations of more universal applicability, and the further problem of attracting a critical mass of material to address comparable research questions rather than having individual insights stand as isolated cases without the kind of follow-up that moves the debates forward in productive ways.

We contrast the work of two archaeologists, Fred Plog (1974) and Peter White (1977), who are interested in how increased adaptiveness is reflected, over time, in cultural change. Among the aspects of this topic with which both are concerned is the question as to what should count as evidence of a culture's improved adaptation. In particular, both have considered what kinds of inference can be drawn from stone tool data. We should note that Plog's views, especially, have been modified since the publication of the book used in this case (cf. Plog 1979). The reason for confining our remarks to this single representation of Plog's views is that this particular set of ideas provides an interesting contrast to White's position. In juxtaposing the two positions in a selective fashion, we are interested in the contrasts that emerge and in their philosophical and substantive implications.

In Plog's view, economic transformations—which he assumes may move a culture in the direction of greater adaptation—are accompanied by technological changes expressed archaeologically in discernible changes in stone tools over time. He has argued, for example, that technological changes occurred during the Basketmaker-Pueblo transition (Plog 1974:140) and that those changes "involved evolution in the direction of increased efficiency" (Plog 1974: 142).

Plog notes that "if the Basketmaker-Pueblo transition was an economic transformation then technological innovation should have occurred during this transition" (Plog 1974: 137). Plog appeals to the literature on economic development as to why such a relationship should hold; he does not say how significant a technological change must be before it would count as an economic transformation of this sort. Arguably, some degree of economic transformation could occur independently of technological change, however the latter might be defined or measured—for example, through reorganization of production activities or intensification of an activity (or the reverse). If adaptive changes can occur without technological ones, clearly improved adaptation need not be manifest in any particular technological category of artifacts such as stone tools.

Peter White offers a counterexample from New Guinea. Although he uses the case to "exemplify the lack of necessary correlation between stone tool morphology and efficiency in harnessing energy" (White 1977: 23), it suggests equally well that economic developments need not be manifested in changes of stone tools. White argues that over a long period of time in a certain part of highland New Guinea there occurred a marked shift from what was largely a gathering and cultivating economy to one involving complex forms of agriculture which supported higher population densities. This level of economic change would seem to be of a larger magnitude than the economic transformation Plog sees as occurring during the Basketmaker-Pueblo transition in Arizona. "On our current data," White writes, "this change is largely unreflected in the wooden tools and negatively reflected (if I may use such a term) in the stone tools" (White 1977:23).

In White's view, while archaeologists may be able to say something about the economy of stone tool production, they are hardly in a position at this point to comment authoritatively on "efficiency" or similar concepts. According to White, pride in craftsmanship and aesthetic considerations, along with any number of other factors, may contribute to the form and complexity of a culture's stone tools. He goes so far as to say that "the majority of stone tool forms were not necessary, in a utilitarian sense, at all" (White 1977:26). He concludes that in relation to other aspects of man's material culture we do not regard increasing complexity as a necessary sign of modernity or progress. "In discussing art, or burial or clothing, we do not demand 'efficiency': we accept elaboration at some times, simplicity at others. Why not with stone tools?" (White 1977: 26–27).

While Plog correlates developments in projectile points, among other things, rather directly with changes in adaptation, White rejects this sort of technological determinism. In his view, comparable levels of energy exploitation and population density could have been achieved historically with stone tools of varying degrees of morphological complexity; he quotes favorably (White 1977:24) Marshall Sahlins' remark: "For the greater part of human history labour has been more significant than tools, the intelligent efforts of the producer more decisive than his simple equipment" (Sahlins 1972: 81). White shares with Plog an evolutionary perspective concerning culture change, but the two have quite different views about the significance of stone tools to that theoretical framework.

On the surface, this comparison of Plog and White has little significance to a work on the philosophical issues in archaeology, for the two archaeologists disagree more on substantive than on theoretical matters. That is to say, they do not hold significantly different notions about, for example, what it means to say that a piece of evidence confirms a hypothesis—or if they do, that is not what is revealed by the above comparison of their views. Rather, they have developed quite disparate ideas of what counts as evidence of a culture's greater adaptation, and their evolutionary frameworks are operationalized along quite different lines.

The comparison is nevertheless of methodological interest, for it underscores the persistent problem for archaeologists of discovering truly general and, at the same time, nontrivial laws, a difficulty we discussed in some detail earlier. Even if Plog is right that, at least in certain parts of the world, economic transformations are accompanied by technological developments, if they are not *always* so coupled then Plog's hypothesis is not a universal generalization but must be, at best, a generalization of limited applicability. The worry is not that one archaeological proposal has been met by a counterproposal; this is an everyday occurrence in any enterprise in which hypotheses are advanced. Rather, it raises a number of concerns about the scale or level of magnitude of applicability of many archaeological hypotheses, ideas, generalizations, or inferences. Does Plog's notion have geographical dimensions? Is it specific to particular regions? What variables underlie the cases in which it is applicable and those in which it is not? Could a more comprehensive generalization accommodate both kinds of cases? Will either notion be systematically pursued, or will new problems claim the attention of both investigators? The high value placed on novelty may subvert systematic follow-ups.

Binford's ideas on the relationship between general and middle-range theory (the latter a concept borrowed from Robert Merton: cf. Raab and Goodyear 1984) are relevant in this context. As a scientific endeavor, archaeology ultimately seeks to illuminate "the conditions that brought about change and modification in the organization of dynamics occurring in past living systems (Binford 1977:7). Archaeologists "seek understanding of the *processes* responsible for change and diversification in the organizational properties of living systems. In approaching this problem, we seek the development of general theory" (Ibid.). As Binford notes, not all archaeologists agree that development of general theory is an appropriate aspect of their task, and, even among those that do, few results of this kind have been produced.

In Binford's view, one of the reasons for this low productivity is the lack of "middle-range theory," that is, theory which relates the observed data, the archaeological evidence, to the processes that gave rise to them: "we must develop ideas and theories (middle-range theory) regarding the formation processes of the archaeological record" (Ibid.). "The development of general and middle-range theory must proceed hand in hand," because

> in the absence of criteria of relevance, we may waste much time in developing middle-range theory concerning the dynamic significance of certain static facts that prove to be irrelevant to the evaluation of our ideas about the general determinant processes that promote change and diversification in living systems. The field must advance as a whole. Advances in middle-range theory divorced from general theory may prove to be a waste of time (Ibid.).

If Binford's claims about the need for "middle-range theory" progressing concurrently with the more general are correct—and they seem reasonable, then, since attempts to build a general theory for archaeology seem to be nonproductive at present, Binford's advice to strengthen middle-range research would appear well taken. The two cases juxtaposed here seem to offer the basis for just such an exploration of "middle-range theory" if the topic of the role of technological change in human adaptation were systematically pursued in multiple locations with hypotheses being tested against empirical data. Higher level accommodations of multiple hypotheses could result.

White raises another issue related to topics discussed in earlier chapters—namely, the effects of the cultural heritage of archaeologists upon their archaeological world view. Many ideas in archaeology, including the idea of "progress," are rooted in the nineteenth-century European origins of the discipline. White argues that these are parochial and not readily transferable to other parts of the world—at least not the part of the world where he works. Not only was the notion of "progress" per se a function of nineteenth-century European values, but European Paleolithic archaeology set a historical precedent in building such values into empirical analysis. The European Paleolithic does exhibit progression in stone tools with increasing diversification of tool forms through time as well as a tendency to get more tools out of a given volume of raw material.

Juxtaposing two works about such different parts of the world thus allows us to direct attention to the questions of scale of generalizations, to the need to define the conditions under which generalizations apply, and to the need for repeated testing in multiple contexts. White has also raised more fundamental questions about the effects of archaeology's Eurocentric background on formulation of hypotheses or inferences.

The confirmation and acceptance of controversial theses in archaeology

Contemporary debates about the possibility of an earlier Early Man in the New World than is presently accepted into the body of core beliefs illustrate several points made in earlier chapters and in the preceding cases. Using the spherical model of the discipline presented in Chapter 4, we have a situation analogous to the earlier Early Man debates of the 1920s (and before), in that challenges to the received view are being advanced but, at the present time, they are placed in the outer parts of the sphere. Some of the strategies being employed in an effort to replace the current core belief with another offer insights into both sociological and methodological issues. Specifically, strategies of opinion changing, the higher demands of confirmation in threshhold and/or controversial cases, and analogies to inference in other areas arise in connection with this topic. Thus, the recurring themes of the role of core beliefs relative to the community base that supports them, the role of core beliefs in acting as precedents in decision making, the persistent problems of what counts as evidence in which contexts, and the creation of effective tests of hypotheses and evaluation of conflicting knowledge claims are rendered more visible by the controversial nature of this case. For our purposes, it does not matter whether or not the challenges are successful in overturning the currently accepted view.

Few topics have evoked consistently stronger responses from both innovative and conservative archaeologists than suggestions of major increases in the estimation of the antiquity of human occupation in the New World. Transoceanic contacts between the Old and New Worlds evokes similar responses, and most archaeologists hold quite partisan opinions on these matters. That views about the age of man in the New World are deeply entrenched in the core system is perfectly understandable. All disciplines have and require basic tenets (cf. Chapter 4). Among other things, accepted views of the age of man in the New World serve to bracket the geological strata in which archaeological field investigations can be expected to be productive. As older ages are accepted, a different range of geological situations becomes of interest to archaeologists. An impressive number of finds have been advanced in recent years as candidates for pre-Clovis occupation of the New World. Remarkable parallels exist to the buildup of possibly early finds prior to the Folsom discovery.

Workers in the Early Man field are especially sensitive to problems of acceptability of their results. At a 1975 symposium on this subject (Bryan, ed. 1978), a number of the participants commented on the nature of the evidence required to change received views as well as on the recalcitrance of the discipline at large in accepting what the individual participants regarded as acceptable evidence. The authors of the Meadowcroft Shelter report at that symposium note that "(P)oorly substantiated data will not suffice to document Pre-Clovis occupation in this hemisphere. Further, proper documentation *must* include *excellent* stratigraphy, coupled with multiple radiocarbon determinations of artifacts of *indisputable* human manufacture in *direct* association" (Adovasio, et al. 1978:178).

One can hardly dispute the desirability of the level of documentation these authors advocate. Nor can it be plausibly suggested that the work presented to date that purports to establish pre-Clovis man south of the Yukon in the New World has satisfied the above standards. As these writers observe, "When these criteria are applied to the vast majority of allegedly Pre-Clovis sites, one or another deficiency is inevitably apparent" (Ibid.). But such considerations are not really relevant to the issue we are concerned with here, which is that few hypotheses concerning the age of man anywhere in the world that have ever been advanced or accepted have been as rigorously supported as these authors advocate. Certainly the Folsom find, on the basis of which the currently received view of the age of humans in the New World became established, was not nearly so well documented (cf. Figgins 1927), though the available evidence was both relevant and convincing.

We suspect that even if evidence of the kind specified by the authors of the Meadowcroft Shelter report was mounted in support of an early Early Man thesis, acceptance of that proposition would not necessarily follow, and this because the established view is so well entrenched. To draw a comparison with Haury's migration hypothesis, we believe that although migration is also controversial, it is not as controversial as challenges to received reviews of Early Man. No one denies that migrations have occurred. Migration is not a core belief, and a suggestion of a fairly short migration supported by reasonable methodology and evidence does not conflict with core beliefs in ways that impede acceptance, although there may be some question as to its significance. Haury did not face the additional task of changing a core belief, with the consequent increased evidence requirements. Archaeology's position with regard to early Early Man might well shift over time as the arguments indicating an earlier age for man in the New World increase in number and strength. In fact, change may well come about as the cumulative result of considerably weaker evidence than the Meadowcroft writers claim is required. In any case, the accepted age of New World man is unlikely to increase substantially in a single jump—say, by 100,000 years—no matter how well documented, although such temporal ranges are being suggested (Reeves 1983a; Irving, Jopling, and Kritsch-Armstrong 1984: 19).

Jose Luis Lorenzo, who has also studied the pushing back of the accepted age of man in the New World, has noted the sometimes less-than-adequate reasons which have underlain the reluctance to change the received view. He has observed, in this context, that "this point has been pushed back slowly and inexorably from the 11,000 year mark of more or less a decade ago to 14,000 years now, but for no logical reason. Once again it is a matter of arbitrary definitions" (Lorenzo 1978:2). He has commented, too, upon the extreme rigidity with which archaeology has protected the received view of the matter: "it is surprising how such an inflexible position is able to obscure the scientific approach under a layer of pettifogging legal phraseology: proof, testimony, evidence, and the like" (Ibid.). While lamenting the legalistic overtones to much argument in archaeology, he suggests that archaeologists, "under threat of ostracism" for having found remains of greater than permitted age, add legally trained spokesmen to their research teams to advocate acceptance of their views (Ibid.). Although not addressing the Early Man question directly, De Boer (1982) has argued along similar lines in his article, "Archaeology as a Myth-Making and Myth-Serving Profession," and Wobst and Keene (1983) have presented socio/political/economic arguments for retaining the status quo in "Origins" research, as was discussed more fully in Chapter 4.

As frustrated as Lorenzo by the archaeological establishment's reluctance to accept earlier-than-Clovis man in the New World, Alexander insists that "we have the evidence. . . , we have had it for a long time, and we should insist that it be used" (Alexander 1978:22). Alexander echoes Lorenzo's somewhat bitterly made recommendation that disputes in archaeology over Early Man issues be decided legalistically, saying that "if a dispassionate approach is to be obtained through a legalistic system, we must provide our advocates with guidelines on burden of proof, and give them sufficient information so that the proper questions may be asked" (Ibid.). Pursuing this idea further, he has demanded answers to questions such as the following:

> What are the rules of the game? Is the burden of proof the same for all archaeological sites, or for all archaeological problems? Do the rules change for acceptable proof of Mesoamerican ceremonial centers, as opposed to eastern longhouses, as opposed to Eskimo tent rings, or Clovis campsites? Is the criterion of "importance" to be overriding and, if so shall we have a supreme court to decide importance"? (Ibid.).

He has considered the role a jury might play in such a aystem of adjudication as well:

A legal system with advocates, testimony and judges will almost certainly require a jury and it is here we may find an impossible hurdle that says: "Go back to the scientific method." For a jury within the English common law system, is made from a group of peers, and the first question asked a prospective juror is "Have you already made up your mind in this case?" (Ibid.).

Two such pointed and not altogether facetious references to legalistic procedures for settling major differences of opinion in archaeology remind us, irresistably, of Spaulding's caustic comment in his review of J. Ford: "Ford's propositions carry the logical implications that truth is to be determined by some sort of polling of archaeologists, that productivity is doing what other archaeologists do, and the only purpose of archaeology is to make other archaeologists happy" (Spaulding 1953: 590).

We would like to consider Lorenzo's and Alexander's comments dealing with legalistic analogy for reasons which would probably be unacceptable to both those authors. This being the case, let us preface our look at the law as an analogy to archaeology by saying that we understand their frustration at dealing with materials they regard as having met reasonable standards of evidence that are disregarded by the profession. Nevertheless, it serves our purposes to take their remarks somewhat more seriously than they intended, for they provide the basis for some illuminating comparisons of decision making in archaeology and the law.

First, several points should be made about the parallels Alexander would draw between dispute resolution in the law and archaeology. Alexander suggests that since jurors must consider the facts of a case before them with an open mind, jurors—and for that matter judges—are disqualified from hearing a case on which they have already reached a decision. But the problem in archaeology that we are considering here is not the flat refusal by the archaeological community to look at evidence which challenges received beliefs. Such dogmatism may occasionally occur, but the problem which archaeologists see arising much more frequently, and which therefore seems to warrant discussion, is, instead, the uniquely rigorous standards that are imposed on the documentation of hypotheses with appear to contradict established doctrine.

It might be objected that our legal system even rejects judges and juries disposed to certain views, not only those already firmly persuaded of a particular outcome. While this is accurate, at least in theory, it cannot be carried out in practice, and it is hard to see how such stringent standards could hold in archaeology or any other branch of inquiry. The requirement of absolute impartiality is impossible to achieve, even in the law where it is especially desirable. In fields such as archaeology, however, it is both unrealistic and unnecessary to expect complete neutrality. What is important, it would seem, is that an archaeologist hold beliefs rationally—and that is not precluded by being simply inclined, or predisposed, to accept one school of thought as opposed to another.

Taken literally, Alexander's remarks on burden of proof reveal another misunderstanding of the law. In civil cases, the party on whom the onus of establishing a particular proposition falls must satisfy the court on a balance of probabilities of that point. This remains the quantum of proof required even where that which must be shown amounts to the commission of a criminal offence which, if prosecuted, would have to be established beyond a reasonable doubt. Notwithstanding these legal principles, however, there is no question that, as a matter of practice, the standard of proof a litigant must meet depends upon what is at stake in the legal action. Nor is this aspect of litigation in any way embarrassing to those involved with it. The degree of proof required in any case varies with "the gravity of the consequence of the finding," and this is readily admitted in the common law courts.

Archaeology's tendency to demand particularly rigorous documentation of hypotheses viewed as especially threatening to received doctrine is rather more like, than unlike, the weighing of competing positions in the law. The more important an issue in both the law and archaeology, the more carefully arguments addressed to it are scrutinized. This approach is only reasonable, reflecting, as it seems to us, a principle constitutive of rational decision making.

Two further points in connection with Alexander's extension of legal procedure to the adjudication of archaeological disputes concern his already quoted remarks on "importance," which we are taking at face value for our present purposes. First, constructive consideration of dispute resolution in archaeology is not noticeably advanced by Alexander's suggestion that, if the concept of importance was somehow the basis for determining the degree of support a hypothesis needed, then decisions as to relative importance would have to be made by a formal institution such as a Supreme Court. If we are correct in our above discussion of decision making in general, and particularly in the law, then in fact we make, both individually and collectively as practitioners of various disciplines and professions and members of a common culture, such judgments as to importance all the time. We do so, when we are performing at our best, on the basis of common sense and in consideration of the facts of the particular situation before us. This is not to say that there are not disagreements over questions of importance—for there are. But we seem to be muddling along quite well without a National Standing Committee on Importance, or the like.

Second, if we continue to take Alexander's remarks at face value, we discover another misunderstanding of the law. In the law, issues of importance or relevance are treated virtually uniformly as questions of fact, not law. That is, while questions of law are decided on the basis of principles extracted from previous cases—or so the theory runs, questions of fact are said to be resolved on the circumstances of the individual case. Thus, on the legal model at least, questions of fact are never the subject of binding precedent. But a problem immediately arises, for—as is widely recognized, at least among philosophers—the theoretical underpinnings of the fact/law distinction are far from strong enough to bear the weight most lawyers want to impose on it.

The general philosophical point to be made is that no precise fact/law distinction can be drawn; the law is modified to respond to the facts of particular cases, and what the facts are, in any case, is in large part a function of the body of law that is applicable to them. When lawyers speak of "the facts of a case," they are not, of course, referring to all the facts of the case (whatever that could mean), but only to the features of the situation which are relevant to the resolution of a particular dispute. The key concept here is "relevance," for it is precisely the question as to which facts are relevant that cannot be answered independently of the body of law applicable to the case.

The above point has its analogy in science, of course. It is commonly thought that, while scientific theories are somehow devised by scientists, the facts they are meant to account for exist independently of those theories—that those phenomena are simply "out there" waiting to be discovered or explained. This view is a misconception, as we have discussed earlier and have argued with respect to legal facts. There is no clear dividing line between facts and theory. The scientist's perception of the facts is as theory laden as his development of theory is fact influenced. Nevertheless, in spite of the fact/law interrelationship, legal questions are categorized—and usually satisfactorily for the purposes of the law—as being of either one or the other type.

Alexander goes on to suggest, as we have indicated above, that archaeologists mend their ways by returning to the scientific method, and this recommendation—unlike his proposals to impose a legal format on dispute resolution in archaeology—he intends to be taken quite seriously. The scientific method, as Alexander sees it, demands the acceptance of a hypothesis even when it contradicts the received view, if it offers an explanation that is better than the current explanation of a phenomenon. In a sense, this construal of science is accurate. In another sense it is not, for there is *always* the question as to what is to count as a better, or best, explanation, and, in the Early Man case, there are obviously serious differences of opinion as to what constitutes evidence either necessary or sufficient to overturn the accepted views. Challenges to fundamental tenets must be better substantiated than less adventuresome hypotheses.

Also, as we noted earlier, scientists don't always abandon a position simply because evidence against it is presented. The history of science reveals numerous examples of theories retained in the face of considerable contrary evidence. Of course, some of those propositions were protected for inappropriate reasons—for example, simply because their proponents could not grasp, for whatever reasons, the concepts involved in the alternative hypotheses advanced. As we noted earlier, according to Kuhn, it was to a large degree because Lavoisier's contemporaries could not see the inadequacies in the phlogiston theory that they failed to recognize combustion for what it was. The oxygen theory of combustion was eventually accepted, but the point remains that the phlogiston account endured for as long as it did at least partly because its advocates did not understand the nature of the differences between it and its most important competition. Those who criticize archaeology because it, or some of its practitioners, have been guilty of the same should be aware that such occurrences, however unfortunate, do not alone set archaeology apart. The more significant point here is that a certain degree of conservatism is perfectly reasonable in science. Indeed, as we noted in discussing Kuhn's concept of science in Chapter 3, without some shared and more-or-less stable foundation of theories or beliefs on which scientists can base their work, progress in science would be impossible.

We have argued that archaeology shares more rather than less with law than Alexander suggested. As a postscript, we note again that archaeology does seem to protect core beliefs in a way that may be unique or extreme among the sciences.

A controversial thesis

The subject of transoceanic contacts is used here to illustrate certain sociological and methodological issues such as the evaluation of conflicting knowledge claims, the nature and context of evidence, and the role of core beliefs in acceptance or rejection of hypotheses—topics that appear again and again in these cases. This case also allows us to explore the less common problem of the effects of unusual combinations of expertise and data.

Long-distance and especially transoceanic diffusion has emerged as one of the most controversial topics in all of archaeology, with most archaeologists adopting a rigid position on one side of the debate or the other. In an effort to overcome this polarization and examine the issue more rationally, Paul Tolstoy (1972) has drawn a distinction between diffusion "as event" and diffusion "as explanation." In the latter, occurrences of diffusion, where established, are seen as useful to the explanation of culture change. With respect to that view, Tolstoy, responding to earlier critiques of diffusion by John Rowe (1966), has written that "diffusion-as-explanation can have only modest pretensions and can hardly be offered, for instance, as 'a substitute for archaeological theory'" (Tolstoy 1972:823).

Our purpose here is neither to advocate any particular example of diffusion or to argue that diffusion does or does not have explanatory power. We do want to note, though, that the resolution of such an issue would depend in part on what one was trying to explain in any specific case, and that, unless one is wedded to single-cause explanations, diffusion could play a useful role in the analysis and understanding of past events.) Instead, we want to look in some detail at a study done by Tolstoy which purports to demonstrate an occurrence of diffusion and to consider why the results of that study have not been widely accepted. We argue that Tolstoy's conclusions have been ignored, notwithstanding the commendable rigor of his research, in large part because they conflict with certain views that are deeply entrenched in contemporary archaeological beliefs.

Tolstoy has been much concerned with methodological issues related to diffusion, and in the particular study with which we are concerned he has examined the parallels between the manufacture of bark cloth in Southeast Asia and Mesoamerica (1963 and 1966). In concluding that the Mesoamerican industry is of Southeast Asian derivation, Tolstoy (1966: 834–836) went beyond drawing the usual comparisons and formulated ten criteria against which he evaluated the evidence collected from the two areas under study. These are paraphrased below:

1. *Number* of descriptive features shared by Southeast Asian and Mesoamerican industries: of 119 Mesoamerican traits, 92 were found in Southeast Asia.
2. *Distribution*: 88 of the 92 bark-beating traits shared by Mesoamerica and Southeast Asia were focussed in a specific group of Southeast Asian industries. In contrast, only 34 of the 92 traits were found in Central and South American industries, of which only 9 are common to those two areas.
3. *Exhaustiveness*: Traits shared by Mesoamerican and Southeast Asian industries include all but 4 of the Mesoamerican traits that also occur anywhere else.
4. *Degree of similarity resulting*: The Mesoamerican and Celebean industries are more like each other than either is like any other. They are closer to each other than the Celebean is to Javanese, Mainland, or Polynesian, or than the Mesoamerican is to the Central and South American.
5. *Local relationships*: The shared traits seem to have developed gradually in Southeast Asia. No such developmental continuum is known for the New World.
6. *World infrequency*: Some 67 of the 92 traits shared by Mesoamerican and Southeast Asian industries are very rare on a worldwide basis; 31 are found in no other industries.
7. *Nonessentialness of shared traits*: Many (44 of 92) of the common characteristics are not required for any other manufacturing step. Thus, a functional explanation cannot be given for these traits.

8. *Plurality of alternatives*: Even where essential, many of these shared traits represent but one of several known, alternative solutions to the problem for which they were required.
9. *Redundancy*: 37 of 92 traits are redundant in the sense that they co-occur with alternatives.
10. *Convergence of evidence*: The same criteria tend to be satisfied by the same traits. (Tolstoy considered this coincidence of the various lines of evidence itself to be significant.)

As a final point noted elsewhere by Tolstoy, there was a temporal overlap of the Southeast Asian and Mesoamerican industries.

In examining the reception given Tolstoy's work by the archaeological community, a number of possibilities must be considered. First, was his evidence not adequate to justify his conclusion that the Mesoamerican bark-cloth industry is rooted in that of Southeast Asia? It must be conceded that the basic archaeological data upon which Tolstoy relies are patchy and often have poor contextual or chronological control. As well, little information is presented as to which traits are functionally similar (and why) and, more seriously—because the claim that many of the shared characteristics cannot be accounted for by "functional imperatives" is an important aspect of his thesis, we are not fully informed of Tolstoy's grounds for rejecting functional or simply commonsense reasons for similarities among the remaining parallel traits.

There is, in addition, another very real difficulty with the evaluation of Tolstoy's data which stems from the fact that few archaeologists are sufficiently familiar with the archaeology and ethnography of both Mesoamerican and the Celebes (not to mention the other relevant areas), or with the technology of bark beating, to comment critically. This illustrates a significant problem for assessing any study which involves an unusual combination of expertise or a radically new organization of data.

Since transoceanic diffusion is one of those subjects that has not gained acceptance in the core system over a very long period, it still rests lightly on the surface of the sphere—to harken back to our spherical model of the discipline. This means, among other things, that the subject has failed to develop a critical mass of researchers who can provide the kind of critical evaluation that leads to improved assumptions and hypotheses and which increases the number of cases under consideration. Tolstoy's arguments are not debated and evaluated as much as they are ignored.

Is it, then, the quality of his evidence that has prevented general acceptance of Tolstoy's conclusions? We think not. Data supporting a thesis of such a broad geographical and temporal scale are inevitably going to be incomplete. The patchiness of Tolstoy's evidence is not in itself sufficient cause for rejecting his idea, especially since Tolstoy seems to have been quite sensitive to the limitations inherent in his data. It would appear that his evidence would have been convincing to the archaeological community had the two areas exhibiting the parallels Tolstoy uncovered been Mesoamerica and Central America, for example.

Another possibility is that the approach Tolstoy took to his study was in some way substandard. This, too, seems very unlikely, for Tolstoy's methodology was without doubt superior to that used in many cases that have not involved controversial topics and which have generally been accepted. Tolstoy, for example, considered three alternative explanations for the similarities he perceived between the two bark-cloth industries: chance correspondence, functional imperatives, and historical relationship. The ten criteria he developed were intended to form a basis for assessing "the likelihood that such similarities as occur may have arisen either by chance or from functional demand, operating independently in each of the two areas compared" (Tolstoy 1972:836). His reasoning led him to conclude that the correspondences were too high for chance and that the functional imperative was too weak to account for the data. Thus, he was left with historical relationship as the cause of the phenomena suggested by his evidence and with the methodological problem of meeting the demand for direct evidence of the mechanisms involved, which represent mechanisms of relatively low probability of occurrence. The fact that long-distance diffusion and/or migration is presumably an uncommon event and therefore low in probability raises a series of questions that are addressed in philosophy of science (*see* Chapters 5 and 6; *see also* W. Salmon 1971).

Tolstoy's methodology can be seen as an application of inference to the best explanation, because it involved selection of a hypothesis from among a number of plausible alternatives through a process of elimination. We have argued earlier that the most productive approach to the confirmation of hypotheses, and in connection with this, the acquisition of knowledge, is likely to be by means of just such a process of reasoning and data handling. Some archaeologists have advocated the consideration of several competing hypotheses as a methodology vastly improved over the usual ones employed in archaeological investigation (cf. Dumond 1977). It seems clear that the structure of Tolstoy's work cannot be what has prevented the acceptance of his proposals.

It is important to note here that there is more to selective confirmation, or inference to the best explanation, than just elimination. The hypothesis that is left is not necessarily a good one simply because all the others considered have been rejected for one reason or another. It is necessary to emphasize that in order for us to be able to say a surviving hypothesis is the best, all the reasonably plausible alternatives must have been included in the initial set of hypotheses. Tolstoy certainly did not exhaust the alternative hypotheses concerning, or possible explanations of, the evidence he uncovered. Levi Strauss (1963), for example, favours an explanation of the similarities between art styles from Amazonia, the Northwest Coast, and Oceania that appeals to the hierarchial nature of the several cultures involved in his comparative study. Others have suggested that Old and New World parallels can be better explained by more remote common ancestry, with the traits in question forming part of the intellectual, or other, heritage of early migrants (cf. Mundkur 1976). Tolstoy's work is important nonetheless, for he has employed a methodology that is more sophisticated than that used typically in archaeology, and few individual research efforts ever canvas *all* the plausible alternatives.

What, then, does account for the archaeological community's reception of Tolstoy's work? Tolstoy himself suspects that

> most of the skepticism directed at [his] and similar claims stems not so much from their methodological weakness as from the difficulty most of us have in visualizing concretely the process of diffusion. Thus, isolationists are evidently unimpressed by ocean currents. . . , recent raft traffic, or the capacity of water-borne intruders to do anything but clean fish and keep quiet (1972:839).

Tolstoy's proposal that transoceanic contact occurred also threatens the view, now deeply pervasive among archaeologists, that the New World is an independent laboratory for evolutionary theory; his suggestion that certain traits found in the New World may have been introduced from the Old challenges a well-entrenched account of culture change. It thus seems clear that core archaeological beliefs figured prominently in the discipline's assessment of Tolstoy's diffusion-dependent inferences.

A counterexample

In view of the methodological points raised in the last section, it is useful to follow that example with another dealing with the same general topic. Here, again, we have conflicting knowledge claims. The problem of a less than adequate data base is rendered even more severe because of the scope of the problem. Although one alternative explanation is favored over another at the conclusion of this study, the nature of the evidence does not provide a clear-cut choice, illustrating that even thoughtful and critical evaluation of a hypothesis can give ambiguous results. This case also speaks to the point that empirical testing beyond explanatory power of a given hypothesis is necessary and raises the issue of the difficulties of application of probability theory to archaeology and anthropology.

Earlier in this chapter we noted that most archaeologists dismiss long-distance diffusion and do not enter directly into the process of critical evaluation of such hypotheses. Although Charles Erasmus is an anthropologist and not an archaeologist, he is a striking exception to our general observation in that he has adopted a more open-minded view of such accounts and has attempted a critical evaluation of the classic case of patolli/pachisi similarities and the reasons for these similarities. Erasmus chose a classic case for methodological reasons; indeed, he chose the patolli-pachisi case that Tylor and Kroeber had previously considered *because* it was a classic study. Erasmus was not primarily interested in obscure games.

Many marked similarities between the various board games found in the New and Old Worlds have been noted, and particular attention has been paid to the common characteristics of the Aztec game of patolli and the Hindu game of pachisi, which, according to Erasmus, include the following:

In both [games] counters were moved along tracks which were cross-like in shape and drawn upon mats or "boards." In both, the movement of the counters was determined by the throw of lots—six cowrie shells in the Hindu version and five black beans with white dots on one side in the Aztec version. Although the scoring was arbitrary in both cases, there was a tendency in both to give greater weight to more difficult throws (Erasmus 1950:369).

In light of these and other similarities, the question has arisen as to whether the New World game developed independently of the Old World game or was a product of diffusion from the older culture. Discussion of this question touches upon the general controversy concerning whether there was pre-Columbian contact between the Old and New worlds.

One of the first writers to stimulate significant interest in the relationship between patolli and pachisi was E.B. Tylor, who was profoundly struck by the number of features common to the two games. In a paper published in 1878, he expressed the conclusion that "the American lot games, like certain developments in metal work, architecture, astronomy, and political and religious institutions, had come somehow from Asia prior to the time of Columbus" (quoted in Erasmus 1950:369–370). In a later paper, Tylor attempted to strengthen his view that the New World games had an Asiatic origin with probability theory. The gist of his argument was that, given the complexity of patolli and pachisi, the likelihood of the younger game's having developed independently of the older was just too low for rational acceptance. As Tylor explained:

> I have found it useful at any rate as a means of clearing ideas, to attempt a definite rule by analyzing such phenomena into constituent elements showing so little connection with one another that they may be reasonably treated as independent. The more numerous are such elements, the more improbable the recurrence of their combination (Tylor 1896:6).

A.L. Kroeber is another who has used the patolli-pachisi case to illustrate the difference between independent parallel origin, or convergence, and the relationship by common descent and spread, or diffusion (*see* Kroeber 1931). He has likened this distinction to that between homologous and analogous similarities in biology, homologous traits being those which are structurally, and not merely superficially, similar. These correspond in anthropology to similarities produced by diffusion, while analogous characteristics are just that, and correspond to those developed independently. Applying this approach to the patolli-pachisi example, Kroeber has said:

> [I]t is difficult to see only a superficial analogy between the Aztec patolli game and the Hindu pachisi game, long ago analytically compared by Tylor. Their specific structural similarities in two-sided lot throwing, count values dependent on frequency of lot combinations, a cruciform scoring circuit, the killing of opponents' counters that are overtaken, etc., make out a strong case for a true homology and therefore a genetic unity of the two game forms, in spite of their geographical separation (1931:151).

Kroeber concluded that the origins of patolli were to be found in the Old World board games, such as pachisi.

Kroeber's confidence in this position later wavered, however. He remained impressed by the similarities between patolli and pachisi, and used probability theory as Tylor had to support the view that "[t]he mathematical probability of two games invented separately agreeing by chance in so many quite specific features is very low" (1948:551). But he was bothered by the question why only pachisi, and not other elements of the culture from which it was taken, would have been adopted: "If pachisi was anciently imported from India to Mexico, as it was later carried from India to England and America, it is extremely unlikely that the people who brought it would have brought that and nothing else; or that only pachisi survived as patolli, but practically everything else brought with it failed to be accepted in Mexico, or died out, or was so altered as to be unrecognizable" (1948:551–552). Kroeber's ultimate conclusion was that the issue had to remain open—that the evidence was too evenly balanced to warrant acceptance of either position.

Erasmus stepped into this classical anthropological stalemate with the publication in 1950 of his article, "Patolli, Pachisi, and the Limitation of Possibilities." In criticism of both Tylor's and Kroeber's work, Erasmus made two related points. First, he compared the rather meager data upon which they based their conclusions with the extensive information upon which Stewart Culin (1903) relied in his study of the games of the North American Indians. In Erasmus' view, the work of Tylor and Kroeber suffered from sampling difficulties—they both worked with too narrow a data base and thus overlooked hypotheses they might otherwise have thought plausible. The problem of insufficient and uneven data is commonly encountered in anthropology and archaeology, and this becomes painfully apparent when data collected in other contexts is used for evidence in support of different kinds of hypotheses. Also, it is inevitable that hypotheses or theories on the scale of the patolli/pachisi debate will be of exceedingly uneven quality as well as having gaps in the data. Tylor and Kroeber, therefore, should not be singled out in this regard. But Erasmus' criticism nevertheless points out a weakness in their research.

Second, and more fundamentally, Erasmus argued that Tylor and Kroeber misapplied probability theory to the data they did have. They failed, according to Erasmus, to recognize that the constituent features of the games they studied might not have been independent of one another. Of course, if they were not, then the appropriate probability calculations—typically called "conditional probabilities," could have yielded dramatically different results. As Brian Skyrms has said: "Conditional probabilities allow for the fact that if a certain statement, p, is known to be true, this may affect the probability to be assigned to another statement, q" (1975:40). He illustrated this feature of the probability calculus in the following way:

> In the example with the die, we found that the probability of throwing an even number was 1/2. However, the probability of getting an even number *given that* a 2 or a 4 is thrown is not 1/2 but 1. And the probability of casting an even number *given* that a 1 or a 3 is thrown is 0. To take a little more complicated example, suppose that the die remains unchanged and you are to bet on whether it will come up even, with a special agreement that if it comes up 5 all bets will be off and it will be thrown again. In such a situation you would be interested in the probability that it will come up even *given that* it will be either a 1, 2, 3, 4, or 6. This probability should be greater than 1/2 since the condition excludes one of the ways in which the die could come up odd. It is—in fact, 3/5. Thus the probabilities of "even," given three different conditions, are each different from the probability of "even" by itself (Ibid.).

The notion that some traits ordinarily occur in combination with certain others, and thus that the presence of one increases the likelihood of those to which it is so related (and correspondingly decreases the likelihood of others), is referred to, in anthropology, as the "limitation of possibilities" principle. For example, Erasmus notes that it has been suggested by Boas that similarities among languages may in part be the result of physiological constraints that "limit the range of possible sounds which are sufficiently distinct to the ear for clear understanding" (Boas 1920:216), and this raises again the usefulness of some of the criteria Tolstoy used in his study relating to alternatives such as functional explanations. The strength of arguments that identify alternatives, as well as the strength of applying probability theory to archaeological and anthropological problems, is that more variables are made explicit and guidelines for rational evaluation are offered. As the explanations which are attempted become increasingly sophisticated, particular limitations on culture change will emerge more clearly (Erasmus 1950:386).

With respect to questions about the development of games, Erasmus saw plenty of room for application of the limitation of possibilities principle:

> There are in America today individuals and companies who do nothing but apply their ingenuity to the construction and invention of games with which to amuse the public and thereby derive their sustenance. Yet, despite this concentration of effort, the dice-and-board games of today do not differ in principle from those of the Aztecs and the Hindus . . . [t]he limitations still there are so great that a manufacturer cannot patent such a toy. Even his copyright extends only to certain superficial aspects like the name and the art design (Erasmus 1950: 386–387, footnote deleted).

In connection with the patolli-pachisi controversy in particular, Erasmus has said that he "sees no reason why patolli and pachisi could not have developed independently from a common widespread trait like the use of lots. The features which [the two games] share in common are not independent multiples which put their chances for recombination beyond the realm of feasibility" (Ibid., 382). In Erasmus' view, "there is no proof or incontestable evidence for either diffusion or independent development in this [the patolli-pachisi] case, nor is there likely ever to be" (Ibid.). He found the parallel development thesis, on balance, the better supported.

It is perhaps worth noting that earlier willingness to accept diffusion as an explanation of the patolli/pachisi similarities may be due, in part, to assumptions current in earlier core beliefs. That is, because of the content of earlier archaeological beliefs and their position within the archaeological sphere (to refer again to our spherical model), there was a tendency to focus on cases that were uncritically presumed to fit a diffusionist model just because diffusion was regarded as plausible. There was, as well, a tendency to dichotomize into diffusion versus local invention things we now tend to regard as the result of multiple and complex processes. In part, then, the current questioning of diffusionist inferences is due to rejection of beliefs and assumptions that held a different position in the sphere of earlier researchers.

Erasmus is among the many anthropologists and archaeologists who have rejected long-distance diffusion accounts of various phenomena. Nevertheless, Erasmus' rejection of the Asiatic origin of patolli differs significantly from that of most of the others dealing with long-distance diffusion because it was the product of a *reasoned* consideration of the hypothesis' plausibility and not an ideological stance against diffusion. Erasmus should be credited with a rather sophisticated understanding of the theoretical foundations of the controversy, for he realized that the competing ideas—diffusion and parallel development—must be carefully formulated if the debate is even to make sense. Even if the parallel development view is accepted, as was his inclination, the problem remains "of defining in this case the degree of 'independence' between the two games" (Ibid.: 382). This is essential because

> The decision as to where convergence replaces parallelism in any case of cultural growth, then, might depend entirely on how far back we wish to push our explanations (Ibid.: 382–383).

Erasmus' work on the development of patolli is valuable because it draws attention to the pitfalls of taking a cavalier approach to the application of probability theory, an approach Erasmus detected in the writings of Tylor and Kroeber. As we argued in Chapter 4, archaeologists and anthropologists should be encouraged to use techniques developed in other disciplines, including statistics, where appropriate. For such borrowing to yield productive results, the limitations of those techniques and the conditions necessary to their correct use must be thoroughly understood. Probability theory is a powerful analytical tool, but its misapplications are correspondingly misleading and thus must be assiduously guarded against. In all fairness, it should be said that archaeologists and anthropologists are hardly alone in their frequent mishandling of probability, for no widely used mathematical technique is less well understood (*see* Tribe 1970–1971 for a discussion of misapplications in law).

Marshalling evidence for a controversial hypothesis

The effect of a new kind of evidence on research strategies is examined. Bone technology in the Yukon, itself controversial, is also key evidence in claims of great antiquity for humans in the New World. The research strategies employed offer unusually good examples of problems in confirmation, illustrating again the close relationship between hypotheses and data. Some of the research has taken the form of establishing the plausibilities of human alteration of some of the bone in question through replication and experimentation. Differentiation of human from other causes of bone alteration provided another rich line of research. Multi-disciplinary studies have been aimed at the geological and environmental context of the finds. Searches have been made for new relevant data. Conflicting knowledge claims remain. The effects of sociological factors are apparent.

The Old Crow locality in the Yukon has been and continues to be the topic of heated debates. Of interest here are the numerous *new* avenues of research that have opened up as a result of inferences about bone technology and the sequence of controversial hypotheses or ideas that are being generated by the Old Crow debates. Such opening up of new research areas as a result of attempts to marshall evidence for controversial findings is one of the ways in which science grows and/or changes.

The most important line of research was probably paleo-environmental, geological, and palaeontological (*see* Hughes et al. 1981). This is not what we wish to highlight, however. Instead, we look at the work involving bone technology—first, because archaeological evidence in the form of a bone tool triggered the rest of the research, but more importantly, because bone technology, as developed in the Old Crow case, constitutes an essentially new category of information in New World archaeology. Although the sort of bone technology involved in the Old Crow debates represents a substantive addition to archaeological data categories, the methodology for coping with such introductions is well-known to science.

Research undertaken to document the presence of pre-Clovis man was begun at Old Crow Flats in the Yukon in 1966 when the first major find in the area was made by C.R. Harrington in the course of his palaeontological research. He discovered a flesher made from a caribou tibia and several other bones he believed to have been modified by man. Harrington brought these artifacts to the attention of W.N. Irving, and both men were led by the geological context of the finds and degree of fossilization of the bones to believe the bones were considerably older than 11,000 years, the accepted age of man in the New World at that time. Irving and Harrington reported the initial findings at the 1967 Society for American Archaeology meetings; Irving also gave a paper on the subject at the Alaska Science Conference in 1967; the 1973 article in *Science* by Irving and Harrington represents the first widespread announcement to the world at large. Harrington's finds were subjected to Carbon 14 analysis, which indicated an age of 25,000 to 30,000 years for them (cf. Irving and Harrington 1973; Morlan 1978).

When they were conducting the research, Harrington and Irving were not aware of any fleshers having been reported as "ancient" although similar fleshers were known ethnographically (Irving and Harrington 1973; Irving 1978). Their find suggested that an apparently recent artifact form was quite old. They made this suggestion in the absence of the sort of evidence archaeologists are accustomed to dealing with in such contexts—namely, stone tools. At the time, they felt it likely that stone tools would eventually be found. The two investigators were nevertheless of the view that the early dates should be seriously entertained. They took the position that the Carbon 14 data, along with the other available evidence—geological, palaeontological, and degree of bone mineralization—and their belief that the flesher had been made on fresh bone prior to mineralization warranted the conclusion that "[M]an lived in the eastern part of the Bering refugium before the peak of the late Wisconsin glaciation" (Irving and Harrington 1973:339).

It had previously been established that Old Crow Flats was part of a large area (including nearly half of the Yukon and much of the adjacent areas of interior Alaska) that remained ice-free during the Pleistocene and which formed the eastern end of the Beringia platform during glacial periods. It was thus not surprising that if evidence of pre-Clovis man was to be found in the New World, it might be found in the region of which the Old Crow Flats had formed a part. Harrington and Irving's work did much to establish the Yukon as an area worthy of major research, and that research had to be multidisciplinary, cutting across, as it did, the boundaries of botany, ecology, palaeontology, geology, and archaeology. In 1975, in the hopes of tapping the area's potential, two substantial research projects were begun: the Northern Yukon Research Project, directed by Irving, and the Yukon Refugium Project, directed by Richard Morlan (the latter combined the efforts of the National Museum of Man and those of the National Geological Survey). The former concentrated on Old Crow Flats; while the latter worked there as well, its charter was not confined to that area but encompassed the entire Yukon.

As of 1977, the only evidence of human activity uncovered in the Old Crow area was secondarily deposited, although both research teams were confident that they were close to finding palaeontological or archaeological evidence in the contexts where it would have been originally deposited. Morlan and Matthews optimistically noted: "We believe we have indeed found that level in the stratigraphic haystack in which the archaeological needle is hidden" (1978:5). Similarly, Irving (1978) reported finding *in-situ* bones "fractured while still fresh" at two localities during the 1977 season. Blue Fish Cave was hopefully regarded (Cinq-Mars 1979).

To digress for a moment, we should note that the idea that humans were present in the Yukon 25,000 to 30,000 years ago now appears to be cautiously accepted by a substantial number of Early Man specialists, although recent challenges to the Carbon 14 dating of the famous cariboo flesher may again shift opinions. The substantive debates about man's antiquity in Old Crow Flats have shifted to much earlier dates, in the range of 60,000, 80,000, 100,000, or 120,000 years ago, and these debates have polarized the participants involved in the Old Crow research. We also note that stone tools have not been forthcoming in the manner expected, and this too continues to raise critical questions. Recent claims by Irving 1982) and his associates (Irving, Jopling, and Kritsch-Armstrong 1984; Jopling, Irving, and Beebe 1981) about stratigraphic evidence of human-produced bone artifacts under Sangoman deposits are apparently rejected or ignored by other workers in the area (cf. Morlan 1984). As is customary in these case histories, we refrain from commenting on the substantive issues, focusing instead on methodological matters. It is necessary, however, to separate the several issues involved (antiquity, geological interpretations, bone technology, the absence of stone tools) before turning to our particular interest—the research directed to bone technology.

The question that these Early Man archaeologists are asking themselves is, "What constitutes valid evidence that can be used to demonstrate the antiquity of man?" (Bonnichsen 1978b:102). As Bonnichsen notes: "Most scholars have been looking for deeply buried sites, clearly of Pleistocene age, which contain stone tools and extinct faunal remains. [But r]ecently a considerable body of new information has been assembled from the Beringia Refugium located at Asia's gateway to North America which does not meet either of the above criteria" (Ibid.). The crucial evidence is bone, and this has presented serious problems because methods of distinguishing human from natural bone modification is controversial, bone technology of the sort postulated is poorly understood, and it has to be shown that the bones in question were not altered by subsequent "formation processes." As a result of these problems, archaeologists working in the area set about strengthening the data base upon which they rely. Their efforts in this regard have taken them in three main directions, and—in large part because key investigators have maintained close enough contact for their findings to be systematically integrated into a common working framework—they have been successful to a significant extent in illuminating a previously underexplored research area in the New World context. One of the lines of work undertaken aimed at showing that the altered bone material from Old Crow was not unique, as had been thought. Comparative studies of other northern palaeontological collections have shown that "altered" bones are found elsewhere. These results have caused archaeologists to begin studying a data base that had previously been regarded as irrelevant to archaeological purposes.

A second major research direction was experimental (Bonnichsen 1973; Biddick and Tomenchuk 1975; Bonnichsen and Will 1980; Bonnichsen 1978b and 1983). It addressed the fracture mechanics of bone and sought to identify the fracturing properties of different kinds of bones. It was important that the investigators at Old Crow be able to show that the bone artifacts they had discovered were not made from fossilized bone, for, if they were, no clear cut conclusions could be drawn about the age of the culture which produced them from the age of the bones themselves. Experiments indicated that green (fresh) and fossil bones break quite differently and, thus, provided a basis for distinguishing artifacts made from the two.

If the first breakthrough in Old Crow Studies occurred in 1966 with Harrington's find, the second was achieved in 1972 when Bonnichsen (1973) recognized a flake production system in the altered bones. Subsequent experimental work and careful observation of the bone artifacts indicated that heavy bone, especially that of proboscidians, can be flaked very much like stone and that bone cores and flakes exhibit morphological characteristics similar to those of stone. These data are significant, because they established that bone can be deliberately shaped into useful forms and increase the likelihood or plausibility that early humans shaped the altered bones found at Old Crow. They do not show that this definitely was the case, however, and considerable variation exists in professional responses as to the significance of Bonnichsen's work for the central question at issue.

Included in the Old Crow bone remains are those with spiral fractures. It is known that such fractures occur as a result of both human and nonhuman actions, and also that those caused by humans cannot always be identified as such. Both research teams working at Old Crow conducted a great deal of research into this problem, but it remains the case that while some spirally fractured bones may be attributable to human production, a substantial number of other processes produce similar fractures.

A third research direction involved assessment of the functional qualities of the altered bones as tools; the results of this have shown that for at least some tasks they are more than adequate. Bone points have been found to penetrate the skin of elephants, for example, as easily as those made of stone. "Tools" produced by both flaking and spiral fracturing have proved useful as butchering implements; in fact, the greater "give" of bone tools makes them superior to lithic tools for levering. Bone tools are not adequate for cutting through elephant hide, however, and even partisans of the idea that humans produced the Old Crow bone "tools" admit that the initial butchering phase of cutting through mammoth hides is not explained with the evidence in hand. Bone tools, because they are relatively soft, lose their edge quickly; however, replacements can be quickly and easily made. Indeed, Morlan suggests that this "throw away" feature may account for the apparent abundance of such tools in the Old Crow area.

Since the initial cluster of new research directions was undertaken, still other avenues of investigation have been initiated as the search for means of better evaluating the evidence has expanded. Some work involves Old Crow materials and personnel still dedicated to the Old Crow project, but many other researchers have become involved in allied research on something approaching a global scale. Faunal collections from various parts of the world (across North America, Siberia, China, Europe, Africa) are being reviewed in an effort, first, to ascertain if evidence of bone technologies comparable to the Old Crow case exist which have been overlooked, and, second, to enlarge the "laboratory" used to differentiate human alteration of bone from bone altered by other means.

Sites in Northwest China have been reported with bone evidence and little or no stone technology. Similar controversies occur for the temporally more remote Australopithicine/Homo habilis-related materials from Africa, where Brain (1969) effectively demolished the Osteodontokeratic culture some years ago (but *see* Brain 1984). Controversies continue to surround East African early hominid finds, as is evident from the Binford/Glynn Isaac debate (cf. Isaac 1978, 1983a, 1983b; Binford 1984). Interestingly enough, the Old Crow research appears to have developed independently of the African debates, but now that the research spawned by the Old Crow finds has broadened so dramatically, there is an interchange of ideas between these two geographically separated arenas of archaeological inquiry.

In another research direction, a great many butchering experiments are being conducted; those most analogous to the Old Crow finds are being conducted on elephants—the "Ginsberg" butchering experiment being a case in point (Stanford et al. 1981; Bonnichsen 1984). Such experiments involve the evaluation of different kinds of tools as well as a close examination of the effects of butchering on bone tools or whatever other kinds of tools are being employed.

In Toronto, bones are being frozen at various temperatures and for varying lengths of time to observe the effects of low temperatures on their physical properties, since the kinds of properties of "fresh" bone which can be maintained through freezing is a topic of some significance. Within the same research group, an ambitious attempt is underway to devise a computer-based means of handling multiple variables relevant to distinguishing human from other bone-modification agencies. Still other discussions center on the necessity of better understanding natural conditions that might produce the appearance of human-produced fractures. For example, could elephants stepping on elephant bones supported on stones or other bones produce an anvil effect, with fracture patterns that duplicate experimentally produced fractures? Larry Agenbroad (1984), for example, has reported on nonhuman bone modification of mammoths at the Hot Spring Mammoth Site, South Dakota. Animal behavior studies are also considered relevant to the debates. Carnivore activities have become a major research focus, with several kinds of investigation in progress. Detailed analysis of carnivore gnawing or tooth marks on bones represents one line of investigation. For Old Crow, for example, Beebe (1983) has identified carnivore marks on Blue Fish Cave bones. Horses recently killed by wolves in the Yukon are being studied by Irving's research group.

Binford (1978b and 1981, for example) has worked intensively in carnivore-altered bone studies and in so doing has called into question many of the inferences made by other investigators. His often pointed criticisms have, in turn, created a second or third generation of rebuttals, as was demonstrated at the Archaeological Symposium of the University of Calgary on Bone Technology in November 1982 (Le Moine and MacEachern, eds., 1983).

Paradoxically, while the many and various results of the massive research projects centered in the Yukon are considered problematic by many archaeologists, the initial find of the flesher and its date have been until recently widely accepted as valid. We noted in Chapter 4 that a measure of the corpus of received views within a discipline can be gauged in part by the content of textbooks. The Old Crow flesher has been cited in texts as evidence for Early Man: "The earliest of the North American locations is Old Crow Flats . . . The flesher is remarkably modern in appearance, and there is no doubt that it is a tool created by human hands. Its age is 27,000 ± 3000 B.P. . . . Although no stonework was present, one has no hesitancy in accepting this find as authentic" (Jennings 1978:22). Jennings (Ibid.) accepts not only the flesher and its date but the presence of humanly modified mammoth bone. Some skeptics argued that although the tool is a tool and the bone may be 25,000 + years old, it does not necessarily follow that the human alteration occurred when the bone was new. The "green" bone properties retained during freezing became a crucial question for such skeptics. As noted above, the initial C_{14} date has now been challenged by a second C_{14} assay, and so the antiquity of this critical tool is in doubt.

Research subsequent to the initial finding of the flesher has had at its methodological core questions of the nature of evidence, what is to count as evidence in which context, multiple confirmation problems, and conflicting knowledge claims. The entire subject of bone alteration has become an expanding area of inquiry which is currently undergoing deep, critical evaluation. The problems of evidence and confirmation have resulted in the exploration of many new avenues of research.

It is, perhaps, premature to expect acceptance or rejection by the archaeological community of the several hypotheses and claims advanced. The topics of bone technology and the alternatives of nonhuman modification do not yet have a corpus of empirical background information equivalent to topics such as lithic analysis, which has a much longer investigative history. No one, including those most accepting of the Old Crow evidence, would claim that all alternatives have been explored. General agreement exists about exploring more alternative hypotheses and firming up those that have been advanced with better empirical foundations. As one example of an alternative hypothesis, carnivore intervention does not seem to explain or account for modification of probosidian bone, since it is unclear which carnivores had the ability to fracture such massive bones. The anvil or trampling hypothesis seems plausible, and Agenbroad (1984) has produced a case study of trampling as an agent of bone modification. Plausibility on any side of an argument cannot count as confirming or refuting particular cases, although it does direct hypothesis choice.

Sociological factors which are clearly apparent in this case include the recurrent theme of influences of core beliefs on the behavior of archaeologists in a threshold or controversial case. More stringent demands were made on evidence and confirmation than would have been required for different kinds of evidence—

stone tools with well-dated Siberian analogues, for example. Of special sociological interest in the Old Crow case is the rapid development of a critical mass of researchers. Not only did the initial finds signal a research area of great potential importance, but the finds were made during an expansionist time in archaeology, when funding for research was relatively abundant and multidisciplinary projects were favorably regarded. "Origins" research was attractive to both researchers and granting agencies. Not one, but two major projects were undertaken.

The research groups attracted to Old Crow experienced many forms of interaction patterns, some acrimonious, but they nonetheless exhibited aspects of the pattern of Big Science as defined by Derek Price (1963; *see also* Chapter 4 this volume), in that ideas and information were exchanged well ahead of publication. Intensive interaction patterns persisted throughout the duration of the major projects and their aftermath. The end of the projects, other demands on researchers' time, and persistent problems centering on the nature of evidence and confirmation have acted to dissolve the critical mass of investigators working on these problems in Old Crow Flats.

Archaeologists across the Americas will pay more attention in the future to indications of bone technology in their faunal collections, and the First International Conference on Bone Modification was held in Carson City, Nevada, in 1984. For many Early Man archaeologists, "normal" studies have been extended to include bone technology, and this will continue to occupy them whether or not the claimed early date of the cariboo flesher that started this research sequence continues to be accepted.

A surfeit of hypotheses

Conflicting knowledge claims are considered in several of the cases included in this chapter, including this one. In contrast to the Old Crow case in which the nature of basic evidence was highly controversial, this case is more concerned with inferences made from a data base about which there is relatively little controversy. The research producing the data base on which inferences are proposed has occurred over a long period of time, and this again is in strong contrast to the Old Crow case, in which key research was of relatively short duration. Problems of conflicting knowledge claims illustrate the recurrent methodological theme of selection from among competing hypotheses and the strategy of creating higher order hypotheses to subsume and reconcile more particular hypotheses. The latter strategy is stressed here. Since the many hypotheses advanced in this case are a product of long-term research over a large geographical/cultural area, problems of missing, incomplete, and uneven data are clearly apparent, as are the problems of fitting such a patchwork data base (as well as new data) into the hypotheses. The relationships between discovery and justification procedures and between data and hypotheses are evident, as is the principle of "inference to the best explanation."

Multiple and sometimes competing ideas of hypotheses have frequently been advanced to account for the same archaeological phenomenon. This is especially easy to see in complex archaeological situations that have attracted the attention of many archaeologists over a long period of time. "The decline and fall of Lowland Classic Maya civilization has provided archaeologists with a wondrous mystery" (Rathje 1973: 405), and it is to this wondrous mystery that we turn for the next methodological case study.

The Maya Collapse is interesting from both discovery and justification points of view, as it illustrates how narrow the line can be that separates these two aspects of science. The formulation of ideas about the collapse are difficult to separate in definitive ways from the confirming or supportive evidence. At times, hypothesis formulation has been as criticized as the supporting evidence. Many of the individual hypotheses or ideas have grown out of "indications" in a particular data base and, to a large extent, derived support therefrom.

For genuinely competing hypotheses, the crux of the matter is that they cannot both be true. Many proposed alternatives are not genuinely competitive. Either they are formulated so that they are not mutually exclusive, or it is found that elimination of one or more is less feasible than combining the "alternatives" into a larger framework. In the latter event, the multiple hypotheses might be seen as addressing partially overlapping problems which can be integrated into a broader account. This is what is happening in the Maya Collapse case.

Earlier accounts of the collapse tended to be rooted in particular data sets, and were, not surprisingly, supported by that data set and perhaps by other evidence. The problem was that other hypotheses had support in other evidence. As these disjointed sets of hypotheses and supportive evidence accumulated, it became apparent that "most of the explanations are not mutually exclusive; recombinations and reformulations are plausible" (Adams 1973:

33). By the 1970s, there was a growing conviction among Maya researchers that the circumstances surrounding the downfall of the Classic Maya varied from region to region, and perhaps even from site to site (Ibid.). Thus, it might be that no one hypothesis would be completely applicable across the board. If this should be the case, it would not be surprising to find hypotheses which are supported by evidence from some parts of the area occupied by the Classic Maya, but which do not gain strength at all from—and may even be disconfirmed by—data from others.

Archaeologists have long argued that the elite classes of the Classic Maya were toppled and elite-related activities ceased at all known centers in the Peten around, very roughly, A.D. 900. It is considerably less clear what happened to the peasants, but major population declines seem to have occurred. Over the years, numerous attempts have been made by a variety of investigators to explain the Maya Collapse, and the resulting hypotheses have been categorized by Richard E.W. Adams (1973:23) as follows:

1. Ecological
   a. Soil exhaustion
   b. Water loss and erosion
   c. Savannah grass competition
2. Catastrophic
   a. Earthquakes
   b. Hurricanes
3. Evolutionary
4. Disease related
5. Demographic
6. Social structure related
7. Invasion related

These hypotheses, as the names of the categories under which they have been subsumed suggest, tended to focus on a particular cause, or triggering mechanism. For example, William Sanders (1962, 1963, and 1973) advances a theory that Adams classifies as being of the "Ecological-Soil Exhaustion" type. Sanders' work reflects a clear appreciation that many factors contributed to the Maya Collapse. Indeed, he describes what he sees as the closely interrelated processes leading up to the collapse as follows:

> Subsistence patterns shift from a bush-fallowing, maize-based system to one of increasing diversification, to heavier yielding

crops, to a dependence on imported foodstuffs, and finally to increasingly tighter control of land use and trade.

Paralleling this process are several closely related phenomena:
rise of population density, decline of per capita income, increasing
local specialization in crops, heavier reliance of the core area on the
periphery for many basic materials and more highly organized
trade.

Territorial levels shift from tribal to two-level chiefdoms, to three-
level chiefdoms, then to three-level and ultimately four-level states.
The process is closely correlated with militarism, possibly involving
a warrior class of Central Mexican type at the end. Correlated with
these events is a shift from egalitarian to ranked to stratified society.
At the same time that this last process is occurring, the older patterns of integration, the *cargo* system and theoretical descent ties,
are losing their strength as integrative mechanisms.

All of these events are correlated with increasing demands on the
labor of the supporting peasantry in meeting their own caloric needs
as well as those of the non-food-producing class, increasing problems of malnutrition, rise in death rates, and a reduction of per
capita income (Sanders 1973:359–361).

It is nevertheless Sanders' view that it was a particular crisis (one
involving overuse of agricultural land with a resultant food production crisis) that precipitated general collapse. Accordingly,
Sanders claims that other considerations are relevant to explanation of the Maya Collapse if they somehow relate to constriction
of the food supply, but not otherwise. In this vein, he says of the
"trade network, and consequently the military hypothesis" that
they "can have explanatory value only if basic foods were involved. . . . [N]one of the craft products derived from exotic
sources were absolutely essential for the survival of the Maya
peasant. The only product that would have presented survival
problems was food" (Ibid.: 363).

Because the event to be explained was a total cultural collapse,
many of the "prime-mover" accounts have included consideration
of secondary factors. We have seen an illustration of this in Sanders' work. Another example of combining prime-mover concepts
with an appreciation of secondary factors is seen in the earlier work

of Sylvanius G. Morley (1946), who pinpointed the spread of savannah grass as the main cause of the collapse; its spread, which was the result of repeated clearing and burning of the forest for agriculture, produced a sod the Mayas did not have the technology to cope with. Morley also postulated a consequent economic collapse which precipitated social unrest, governmental disorganization, and religious disbelief (see also Adams 1973:25). The seeds, then, of a more comprehensive approach to the problem were sown some time ago. It was only after several major projects had been conducted in the Maya Lowlands after World War II that the collapse was made the subject of a symposium held at the School of American Research. General agreement was reached that the breakdown of the Classic Maya was "due to a series of interrelated factors none in itself sufficient to account for all of the observed phenomena" (Culbert 1977:509).

The view that has emerged is that the framework appropriate to studying the Maya Collapse must encompass a number of interrelated variables. The "primary cause" and piecemeal approaches that were previously dominant are now regarded by most people working in the area as too simplistic. As Adams has said: "the explanations of the collapse have generally moved from the 'prime mover' category to the multifactor approach" (1973:33). This development in Mayan studies seems to have been a productive one, for research in the field has recently progressed rapidly. Individual investigators, or teams of investigators, apply their special expertise to the problem, as they always have, but now this is done with a far more serious eye to how their results can be fit into an overall solution to the collapse. Thus, while Frank Saul (1973) focussed on "the importance of the disease burden," he refers to that consideration as only "a potential factor" in any explanation of the phenomenon. People working in the area still have their favorite hypotheses, of course, but enthusiasm for them now seems tempered by an appreciation of the extreme difficulty of the task. As William Rathje has said: "The search for simple causes and effects in the last days of the Classic ceremonial centers has led to an understanding of the complexity of the collapse" (1973:405).

Rathje has developed his own collapse hypothesis, one that is heavily systems oriented and takes the form of what has been called a "market" or trade model. This account emphasized "potential for variety," for in Rathje's view:

One of the crucial nodes of adaptation is the "variety pool." The amount of potential variety in a system's pool (that is, the number of potentially viable alternatives) constrains the ability of a system to construct new "maps" in response to changes in the environment to which it has to adapt. For the purposes of this paper, the amount of variety in a system is equated with "adaptability." If there are few constraints upon variety, a system has the potential to successfully modify its structure in response to environmental variation. If the constraints upon internal variety are severe, a system's potential to modify itself in response to change is low (1973: 406, reference deleted).

According to Rathje, the essential weakness of the Classic Maya culture was a lack of resource variety, which was manifested in a dependence on trade. The collapse was brought about when trade was severely curtailed: "Because of constraints on variety, the constituents of cultural systems without efficient trading networks could not be sustained, and large-scale depopulation resulted" (Ibid.:453). The constraints were categorized as being related to basic resources and ecological and geographical factors. Although Rathje urged that special attention be paid to a particular feature of the Maya culture—trade, this was not to be done to the exclusion of other considerations. In keeping with the appreciation he showed for the multi-faceted nature of the collapse, he advanced his thesis to supplement other hypotheses that have been proposed (Ibid.:405).

A number of archaeologists have proposed a systems approach as the framework for handling multiple variables in an explanation or interpretation of complex phenomena such as the Maya Collapse, the initial domestication of plants and animals, and the origins of civilization. We have argued earlier that a systems approach as a platform-for-viewing or world view can be useful to archaeology; however, as long as a very low-level, generic systems concept is used, it is empirically empty (M. Salmon 1978) and, to paraphrase Popper (1962), it explains nothing because it "explains" everything.

Rathje's work, for example, was expressly premised on a systemic view of culture (1973:406), but this is not what is significant about his contribution. What is important are the *specific* cultural processes he postulated and their consequences in the Classic Maya period. His view that the "Classic Maya cultural organizations are systems of adaptation to ecological and sociopolitical environments" (Ibid.) is reasonable enough, but this is more a world view than an inherently testable hypothesis. What is potentially testable are Rathje's ideas that Classic Maya cultures were heavily dependent on trade, and that when that trade was disrupted, as he claimed it was, this became a major factor in the collapse.

Precisely the same point can be made about the theory Willey and Shimkin advanced in their chapter of *The Classic Maya Collapse*. There they argued for a model which

> attributes to a special development of elite culture a primary role in the generation of the climax manifestations of Maya Lowland civilization. The success of the system produced growths of population and of competing centers which led to increasing rigidity in the system as it was subjected to internal stresses and external pressures. The system failed through inadequate recognition of these stresses and pressures and through inappropriate responses to them. The economic and demographic bases of the society were weakened; the consequences were the collapse of the system, the decimation of the population, and a retrogression to a simpler level of sociopolitical integration (1973:489–490).

The generalities which constitute their model are virtually useless from the point of view of expanding our understanding of the collapse. What *could* cause the failure of a system, according to systems theory, other than "inadequate recognition of stresses and pressures" and "inappropriate responses to them"? Much of the model is true by definition and so empirically empty, although the systems-based world view is a valuable step forward from single-cause approaches. When Willey and Shimkin gave the model empirical—i.e., testable—content, which they tried to do by plugging hypotheses about the Classic Maya into its general formulae, the model took on real significance.

Kent Flannery (1972) has developed a model of cultural evolution that relies upon the systems approach but which is more specific in its content than many of the other applications of systems theory to archaeology. His model has proven useful to the generation of hypotheses for testing. Nevertheless, many of the same questions we have discussed arise in connection with his work. What, for example, is to count as evidence of hypercoherence—a concept central to his theory of cultural collapse? What threshold must be crossed for hypercoherence to lead to collapse, and how do we recognize that threshold? It cannot be answered that the threshold is passed if collapse occurs, for that would be patently circular.

More complex formulations integrating several hypotheses at higher levels appear to be much more promising and conceptually satisfying, making the challenge of testing them entirely worthwhile. At the same time, multivariate hypotheses, whether or not heuristically framed in a systems approach, are inherently difficult to test. It is harder to see how numerous factors fit together than it is to determine how one variable and the phenomenon to be explained interrelate. This is rendered yet more difficult when different hypotheses have differential support on different bodies of archaeological data. The integrated concepts are not only more complex in the number of relevant variables to be considered; they are also more general. The distance between the concepts and the data base is increased in significant ways. This has two consequences that recur in archaeology with great regularity. One is that this increases the difficulty of testing beyond the concern expressed about sheer numbers of variables. The other is that the interesting and most satisfying kinds of explanation must go beyond the data they are meant to explain, a problem seemingly more acute in archaeology than in many other sciences. Direct testing of many interesting archaeological hypotheses may prove to be impossible. This may not, however, be intrinsically different from the evidential problem experienced in some better established disciplines, where increasingly sophisticated knowledge has often been gained through creation of more and more abstract hypotheses. In such cases, similar problems of confirmation have arisen. Einstein's Theory of Relativity, for example, presents serious problems for confirmation, for almost any kind of test of the theory would have to be indirect.

Indirect testing has been useful to all sciences and provides potentially useful directions of inquiry for archaeologists facing difficult problems of confirmation. Willey and Shimkin argue that the development of models is essential to solution of the collapse mystery; at the same time they appreciate the inadequacy of relying solely upon direct archaeological evaluation of those constructs:

The data requirements of these models are large and cannot be satisfied by direct archaeological materials alone. Consequently, a variety of comparative and experimental information must be used.

It is suggested that such data be used at two levels. The first is one of critical suggestions, wherein widespread sources can be used without too much reference to context. The second is that of controlled comparisons, wherein systems that appear analogous are matched after careful examination of their differences.

In all cases, the great importance of protohistorical and early historical information on the Maya area must be stressed (Willey and Shimkin 1973:497).

Other possibilities for seeking indirect confirmation include simulation (cf. Hosler, Sabloff, and Runge 1977; Sabloff 1980) and other forms of analogy (Wylie 1980 and 1985).

Early in the development of archaeological interest in the work of philosophers of science, Binford (1968) suggested that explanation of the Maya Collapse should be subsumed under a law in the D-N model of explanation. The difficulty with this is that when inferences go a substantial way beyond the premises, when the "truth" requirement for the premises is not convincingly met, and when the introduction of new conditions in the justification can alter the conclusion suggested by the premises, the mode of inference being used is inductive rather than deductive. For the case of the Maya Collapse, as for almost all interesting archaeological cases, we shall require a combination of inductive and deductive inference.

We have suggested that inference to the best explanation construed as a particular form of eliminative inference provides one of the most promising characterizations of archaeological inference at the present time. Like many research strategies, this mode of inference is easiest to implement at lower levels of testing where variables can be more tightly controlled—i.e., fairly close to the data base. As individual hypotheses are combined and integrated, the same attention should be given to alternatives and the grounds for elimination. This level of providing alternative accounts is admittedly difficult to achieve. Among other things, as multiple factors are considered and integrated, the range of data to be considered in assessing the new formulations increases dramatically from the results of one project, for example, to the accumulated data base for the Classic Maya in the Peten. This enlarged data base is exceedingly uneven, having been collected over decades for different problems using different methods and techniques. The problems of testing progressively more generalized and multivariate hypotheses are very hard to overcome. These problems are, if anything, compounded by embedding the Maya Collapse into discussions of the collapse of civilizations—a topic that was the focus of the recent seminar at the School of American Research chaired by Norman Yoffee.

It may be, as Achinstein says, that "Given an explanation which does not explicitly invoke a law, if all the assumptions made in such a statement were themselves to be explained by appeal to more basic assumptions, and similarly for this in turn, then at some point it is very likely that laws would be invoked" (1971:97).

Finally, it is worth noting that the group of archaeologists working on the Maya Collapse exhibits some of the characteristics Kuhn and others have suggested are found in the most productive research units in science, and which we have noted as a relatively short-term phenomenon in the Old Crow Case. Key Maya researchers keep in close contact with one another. They share their results and advance the state of Maya studies through informal communication which stays ahead of publishing—something we also noted for Old Crow and bone technology research. Small working conferences facilitate the exchange of information. The Maya Collapse conference held at the School of American research has been followed a decade later by a second conference involving many of the same participants. The body of knowledge available to scholars is growing rapidly in size and sophistication.

Competing hypotheses—regional variations in the Clovis culture

Selected aspects of reasons for entertaining hypotheses (including the role of plausibility) and confirmation are explored in this case, which, once again, involves conflicting knowledge claims. This case shares with the Maya Collapse case the problem of testing hypotheses when the relevant data come from a large geographical area and have been collected for different purposes. Confirmation or rejection of either of the two competing hypotheses involves not only evaluation of the same uneven evidence from the two perspectives, but, in addition, what constitutes apparent support in the two cases overlaps in spite of the fact that both hypotheses cannot be true. The role of analogical reasoning is also important here.

As we noted in the preceding section, not all cases of multiple hypotheses involve conflicting ideas: sometimes it is possible to bring together hypotheses that focus on different aspects of a problem in a more comprehensive account. At other times, this cannot be done. When hypotheses are simply incompatible we are forced to choose among them, and in this section and the following, we consider hypotheses, all in some way relating to the Clovis culture, that are, in this sense, competing.

Hypotheses may be incompatible in a number of ways. Here we have in mind incompatibility in the sense that both (or all, if there are more than two) could not be true. If one is true the other must be false. It is always possible that both are false, for we may not yet have hit upon the correct hypothesis in our research.

From a practical point of view, it is not always easy to be sure that hypotheses are genuinely competing. There may be good reasons to insist, for example, that hypotheses are not really comparable unless they address the same problem, construed fairly specifically. Hypotheses about the social organization of Clovis groups are not necessarily in competition with hypotheses addressing the economic basis of Clovis society unless one hypothesis has clear-cut implications for the other.

Once we have ascertained that we are dealing with incompatible hypotheses, the question then becomes—how should we make the choice? The straightforward answer seems to be that we are after those hypotheses that are best confirmed. As we have previously discussed at length, what is meant by "confirmed" in this context is not entirely clear, for a number of interpretations have been given that term. (*see* Chapter 6, *Confirmation*, for a discussion of the most important of them). We have argued that the most promising concept of confirmation is essentially eliminative. The selected, or best, hypothesis in this approach is the one which survives elimination; it is the one that is best confirmed in the circumstances, because we have instances that select it as against the alternatives, or because we have rules for elimination of competing hypotheses, leaving the one in question uneliminated. It will be crucial that the list of hypotheses from which the selection is to be made include all those that are most probable, for to select the best hypotheses from a group in which none is likely to be true will obviously be counterproductive. In this process, it is important to recognize the tentativeness of any selection. New information may alter the choice or bring some additional candidates to the set of hypotheses to be considered.

It is therefore important, assuming that we want to minimize the possibility of working with false hypotheses, to start with hypotheses that are as plausible as possible. At the same time, we must recognize that there is no formal way of ensuring this plausibility. Indeed, the role of plausibility as grounds for accepting hypotheses for testing is complex. Plausibility may be in part a product of core beliefs which can and do change over time. Plausibility can also be considered as an informal expression of probability, but as Salmon (1971) argues, low probability and therefore less plausible events do occur, a point perhaps more easily seen in the long-distance diffusion case discussed earlier in this chapter. Plausibility as a reason for entertaining hypotheses is also seen in the Old Crow case.

Ron Giere is among the philosophers who have been at pains to stress the importance of hypothesis-candidate plausibility to the success of eliminative confirmation. In one of the examples Giere has used to make this point, he discusses some of the popular literature on UFOs (Unidentified Flying Objects) in which, he says, arguments by elimination figure prominently:

> Popular books and articles on UFOs tend to focus on the people making the reports. There are two reasons for this. One is that by showing that the subjects are honest, stable people, one eliminates several possible terrestrial explanations of the reports. The other is that most popular writers on UFOs are unfamiliar with the wide variety of natural atmospheric phenomena, both optical and electrical, that might have been the basis of the subject's experience (1979:152)

The latter point is, of course, a critical one; as Giere notes, as a result of this ignorance, these authors are not in a good position to argue that what was experienced was

> not some natural phenomena . . . [U]nless one can, with some confidence, rule out all possible known natural causes of the observations, the argument by elimination fails to reach the desired conclusion: the ETV [Extraterrestrial visitation] hypothesis. One

would still be left with a number of other possible hypotheses, any of which might be true (Ibid.).

Further complications arise because, even if one could somehow rule out "all *known* possible natural causes of the reported phenomenon":

> one would still be left with one possible claim other than the ETV hypothesis. The phenomenon might be an *unknown*, but natural and terrestrial, occurrence. This claim is impossible to eliminate. . . . At this point the attempt to justify the ETV hypothesis by elimination breaks down completely. One will always be left with two possible hypotheses—unknown natural cause and ETV—and with no way to eliminate the first (Ibid.).

Giere's purpose in using the ETV hypothesis is not to argue that it is false; indeed he expressly notes that it might be true. His aim is to show that the hypothesis cannot be justified by elimination. In one archaeological context, Binford (1981: 180) has charged that Morlan is guilty of the "classic argument from elimination" (*see also* Morlan 1983).

The aim which underlies the search for confirmed hypotheses is, of course, the acquisition of knowledge. Questions about confirmation seem to turn ultimately into questions about knowledge, and knowledge, for a field like archaeology, can perhaps be most adequately characterized in terms of what Gilbert Harman (1965) has called "inference to the best explanation." Harman (1970:83) has claimed that whether we know something depends on whether our reasoning justifies our conclusions. Reasoning that gives us knowledge then, frequently, is inference to the best of competing explanations.

There is, of course, a difference between warranted inference or warranted assertibility and knowledge itself. If we are warranted in making a certain inference or assertion, we may be equally justified in claiming to know what is inferred or asserted; but to say that someone really knows a certain proposition is to say that the proposition is true (for one cannot know as true what is false), and this is to say more than that we are justified in claiming to know it. It is therefore important to be clear that the move from warranted inference to knowledge *is* a move, and in scientific contexts we can at best be certain of the former, for we can never be absolutely certain that a given proposition is true. It is this latter fact that suggests that it might be reasonable to conflate the two (i.e., warranted inference and knowledge), and for practical purposes this does often seem acceptable.

Some archaeologists, such as Dumond (1977) and Hole (1973), have favored an eliminative, or at least a comparative, approach to confirmation. The problems of how knowledge and reasoning relate to one another do not appear to have been addressed in any detail by archaeologists, however. In this section and the next, consideration will be given to how the notion of inference to the best explanation might play a role in the adjudication among competing archaeological hypotheses. While the criteria for choice of the best explanation that we use below are those put forward by Paul Thagard (1978), it is clear that his approach to the problem is but one among a number that could be and have been advanced. For example, Kuhn offers as reasons for accepting one account over another "reasons of exactly the kind standard in philosophy of science: accuracy, scope, simplicity, fruitfulness, and the like" (1970b:261). Our treatment of inference in this section should be taken merely as an example of a possible structure for inference to the best explanation rather than as a definitive account of that reasoning.

The Clovis culture occupies a unique position in the body of knowledge concerned with the peopling of the New World. Artifacts defined as Clovis or as related to Clovis are thought to have extended from what is now the western to the northeastern United States and into presentday Nova Scotia. The projectile points designated as Clovis, or Llano (Sellards 1952), are usually regarded as being confined to the Great Plains and adjacent localities. Fluted points in the Eastern Woodlands are sometimes distinguished typologically (the Ohio fluted, for example), and there are relatively few of the eastern sites with good geological contexts or sound Carbon 14 dates. The view that these eastern fluted points belong to the same time period and/or "culture" is itself a hypothesis not yet fully supported.

The Clovis horizon is the earliest temporal horizon on this continent to receive virtually conclusive archaeological confirmation. Archaeologists are agreed that human population existed on this temporal horizon and much of the archaeological evidence is indicative of big game, especially mammoth, hunting and the use of fluted points. On the other hand, many of the details about that temporal period are very much in question. One of the Clovis-related controversies is concerned with regional variations within Clovis. Once again, we refrain from debating the substantive issues, using these cases for methodological purposes because such controversies nicely highlight the inferential matters we wish to explore.

While consensus exists about the reality of Clovis archaeological remains and dates, there are at least two strikingly divergent views as to how the differences among the groups subsumed by the labels "Clovis" or "Clovis-related" developed. In the scheme advanced by Vance Haynes, Clovis is seen as the result of the widespread dispersal of a culture brought from the Siberian Paleolithic (1978:130). The variations in the Clovis culture which eventually developed can, in part, be attributed to the variety of geographic areas into which the culture spread.

Robson Bonnichsen presents a contrary idea. In his view, Haynes' "migration hypothesis" is inadequate to account for the pattern of regional variation in the Clovis culture. According to Bonnichsen, insufficient attention has been paid to what he sees as pronounced differences in artifacts from the various Clovis groups. In this context he notes that:

> Analysts have traditionally ignored structured variation exhibited by fluted points. The lanceolate shape and basal flutes have been heavily weighted by most investigators. . . . In fact, a great deal of variation exists . . . For example, the reduction strategies and point shape forms and technologies of Clovis points from the Anzick and Simon Sites are quite different than patterns exhibited by specimens from the Murray Springs Site and Colby Site (1978a:133).

Evidence of this sort has led Bonnichsen to argue that regional differences in the Clovis culture can be better accommodated by a scheme which provides for a number of pre-Clovis populations into which the distinctively Clovis innovations diffused rapidly. The four or five traits he regards as distinctive Clovis technological innovations are heat treatment of stone, pressure flaking, extensive use of bifaces, projectile points, and the atlatl. The fluting of projectile points is seen as a preferred atlatl hafting technique (Ibid.:134–135).

In assessing the comparative strengths of these accounts it is of interest to note that a central element of each is based on what proponents of the alternative would say was woefully inadequate evidence. Obviously fundamental to Bonnichsen's account is the presence of populations in the New World at a time earlier than the Clovis period; Haynes claims that the evidence for such a population base is tenuous at best, pointing out that:

> Where stratigraphic successions permit, correlations on the basis of culture, fauna, vegetation changes, and radio-carbon dating are consistent, and strata below Clovis levels are without archaeological evidence in spite of evidence of suitable game animals in the western successions. After Clovis time the increase in site size and number of sites indicates an increase in both population and cultural diversity. Extrapolating back in time we see few people south of the Mackenzie Valley before Clovis (1978:129).

In the case of Haynes' account, the conception he has developed of Clovis culture is thought by many to be based on too facile generalizations from the data. Specifically, critics charge that some of the traits he attributes to that culture are supported by only single occurrences. An example of the sort of claim they would object to on this basis is Haynes' observation that: "Reverence for the dead and a belief in the hereafter is indicated by the magnificent grave goods and red ocher at the Anzick site" (Ibid.:130).

It should be noted here that the question as to precisely what a given piece of evidence is evidence *of* is one which arises time and again in archaeology, as it does in other disciplines. One aspect of the problem raises sampling issues. When, for example, can data be said to be representative of a class of entitites? Another concerns the proper use of analogy. Can one conclude, as has Haynes, that all groups which used Clovis projectile points, or fluted points of comparable form, during the time span in question also employed mortuary practices such as those exemplified at the Anzick site?

The latter query raises another point. Even if magnificent grave goods and use of red ocher were found to characterize Clovis mortuary practices generally, it would still have to be asked how distinctive these features were and whether they needed explaining. It would not be in the least remarkable to discover that any group anywhere in the world at that time revered the dead or held a vivid belief in an afterlife. Nor do the presence of elaborate grave goods or red ocher seem to be particularly unique. At earlier dates, we have the Old World evidence of the elaborate burial at Teshik

Tash and the "flowers" that may have been associated grave offerings at Shanidar, among other examples that could be cited. Hypotheses concerning less common features would seem more useful, and, in fact, the Anzick site, because it represents a cremation—a somewhat unusual method of disposing of the dead at that date—provides some basis for the development of such an approach. It is thus clear that what will count as an explanation in a given context will depend in part on what can be taken for granted and what, by contrast, seems to demand explaining. This is a function of such variables as the current state of knowledge, and cannot proceed according to some abstract logical formula alone.

At present, support for the two accounts is about equally balanced with neither receiving fully adequate support. When neither one of two competing hypothesis can be eliminated, the most rational reaction may be to suspend judgement. Stalemate situations in archaeology are probably more common than most of us realize. The patolli/pachisi case discussed earlier in this chapter, for example, was effectively stalemated for several decades, and even after Erasmus' insightful review, he can only suggest that one alternative is preferred to the other while noting that the less favored alternative is still possible. In cases lacking clear-cut support for one hypothesis and elimination of others, the next step would be to marshall further support for both accounts and again evaluate for elimination of one or the other. There are ways in which this might be done. The competing Clovis hypotheses case would be advanced if Bonnichsen's analysis of the technological differences among fluted points and other parts of the assemblages were to be applied to more (preferably all) of the relevant artifact assemblages, and the variation documented at detailed levels.

Demonstration of regional variability would not be sufficient, although it would be necessary, to establishing Bonnichsen's account as the preferred one. It must be acknowledged that Haynes' migration account could withstand the discovery of evidence showing extensive regional variations in Clovis, especially in the absence of tight temporal controls. Haynes could postulate, for example, a rapid initial dispersal of Clovis immigrants, equipped with both the technology brought with them from Siberia and the newly acquired fluted points which Haynes claims are New World in origin ("Fluted points did not cross the strait in either direction" [1978:130]), followed by some hundreds of years of regionalization. The danger in this sort of maneuver is that the preferred explanation will be extended to cover every conceivable case and thus no data will be allowed to count against it. As we have noted earlier, it must be possible for data to count against a hypothesis, for if this is not the case, the hypothesis is empirically empty and cannot be tested (cf. Popper 1962, especially 255–258). Such a result must be guarded against by one's being as specific as possible about the scope of each of the two competing hypotheses under consideration and about what kind of evidence is to be taken as falsifying either one.

The most convincing evidence for deciding between the two accounts may not come directly from Clovis age sites at all. Rather, the documentation of pre-Clovis populations in the relevant areas would probably provide the most definitive results. If such evidence were accepted, and if Bonnichsen could show regional continuity of technological patterns in those parts of the assemblages other than the traits he argues represent Clovis diffusion, then we could decide between the two competing hypotheses in favor of Bonnichsen's view.

Because it is improbable that the Haynes-Bonnichsen dispute will be resolved on the basis of Clovis evidence, some archaeologists might be inclined toward the view that the issue no longer constitutes an appropriate focus for research. The selection of one account over the other is likely to rest on evidence that will accumulate slowly if at all. For the moment, the competing hypotheses appear to be incapable of being tested at levels that would allow immediate elimination of one of the accounts. To dismiss the controversy at this point would seem premature, for if what we are after is knowledge—in the sense of inference to the best explanation, then the mere identification of a problem area may be useful and there are bases for choice among theories other than how much each explains.

Paul Thagard suggests two additional criteria, simplicity and analogy: "The explanation of fact F by a theory T requires a set of given conditions C and also a set of auxiliary hypotheses A" (1978: 86). Thagard says, then, that "simplicity is a function of the size and nature of the set A needed by a theory T to explain facts F" (Ibid.); "a . . . theory not only must explain a range of facts; it must explain those facts without making a host of assumptions with narrow application" (Ibid.:87). In Thagard's view, this amounts to saying that the theory must explain those facts simply. It is not sufficient that a hypothesis account for a great deal of evidence if it does so only through the addition to the original account of statements that explain a single fact, or a small number of facts.

Thus, a possible objection to Haynes' migration theory might be that, in order to explain particular regional variations in Clovis artifacts—variations that Bonnichsen's work has brought to light, Haynes is forced to amend his account with ad hoc statements of the sort described above. This is no more than a suggestion, for comparisons of explanatory simplicity are difficult to work out and involve close study of the competing explanations' sets of auxiliary hypotheses. As Thagard says: "The matter is not neatly quantitative . . . Thus the comparative simplicity of two theories can be established only by careful examination of the assumptions introduced in the various explanations they provide. As has often been remarked, simplicity is very complex" (Ibid.: 87 and 88). It is important to understand here that Thagard, and others who take a similar approach to theory choice, are not advocating that scientists study only simple entities, or that they undertake only simple problems. The simplicity these philosophers are interested in is the simplicity of the explanations advanced to account for the subject matter, not that of the subject matters themselves. As Thagard says: "ontological economy is not an important criterion of the best explanation" (1978:89).

Nelson Goodman's observations on this issue are also helpful:

> Philosophers sometimes mistake features of discourse for features of the subject of discourse. . . .
> Obviously enough the tongue, the spelling, the typography, the verbosity of a description reflect no parallel features in the world. Coherence is a characteristic of descriptions, not of the world: the significant question is not whether the world is coherent, but whether our account of it is. And what we call the simplicity of the world is merely the simplicity we are able to achieve in describing it (1972:24).

Analogy is the third of the criteria upon which Thagard claims theory choice is based, and one which will be considered in greater detail in the following section. Thagard acknowledges that "at first sight analogy appears to have little to do with explanation" (1978: 89), but he nevertheless maintains that "analogies support . . .

theories by improving the explanations that the theories are used to give" (Ibid., footnote deleted). He elaborates as follows:

> Suppose A and B are similar in respect to P, Q, and R, and suppose we know that A's having S *explains* why it has P, Q, and R. Then we may conclude that B *has S* is a promising explanation of why B has P, Q, and R. We are not [or may not be] actually able to conclude that B has S; the evidence is not [or may not be] sufficient and the disanalogies are [or may be] too threatening. But the analogies between A and B increase the value of the explanation of P, Q, and R in A by S (Ibid.:90).

Using this approach, then, archaeologists might be able to conclude that either Haynes' or Bonnichsen's account was preferable because it bore greater similarity than the other to better established accounts. Indeed, that adjudication among competing theories might be based on differences in analogical support is a potentially fruitful possibility, and, if Thagard and others who focus on the importance of analogy to scientific reasoning are correct, degree of analogical support is commonly used as a criterion in theory choice, at least in the traditional sciences such as physics.

It might be, of course, that even after thorough consideration of factors other than the evidence accounted for by each theory, disputes such as that between Haynes and Bonnichsen would remain unresolved. After all, it will always be possible that two or more alternative accounts are equally acceptable no matter how many different bases have been used for their evaluation. This is not to say that competing accounts can all be true, for if they are incompatible this is by definition impossible. We may, in the end, discover that they are not incompatible, in which case the various accounts may be susceptible to inclusion in an overall theory as we discussed in the section above. What is acceptable—as opposed to true—is always a function of what we know at the relevant time; thus, in any particular case it may be that we are aware of no basis on which to choose among alternative explanations. This may come about, for example, because—as so often happens in disciplines such as archaeology—we have insufficient data in particular cases to draw the distinctions necessary for selecting the best explanation. The general point is that there is more to the identification of hypotheses for acceptance in science than comparison of the number and variety of facts explained by competing theories;

factors such as the simplicity of and the analogical support for those theories must also be examined. Furthermore, where these are competing hypotheses and there are no clear criteria for choosing among them, we may even be justified in particular circumstances in behaving as though each was acceptable in order to pursue the research further.

The accounts of Haynes and Bonnichsen are only a small part of Early Man research. Both Haynes and Bonnichsen have heavy professional commitments to other research and other employment-related duties. The scale of the task of eliminating one of these inferences is enormous. From the perspective of Early Man studies generally, decisions are routinely made to concentrate on specific problems of more manageable scale with greater potential payoff for investment of resources and effort. It is nonetheless important to have broader alternative accounts made explicit and placed before investigators working in the Early Man area. One reason for this is that it prevents debatable beliefs from becoming too deeply entrenched in the core belief system prematurely.

In concluding this section, we offer a few comments on the concepts of migration and diffusion which figure prominently in Haynes' and Bonnichsen's respective accounts. Many archaeologists view these concepts as nonexplanatory—as historical events themselves in need of explaining rather than as processes, the study of which aids understanding and extends knowledge. Archaeologists holding such a view would be inclined to dismiss both Haynes' and Bonnichsen's accounts as beyond the realm of a scientific archaeology.

The Clovis case perhaps brings into better focus some of the points made earlier in this chapter about underlying sources of credibility. In those cases migration and diffusion were regarded as having low probability or even as being totally implausible. This is reversed in this case as a function of received views about prior events. If one views the New World as previously unpopulated, migration on the grand scale becomes a noncontroversial factor, which contrasts markedly with views of migration after the basic population of the New World is presumed to have been in place. This, in turn, illustrates the fact that certain subjects such as diffusion and migration shift positions in the disciplinary sphere according to the substantive context.

Although they use the two contentious concepts in their work, neither Bonnichsen nor Haynes would view resolution of the migration-diffusion controversy as the most important goal, or possible result, of Clovis studies. Instead, both would presumably agree that the advances to knowledge provided by their area of research will be of a much less abstract nature and will emerge from activities such as careful excavation of new sites, more sophisticated technological analysis of artifacts, improved environmental reconstruction, better demographic data, and a deeper understanding of animal behavior and its relationship to hunting strategies. The question may be raised as to the value of attempting, as Haynes and Bonnichsen have, an improved understanding of a great deal of data, of uneven quality, from vast geographical areas. Most archaeological investigations are and will continue to be carried out in connection with rather limited data bases. The broader questions need to be advanced and want answering, however.

The migration and diffusion aspects of Haynes' and Bonnichsen's accounts are held, albeit often implicitly, by many in the discipline, and it is thus useful to have them expressly stated for the purposes of more clearly understanding their implications.

Views about migration and diffusion, whether implicitly or explicitly held, undoubtedly influence the design of research projects and the choice of problems addressed. Hypothesis formulation, research design, and favored approaches are quite interrelated at pragmatic levels, even if it is useful to separate them for methodological clarity. The careful articulation of migrationist and diffusionist views will most certainly spawn new topics for examination. For example, whether or not Haynes' or Bonnichsen's account survives further scrutiny, Bonnichsen has made a significant observation about variability that demands an explanation of some sort. Until other hypotheses are advanced, it is from Haynes' and Bonnichsen's hypotheses that we have to select the best explanation.

Clovis mammoth harvesting

In the following case, a specific site becomes the subject of conflicting claims to knowledge as a result of a reexamination based on a different set of assumptions from those initially used. The role of analogy in hypothesis formulation and testing appears again.

In this section we turn to another set of competing hypotheses concerning the Clovis culture. The specific issue addressed by these hypotheses is how mammoths were harvested by the Clovis populations. The generally accepted account of the matter seems to be that put forward by Vance Haynes, who claims that "big game takes were commonly a single mammoth . . . or a small group of bison" (1978:129–130). The remains of mammoth kills—in particular, those from the Lehner Ranch site in southeastern Arizona—seem to show that the animals taken by Clovis hunters were for the most part young; since mammoths, it is thought, lived in family units encompassing a number of generations, Clovis hunters must have selected the young animals out from the others. In the received view, then: "This age selectivity was . . . no accident. It is explainable on the basis that the younger animals would be easier to isolate from the herd, easier to kill, and tastier" (Haury, Sayles, and Wasley 1959:28).

An alternative account of the Clovis procurement strategies is presented by Jeffrey Saunders (1978), who argues that the Clovis culture may have cropped whole groups of mammoths, not individual animals. Drawing on Laws' (1966) work on the age criteria for African elephants and several sources on elephant behavior (cf. Saunders 1977 where he cites Buss 1961; Winter 1964; Sikes 1971), Saunders suggests that the group of mammoths whose remains were found at the Lehner site may have been quite similar to modern African elephant social units in age range and discontinuous age composition. The oldest elephant at Lehner, for example, is well within the age of contemporary African matriarchs. He concludes that those remains—even if they represent a unit which includes a large number of relatively young animals—may well represent a single kill of a single mammoth group. According to Saunders, one should not be too quick to say that Clovis hunters selected from young animals.

From his conclusion that Clovis hunters may have cropped entire mammoth social units, Saunders goes on to argue—again on the basis of Laws' African work—that such elephant groups can be seen as microcosms of the entire mammoth population. Saunders then looks to these microcosms for clues to the extinction of the species and tentatively concludes that the Clovis method of harvesting through catastrophic sampling was the cause of the disappearance of the mammoths:

> It is suggested here that attributes of health . . . indicate, *as a first approximation*, that *Mammuthus columbi* in the Clovis horizon was in good condition and certainly not manifesting the symptoms of stress which would equate with climatic perturbations (or causes of extinction so based) at the end of the Pleistocene. This could be interpreted to implicate man/animal interactions as a probable factor in the elimination of *Mannuthus columbi* from the Pleistocene magafuna (Saunders 1978:101–102, references deleted).

Saunders further supports his theory about the Clovis hunting strategy by pointing to the reluctance of elephants of a matriarchal social unit to abandon wounded or dead group members, and hypothesizes that if mammoths resembled African elephants in this regard, then mammoth behavior may have encouraged, or in some cases even forced, hunters to take whole groups. Saunders, indeed, suggests that "the evidence from the Lehner Mammoth Kill Site indicates that Clovis Paleoindians knew *and in fact exploited* behaviour patterns of *M. columbi* in subsistence procurement" (Ibid.:99).

The new approaches that are being taken to the study of Clovis hunting strategies, of which those such as Saunders' which focus on elephant behavior constitute an important part, are challenging received views and giving rise to hitherto uncontemplated ideas. For example, hypotheses that the animal remains from one site represent a kill locality become vulnerable to radically different hypotheses when the possibility is introduced that wounded elephants wandered some distance away from where they were attacked, or when the argument is made that the clustering of a number of mammoths in one place is the result of animals, wounded elsewhere, having made their way to water before they died (although the latter hypothesis has not been defended seriously).

Indeed, if the concept that specific sites represent discrete and unrelated events is questioned (ignoring for the moment the question of whether multiple animals in a single site represent discrete kill episodes), it becomes difficult to *disprove* a hypothesis—for example, that the several southeastern Arizona Clovis sites represent more than a single hunt or the activities of a single band over a season, to suggest one extreme. Even to entertain such a possibility forces us to reexamine available evidence in a new light, for a single band of Clovis hunters engaging in catastrophic sampling over a relatively short period of time is a very different matter from multiple bands taking single animals over a period of five hundred years.

To return to our main concern here, the question is, of course, how to adjudicate between Haynes' and Saunders' accounts of Clovis hunting strategies. It should first be noted, in this context, that the remains of Clovis mammoth kills are not likely to resolve the issue, at least not at this point. It is fundamental to Saunders' arguments, for example, that Clovis populations cropped whole groups at specific sites, and if the evidence from such sites could be dated very precisely then the problem might be solved. But demonstrating that the multiple animals at sites such as Lehner are the product of a single kill is not easy, because Carbon 14 dating is not yet sufficiently refined.

Saunders (1977) identified the remains of a single mammoth as being widely dispersed across the Lehner site and has relied upon this fact to bolster his argument for a group kill and against discrete kills by Clovis hunters. But this evidence is also inconclusive, for other, equally plausible, explanations of the scattered remains are available: disturbance by scavengers or post-kill human activity at the site are two possibilities.

Is there, then, a basis upon which a selection between the two accounts can be made? It would seem here that Thagard's work on criteria for theory choice can suggest some options with regard to selection from among alternative views. While his simplicity factor may not cut either way in this case, consideration of the third of his criteria—analogy—may prove more fruitful. Thagard has argued that "analogies support . . . theories by improving the explanations . . . the theories are used to give" (1978:89). Analogies do this, he has said, in two ways. First, they lend credibility to the explanation being put forward; and, second, they increase the understanding that explanations are meant to produce.

Saunders' group cropping theory relies heavily, as we have seen, on two facets of an analogy he draws between present-day African elephants and Clovis-period mammoths. Saunders challenges Haynes' and others' interpretation of the age range of the Lehner group on the basis of data about those elephants and claims, as well, that mammoths may have defended their wounded and dead, as do African elephants. Haynes' account of Clovis mammoth harvesting involves no such analogical support. So, on Thagard's model of theory choice—since neither the available evidence nor the relative simplicity of the competing explanations is, it would seem, determinative, perhaps Saunders' account is to be preferred because of the analogy upon which it is based. But this move would be premature, for there is still the question, which a number of archaeologists have raised, as to how convincing the evidence for that analogy is.

Various penetrating criticisms of that evidence have, in fact, been voiced. The first concerns the level of our understanding of elephant social organization and behavior. Animal behavioral studies are becoming increasingly important to archaeologists concerned with hunting societies, for hunting strategies must, of course, be adapted to the behavior of the pursued species. Thus, the behavior of cariboo, musk-oxen, bison, and other animals have figured in archaeological interpretation with varying degrees of sophistication. But data on elephant behavior is not yet well documented. Indeed, archaeological familiarity with elephant behavior is at such a primitive stage of development that at academic conferences on the subject, personal anecdotes about human/elephant interaction still provide much of the material for discussion. Each of the papers Saunders has presented on this topic (the first at the Paleoindian Lifeways Symposium held at Texas Tech University in October, 1975; the second at the American Quarternary Association Meeting in Edmonton in September, 1978) provoked extensive discussion in which personal experiences and elephant stories were recounted; other individuals raised additional information from animal behavioral studies. It has been suggested that it is dangerous to draw conclusions about Clovis-period mammoths on the basis of supposed parallels between that species and present-day elephants because so little is known about the latter. That is, one side of the data sets being used analogically is not strong enough to support the inferences made on the other side of the analogy.

A second significant misgiving that has been expressed about the analogy upon which Saunders relies is that that analogy depends upon evidence drawn from studies of African elephants. While there are more data available on that species than on the

Asian elephant, it has been suggested that using what is known about Asian elephants would provide a different behavioral model—although *how* different is not at all clear. Selection from among alternative analogues therefore presents another level of judgement.

The third and perhaps most serious criticism is that Saunders has emphasized animal (elephant) behavior without giving enough weight to human activities in the human-animal interaction of hunting. It is apparently true that African elephants will stand in a defensive group and defend members of the group upon some occasions. African elephants also can be stampeded, in which case they spread out. To introduce a further analogy, this one from North American bison behavior, a bison herd at rest may allow a mixing of different ages and sexes; however, if the group is stampeded or begins running, the herd often segregates into discrete groups of cows, bulls, and calves. This behavioral pattern may help explain age/sex selectivity in bison kills. The point is that any animal group has more than one mode of behavior, and human intervention can sometimes initiate animal behavior of particular kinds. There may be more variability in elephant behavior under different conditions than Saunders has indicated.

The strength of the analogical support for Saunders' account is uncertain, and it is too early to say on the basis of Thagard's criteria for theory choice that Saunders' approach should be favored. It does not follow that the application of Thagard's criteria is, in these circumstances, futile. The analogy upon which Saunders relies is not yet very well established, but additional elephant studies may confirm it further, strengthening Saunders' account of Clovis mammoth harvesting as against that of Haynes, or may cause it to be rejected. While an inference to the best explanation analysis may not at this point answer the Haynes/Saunders dispute, it suggests an approach—through application of Thagard's analogy criterion—which may in the end resolve the stalemate produced by consideration of only the amount of evidence explained by each hypothesis.

Concluding remarks on cases

Having discussed a number of examples of archaeological research and having tried to draw out the various sorts of inference they involve and the problems with reasoning in archaeology they illustrate, it remains to make some general remarks about what these cases show. Perhaps the most obvious point that needs to be made explicit is that the models of reasoning, and even the issues so beloved of the positivists and thus so central to the early efforts to make archaeology philosophically self-conscious, are not necessarily appropriate or germane to practice. Time and again we have seen that the formal models of explanation and confirmation adopted at first by the New Archaeologists frequently do not represent very accurately what we take to be quite sound inference in actual cases. These cases, then, serve to substantiate our assessment in earlier chapters of the extent to which the positivist models (as appropriated by the early New Archaeology) are relevant to the actual practice and theory of archaeology.

The cases discussed in this chapter illustrate a number, but not all, of the philosophical issues discussed in previous chapters. A number of methodological problems are confronted again and again in archaeological practice. Mainly, these problems concern the resolution of real or apparent conflicts between knowledge claims, especially where the data are incomplete or ambiguous or where the interest in a particular problem has not, for whatever reason, reached a certain threshold that would allow for sufficient information and discussion to resolve disputes. More specifically, it is possible to identify at least four issues illustrated by these cases:

1. the role of core beliefs, and of the community base that sanctions them in archaeological reasoning, is addressed in the migration/diffusion examples and in the games case;
2. the import of critical discussion and of reaching a threshold in community involvement for the profitable development of a line of inquiry is exemplified by the Clovis and Early Man debates;
3. the import of recognizing the need for testing, beyond assessments of plausibility and explanatory power, is illustrated in early stages in the games case; and
4. perhaps most centrally, the importance of appreciating that what is sought is knowledge in the sense of "best explanatory hypothesis," given the available background information and the evidence, can be seen in the diffusion and migration case and in the games case.

The cases are thus tied to our discussion in earlier chapters of such matters as the social and political context of archaeology, and, especially, the way in which philosophical accounts of explanation and confirmation can be adapted to apply to archaeology.

It seems worthwhile to note that in a number of these cases the way in which considerations of evidence, plausibility, and explanatory power are handled illustrate quite nicely some of our points in this chapter concerning the legalistic nature of some archaeological reasoning. The idea that absolute and incontrovertible proof is required before we can rationally accept a hypothesis seems to be common among scientists. In archaeology, this view is perhaps partly a version of this general notion and partly a function of the reluctance to disturb accepted theories and core beliefs. But it should be clear both from these cases and from our discussion in previous chapters that archaeological inference, like inference in any field, is a complex matter that cannot be represented by some simple model without serious loss of accuracy and, thus, applicability. At any given time we must draw conclusions on the basis of a number of factors which include not only the immediately available evidence but also our assessment of the plausibility of the hypothesis in question, and this will often depend upon core beliefs.

Clearly, it would be a mistake to demand a degree of proof that is in practice unattainable before accepting any new hypothesis, just as it would be a mistake to accept a hypothesis on insufficient evidence just because it is new and interesting. There is no general nontrivial answer to the question of how much evidence is enough; this will depend upon the background information, the evidence for the hypothesis under consideration and for competing hypotheses, the results of testing, and various other factors. That greater quantities and higher standards of proof might be demanded of hypotheses that disturb well-entrenched views is only natural; but, of course, we must take care that we do not set our standards of proof so high that no revision is ever possible. In all of this the legal analogy is interesting and helpful, and inference in law as much as inference in natural science can provide a model for helping us to understand, to assess, and even to develop inference in archaeology. Conclusions are acceptable, though not immune from revision, if they provide the best explanation of the data, taking account of all available relevant information; but it must be remembered that such explanations are only contingently best, and what will count as best may change with new information, new perspectives, or changed emphases on existing data.

Finally, we want to call attention to a methodological area not sufficiently highlighted in the discussion of the preceding cases. We call this the second level of testing or evaluation. Whereas much of methodological awareness in archaeology has focused on primary research, the second level of testing is concerned with ongoing critical evaluation of a topic, open debates about a particular hypothesis or research area, and attempts to duplicate results. We have earlier noted that, in Kuhn's view of science, certain kinds of criticisms have no place in "normal" science. Not all challenges are fundamental, and Kuhn's view of science allows for a number of forms of critical evaluation. The checking and rechecking of one another's work would seem to be an integral part of the groups forming the heart of normal science.

There are good sociological reasons why the second-level evaluation is more productive than the testing of individual hypotheses in primary research, and these are somewhat parallel to the ones that make small working conferences potentially successful. The topic debated is focused enough to involve shared concerns; the range of expertize and problem orientation is sufficiently diverse to provide useful new perspectives. When several researchers focus on a topic, the possibility for self-correction and for exposing the strengths and weaknesses of a position are enhanced. This sort of long-term scrutiny of a topic is less common than one might wish, largely because of the high value placed on novelty and innovation. Substantial numbers of new ideas are not applied to new contexts, nor are they retested in the original one. Some noteworthy cases in which second-level evaluation has proven to be extremely enlightening include the attempts at social reconstruction, the Mousterian debates, the question of hominid hunting versus hominid scavenging, and the problem of bone technology.

Recent trends have seen the growing importance of historical and sociological studies of science in the broadened practice of philosophy of science. Just as philosophy profits from case studies, so too do scientific disciplines gain from a consideration of their substantive and contextual factors deployed against a philosophical background. In this chapter we have looked at a few controversial cases in order to clarify the ways in which substantive archaeological issues employ or might employ philosophical concepts. In this process, certain traditional philosophical concerns emerge as not being very important to archaeology while others are encountered frequently.

8 / The Status of Theorizing in Archaeology

Breaking down disciplinary boundaries

One fundamental issue underlies all the employing of philosophical concepts in archaeology—the relationship between theories and concrete work within the discipline. One could say that what is at stake is the relationship between theory and that which is being theorized about. When the issue is viewed this simply, it would seem that even those who advocate turning away from philosophy and getting on with the business of doing archaeology would have to accept the importance of making the links between theory and practice. It is precisely in making these links that philosophy can be of value to archaeologists.

The preceding chapters have been devoted to looking at a number of aspects of the relationships between archaeology and philosophy. We have examined some of the main philosophical concepts imported into archaeology for the purpose of placing them in their philosophical context, sorting out some misconceptions, and pointing out something of their relationships to archaeological problems. We have discussed such concepts as explanation and confirmation, induction and deduction. It is important to remember that many philosophical debates of the twentieth century have concerned problems that are echoed in archaeology today, and archaeologists who are interested in a philosophically-based analysis of their discipline can find much of interest in the debates. For example, it is instructive for archaeologists to consider why the concepts of verification and operationism were found to be difficult to characterize in unexceptionable ways or why Hempel's formal models of explanation and confirmation did not work.

Besides reviewing logical aspects of science, we have included discussions of the context in which archaeology is practiced as well as actual archaeological cases. Assigning such nonlogical matters a role in understanding how science works is compatible with contemporary trends in philosophy of science. In any case, since the effects of sociological, political, economic, and other nonlogical factors on the practicing of science are undeniably important in archaeology, attempts should be made to understand them. Still, though the logical and nonlogical aspects of science are intertwined, it is useful to try to keep them as separate as possible for analytical purposes.

One of our prime considerations has been to provide a more holistic picture of philosophy than has usually been available in archaeological writings using philosophical concepts. This broader (although necessarily still incomplete) survey of the historical background of key concepts, the debates surrounding them, and alternatives that have been proposed to them is of great importance at the present time because much of the backlash against philosophizing in archaeology is due to overly narrow and often inappropriate borrowing from philosophy in the early stages of self-conscious philosophical awareness. Not only did basic misconceptions become entrenched in the archaeological literature, but archaeologists became discouraged by the apparent failure of certain imported concepts to provide hoped-for results. Such discouragement retarded for a time the exploration of parts of philosophy that offer much more fruitful leads.

The motivations behind early appeals to philosophy cannot be faulted. The conscious use of theory was to be strengthened, explanations were to be sought, and scientific methodology was to be applied. One might say that the link between theory and the subject matter of theorizing was central, as it still is and presumably will continue to be. The New Archaeology was decidedly successful in reorienting the entire discipline toward a greater awareness of theoretical matters, as well as in redefining the topics to be investigated and methodologies to be employed.

Although the motivations were impeccable and the resulting change in the discipline remarkable, a central part of the transformation process was flawed. It is important to understand the nature of these flaws in order to build a case for the continued use of philosophical concepts in archaeology. The problems encountered by the New Archaeology have been discussed in various places throughout the previous chapters. We want to draw together the most noteworthy of these problems here.

One set of problems centers on the borrowing of particular philosophical concepts and the expectations held by archaeologists concerning what the borrowed concepts could accomplish in the archaeological context. With regard to borrowed concepts, many archaeologists have adopted a single philosopher's view—often, Hempel's view on explanation—without considering even the prominent alternative positions. It is curious that archaeologists should have settled so readily on Hempel when Popper has been followed most closely by other scientists interested in theoretical issues, and particularly since the views of the two, at least on the nature of explanation, are substantially the same. In adopting a favorite philosopher, archaeologists have often failed to recognize that important philosophical theses are always controversial. We have earlier acknowledged that most archaeologists, like most other nonphilosophers, are often not equipped to evaluate competing philosophical theories. We nevertheless remain committed to the view that archaeologists should be aware that evaluation of that kind is needed and that it is available in the philosophical literature.

Expectations about what the borrowed concepts could do in the archaeological context have sometimes been unrealistic. Some archaeologists seemed to want straightforward answers to what are extremely complex questions. A desire for neat, simple solutions is entirely understandable, and indeed widely shared. Unfortunately, philosophy—in large part because its tenets are controversial—does not admit of such answers. Philosophical theories of any significance tend to be complex, and to overlook this aspect of the discipline is to miss its very essence.

It can be argued that some twentieth-century philosophers of science have encouraged, however unwittingly, the making of this latter type of error. In advancing accounts, for example, of explanation that are neatly formalizable (that is, completely captured by formal logic), as Hempel did at one stage, it could be said that philosophers conveyed the impression that philosophical problems are amenable to such treatment. The problem is that they are not, or at least that to date all such "solutions" have proven inadequate by failing to capture what it appears we really want to say about the issues in question. We have seen, for example, in our discussion of explanation, that mere conformity to the D-N model does not guarantee explanatory worth. Cases are plentiful that satisfy all Hempel's conditions of adequacy but cannot be regarded as genuine explanations. We have used Hempel as an example, but there are many philosophers about whose work the same could be said. As we discussed in Chapter 1, Hempel was part of a major philosophical movement—logical positivism—which dominated philosophy of science in the 1930s and for some twenty years afterward and shared his commitment to neatly formalizable responses to philosophical questions. The logical positivists sought to articulate an idealized account of science as they thought it would have to be if it was to accomplish the things generally expected of it. Their attempts in this direction were characterized by both a strong emphasis on precision and a heavy reliance upon formal logic. It is really a version of the logical positivists' approach to philosophy of science that archaeologists adopted in the early 1960s.

As was made clear earlier in this study, Hempel is truly a critical figure in philosophy of science, and anyone exploring that discipline must sooner or later encounter his work. The points we want to make about the overcommitment of some archaeologists to his D-N model are:

1. Hempel discussed models of explanation other than the D-N model;
2. he discussed things other than explanation (i.e., confirmation);
3. Hempel's own ideas have changed;
4. Hempel's work has been the subject of criticism and debate, much directly relevant to archaeological problems; and
5. there are other important philosophical positions to be considered.

By the time archaeologists discovered logical positivism and the work of Hempel, philosophy of science had already evolved beyond them. Philosophers had become far less inclined to force substance into rigid formulas, and many had begun to doubt seriously that precise, logical formulations would ever be found for the central concepts of science. Still, there has been a continuity in the development of philosophy of science in this century. Although some of what is being done today in philosophy of science looks quite unlike what was done in the 1930s, much of the best work from the past has been integrated into present views; and of course the tools of logical analysis continue to be used. Ian Hacking's work provides one illustration of the interest of many contemporary philosophers in both philosophical analysis and the history of ideas. In his *Logic of Statistical Inference* (1965), he analyzes the concept of statistical inference. In his later work, beginning with *The Emergence of Probability* (1975), he takes a philosophical look at the history of probability, induction, and statistical inference. Philosophy of science has thus become more varied itself, and in its applied aspects may have more to offer to disciplines such as archaeology.

Another set of problems surrounding the application of philosophy to archaeology centers on the application to archaeology of the concepts treated by philosophers. It seems as though some archaeologists believed that framing their discussions in a Hempelian format bestowed scientific status. In selecting explanation as a primary goal, related concerns such as confirmation and theory building were virtually ignored. Philosophically based methodology at times seemed detached from substantive issues. Polemics within archaeology erected false dichotomies, and polemics from philosophers aroused the ire of archaeologists.

In spite of the highly visible discontent about archaeological philosophizing (cf. Flannery 1982), we would like to argue that much common ground exists between the two disciplines—common ground that can be profitably explored by both. Archaeology offers philosophy the challenge of extending its understanding of science into a new area. Using archaeology as a philosophical case study would have been unattractive to philosophers a decade or so ago. As philosophy shifted to a more flexible stance that allowed the incorporation of substantive and contextual as well as logical aspects of science, however, a broadening of philosophical perspectives to include archaeology became feasible. It is encouraging that a number of philosophers are moving in this direction.

Likewise, we argue that archaeologists can gain from a critical and judicious use of philosophical concepts, especially as guidelines to straight thinking, in conceptual analysis, and in hammering out the theoretical underpinnings of the discipline. While a great deal of archaeological work can be done without reflection on philosophical matters, a complete rejection of philosophical awareness would be counterproductive, especially at the present stage of rapid development when theory and practice must be particularly closely related. Ignoring philosophical concerns carries the risk of having to confine one's attention to relatively routine areas of research, for the more interesting ones often involve philosophical considerations.

Still, the past fifteen years or so have taught us much about the ways of connecting philosophical and archaeological work that are likely to be more or less fruitful. Rather than looking to philosophy for "correct" models of the nature of science and imposing those on what archaeologists actually do, it makes sense to look to the best work in archaeology itself, analyze it, and conceptualize a method from that. This can then be checked against philosophical models and adjustments can be made in both directions. The fact that the process is necessarily two-way may be the most important lesson learned from the excursions of archaeology into philosophy (and the other way round) these past fifteen years.

Increased sophistication

In looking at the relationship between philosophy and archaeology and where that relationship stands today, it is essential to recall how little awareness most archaeologists had of philosophical analysis when the New Archaeology first emerged. That gap is perhaps best illustrated by what is still considered by some commentators to be the fundamental doctrine of the more avant-garde branch of the discipline. As Dwight Read and Steven LeBlanc (1978:307) put it: "Basically, the 'new archaeologist' position begins by rejecting authority as the criterion for judging the validity of conclusions and interpretations in archaeology."

There is nothing wrong—indeed, there is everything right— with refusing to allow an archaeologist's prestige, or status, to determine the acceptability of his or her hypotheses. What is revealing about this statement is that it needed making—that in the view of New Archaeologists this proposition did not go without saying. Conclusions and interpretations are not, of course, to be evaluated on the basis of the credentials of the person putting them forward, but on the strength of the arguments themselves. Recognition of the fallaciousness of attacking (or accepting) a position on the basis of the person who has advanced it seems elementary enough, but the mistake is sufficiently prevalent in all areas of thought to have been given a name: the "genetic fallacy."

Even if propositions such as the one quoted above must still be verbalized, it is nonetheless clear that an appreciation of the theoretical underpinnings of archaeology has deepened dramatically. One of the consequences of growing sophistication in philosophical matters it that the early figures of the New Archaeology now share the stage with a larger contingent of New and "newish" archaeologists. The philosophical discussions have diversified (cf. the exchanges in the June and September 1978 issues of *Current Anthropology* between Binford, Stickel, and Read and LeBlanc).

In attempting to clarify the status of philosophizing in archaeology, an obvious first step is to look at recent reflections on these matters by L. Binford, who remains one of the most articulate spokesmen in archaeology. Binford edited the 1977 publication *For Theory Building in Archaeology*; we would like to hearken back briefly to some of the points he made in the introduction to that book.

One of the most significant themes that Binford developed in the introduction to *For Theory Building in Archaeology* concerned the connection he saw between explanation and understanding. He wrote: "I have suggested that . . . [many of the research activities in which archaeologists engage] will not move us in the direction of *explanation, and therefore understanding*, of the facts observed" (1977:1) (emphasis added). The point has been frequently made that the adequacy of an explanation is somehow dependent upon the understanding to which it contributes. Clearly, understanding depends upon context, and what will be illuminating in one context to a particular audience may not help in the least in another. But the further importance of this link to understanding is often overlooked. If the extent to which a putative explanation of X is a satisfactory explanation turns, at least in part, on whether, or to what degree, that explanation illuminates the processes that brought X about, then the adequacy of explanations cannot be characterized in formal terms alone. The point is—and we have been at pains to make it in this book—that the logic for an explanation is a separate matter from substantive content, and the latter is presumably crucial to understanding.

In Binford's view, the kind of understanding we are after in science is "understanding of the *processes* responsible for change and diversification" (Ibid.:7). He urges archaeologists to "seek the development of general [and, elsewhere he makes clear, middle-range] theory" (Ibid.), which will account for the dynamics of culture change (*see also* Raab and Goodyear 1984 on the topic of middle-range theory). According to Binford, what we want to know is "why the world is the way it appears to be" (Ibid.:5); thus, we must look to discover the processes that have resulted in the phenomena we observe. These remarks are very much like those we made in Chapter 5, where we discussed, among other issues, the role of general laws in explanation.

It is worth mentioning some of the other issues about which Binford had insightful comments in *For Theory Building in Archaeology*. He drew the discovery/justification distinction—often mistakenly characterized by archaeologists—rather nicely, if perhaps a little rigidly:

> Allow me to emphasize the point that scientific methods are designed to evaluate ideas. Science seeks to employ empirical materials in such evaluations . . . Many who have been confused about these points tend to view empirical materials as sources or inspirations for ideas. This may be true in many cases, but the sources for ideas are not the concern of science directly; the primary concern is *only the evaluation of ideas once they have been advanced* (Binford 1977:2).

Although in this *Introduction* Binford did not refer directly to any of the work of philosophers such as Kuhn and Popper on the relationship in science between the source of ideas (discovery) and the evaluation of them (justification) which we discussed in Chapters 2 and 3, he clearly appreciated the difference between the two and the fact that scientific theorizing is concerned mainly with the latter.

Binford also manifested awareness of the fact that all observation is theory laden—that the world is not presented unambiguously for our scrutiny, but that we have to impose meaning upon what we see. Regarding this he said:

> [T]here is a paradox in that the scientist must use conceptual tools to evaluate alternative conceptual tools that have been advanced re-

garding the ways in which the world works . . . There is no way out of this paradox: An evaluation of one set of ideas is dependent upon the accuracy of an assumed set of meanings, not currently under investigation, about what our observations meaningfully imply! (Ibid.:3).

Finally, and related to the above point, Binford's overall analysis of how science is done is accurate. He seems to appreciate, for example, that science does not proceed in only one direction, but that scientists must constantly be prepared to go back and forth between their data and hypotheses, sometimes amending the hypothesis under scrutiny in light of data obtained and often then having to search out new data for evaluation of a changed hypothesis. He is aware, too, that while—in a sense—everything in science is open to question, some things must be assumed constant for the purposes of testing, and, therefore, for the acquisition of knowledge.

Another aspect of current philosophical thinking in archaeology can be seen in "The Prospects for a Philosophy of Archaeology" symposium held at the 1983 Society for American Archaeology meetings in Pittsburgh. Some of the positions argued by participants will be discussed in greater detail in the last section of this chapter. Noteworthy was the remarkable concensus among participants about past problems associated with philosophizing in archaeology. A clear shift away from logical positivism emerged. An unexpected and unpremeditated division of labor evolved in which the participating archaeologists tended to address past problems and prospects with little reference to concrete archaeological data or cases; in contrast, the philosophers for the most part considered archaeological concepts or cases which either served as the basis of focused philosophical analysis or which were subjected to "internal" analysis. In general, it seems fair to say that archaeologists are no longer bulk consumers of particular philosophical programs. The awareness of past flaws, the wider and more critical exploration of philosophy by archaeologists, the substantively grounded analyses by philosophers, the more realistic expectations about what philosophy might contribute to archaeology, and the shift to incorporating substantive and contextual features into the discussions are all portents of growing sophistication.

These portents have been realized in a concrete way with the recent publication (1984) of *Archaeological Explanation*, a second volume by Watson, LeBlanc, and Redman (authors of *Explanation in Archaeology* [1971]), and *Philosophy and Archaeology* (1982) by Merrilee Salmon, which has given the field two interesting approaches to theorizing in archaeology—one from a primarily archaeological perspective, the other primarily philosophical. We have not provided detailed discussions of these works earlier in the text because, especially in the case of *Archaeological Explanation*, the work appeared after the bulk of our manuscript was in the hands of the publisher, but also because, as these two books have aims that overlap with our own, we thought it best that interested readers make their own comparisons and contrasts. Nevertheless, it seems appropriate to provide a brief indication of how we think it all fits together.

Philosophy and Archaeology is a work of applied philosophy, intended primarily for an archaeological audience. It deals, in separate chapters, with the standard problems of philosophy of science as applied to archaeology: the existence of laws, the nature of confirmation, the role of analogy, functional explanation, the structure of explanation, and theory building. In each case, Salmon brings to bear a particular philosophical perspective to attempt to resolve certain long-standing issues within archaeological theorizing. She argues, for example, that there are important laws of archaeology, and that these are often statistical in form. She rejects the hypothetico-deductive method of confirmation for establishing archaeological laws in favor of a Bayesian account, and argues for the use of analogical arguments to establish the prior probabilities of hypotheses that we may wish to subject to testing. She also tries to show that the appropriate model of explanation to be applied to archaeology is the Statistical-Relevance model (developed primarily by W. Salmon), and that this model, unlike other standard models of explanation, can accommodate functional explanations, which are so important for archaeology.

Salmon thus offers a particular philosophical account of various central concepts such as "law," "confirmation," and "explanation" which she believes captures their meaning in the archaeological context. She sees the enterprise largely as one of bridge building between archaeology and philosophy by providing an approach to the genuine philosophical concerns of archaeologists, applying philosophical insights to real archaeological examples, and making evident to archaeologists some of the deeper problems which underlie the study of any science—problems worthy of study for their own sakes whether or not they have solutions that can be directly applied in archaeology. The approach is philosophical through and through, but carefully informed by archaeological examples and sensitive to the applicability of the positions adapted to genuine problems in archaeology.

Archaeological Explanation takes quite a different tack. This is a book by three archaeologists steeped in their own field, but whose work is thoroughly informed by an understanding of the relevant work in philosophy of science and a vision of archaeology as a science with the particular approach to methodology that it entails. The approach could thus be described as the mirror image of that taken in *Philosophy and Archaeology*. Watson, LeBlanc, and Redman begin with a discussion of the concepts that have been thought basic to a philosophical understanding of archaeology—the logic of science, induction, truth, hypotheses, laws, explanations, and the like. The various models of explanation are surveyed, including models of explanation in history and their possible differences from scientific models, but this is essentially as background for a discussion of the practice of archaeology as a science. The conceptual background and the practice of archaeology together inform the rest of the work. Thus we have one chapter devoted to systems theory and another to the ecological approach to archaeology— probably "the two most popular methodological orientations in contemporary prehistoric archaeology" (1984:ix), and each is discussed briefly in relation to relevant philosophical concepts (e.g., functional explanations in the case of systems theory), but primarily in relation to current archaeological investigations and results. There are also two chapters devoted specifically to archaeological method—data collection, research design, sampling, and analysis of various kinds. The final chapter comprises a discussion of archaeological publication, contract research, and various current approaches to archaeology as social science.

Archaeological Explanation is a particularly useful discussion of theory, method, and current directions from the standpoint of three first-rate practising archaeologists who are sensitive to the conceptual foundations of their discipline but who are, quite rightly, immersed in the *doing* of archaeology with all that implies about construction and testing of particular archaeological theories, evaluating claims to archaeological knowledge, and so on. These two recent works thus exemplify two of the possible approaches to a consideration of the relations between philosophy and archaeology—one primarily from the standpoint of the philosopher and the other primarily from that of the archaeologist. It is thus clear that the relationship between theorizing in philosophy and in archaeology that emerged with the New Archaeology has continued to develop, and has become much more refined, particularly in the last five years.

Inference to the best explanation: the last case[1]

It is clear that a major worry on the part of archaeologists who have turned to the philosophy of science has been that their work was not sufficiently rigorous; but in turning to philosophy many archaeologists have labored under a misconception concerning the rewards of methodological rigor. Even the most conceptually astute and methodologically fastidious researcher can reach a false conclusion, for the possibility of such a result is an inevitable feature of inductive reasoning. Unlike earlier writers who held that deduction and deductive reasoning were essential to the archaeological enterprise, and conversely, that induction and inductive reasoning were to be avoided as lacking in rigor, we have maintained that inductive reasoning is a necessary and vital part of doing archaeology, as it is of the scientific enterprise in general. The limits of any body of knowledge cannot be expanded without the use of ampliative reasoning, in which, even if all the premises are true, the conclusion may yet be false.

In G.K. Chesterton's story "The Honour of Israel Gow," Inspector Craven of Scotland Yard finds a number of items loose and unexplained in Glengyle Castle—a hoard of precious stones, mostly diamonds; heaps of snuff on mantelpieces; minute pieces of metal—springs and wheels; and wax candles without any candlesticks. "By no stretch of fancy," he says, "can the human mind connect together snuff and diamonds and wax and loose clockwork." The indomitable Father Brown comes to his aid (Chesterton 1951: 106 and 107):

"I think I see the connection," says the priest. "This Glengyle was mad against the French Revolution. He was an enthusiast for the *ancien regime*, and was trying to re-enact literally the family life of the last Bourbons. He had snuff because it was the eighteenth century luxury; wax candles, because they were the eighteenth century lighting; the mechanical bits of iron represent the locksmith hobby of Louis XVI; the diamonds are for the Diamond Necklace of Marie Antoinette."

. . . "What a perfectly extraordinary notion!" [cries Father Brown's amateur detective friend Flambeau]. Do you really think that is the truth?"

"I am perfectly sure it isn't," [answers Father Brown] "Only you said that nobody could connect snuff and diamonds and clockwork and candles. I give you that connection offhand. The real truth, I am very sure, lies deeper."

The tale proceeds with Father Brown's offering two more peculiar explanations of the odd coincidence of materials in the castle, disavowing each in turn as merely *an* explanation but certainly not the true one. After the third farfetched story Flambeau asks: "Is that all? Have we got to the dull truth at last?" " 'Oh, no,' [says Father Brown]. 'I only suggested that because you said one could not plausibly connect snuff with clockwork or candles with bright stones. Ten false philosophies will fit the universe; ten false theories will fit Glengyle Castle. But we want the real explanation of the castle and the universe.' "

In any science (or, for that matter, any intellectual endeavor), insofar as we are seeking explanations, it is true or real or correct explanations that we want. But how can we tell when an explanation *is* the correct one? We can test it, of course, or at least we can test whatever hypothesis it incorporates, but all that a good test will show is either that the hypothesis cannot be correct as it stands or that it has been to some extent corroborated. Whether it is true or not is quite another matter, and this is something we are often not in a position to know. What we are in a better position to know is whether the hypothesis is the best explanation of the available data, and this is probably what must satisfy us, at least provisionally, as we seek to explain puzzling phenomena. The idea is that, given a set of rival or competing hypotheses, we ought to accept (though not necessarily to accept as *true*) the best from among the alternatives. The question of what criteria are appropriate for hypothesis selection is complex—such desiderata as simplicity and explanatory power are often mentioned; but whatever the criteria, the suggestion is that we reject the hypotheses that are eliminated on the basis of those criteria until one, presumably the best, is left. One obvious problem here, which we treat later in this section, is that it is difficult to be sure that we have before us all of the relevant, plausible hypotheses from among which to choose the best.

The idea of inference to the best explanation is by now familiar to the reader of this book. It is well to remember that the best explanation at a particular time on the basis of the evidence available may not ultimately be the correct one; nevertheless, it may be worthy of acceptance if anything is, although such acceptance need imply no commitment to belief in the truth of the "explanatory" hypothesis. Thus, to accept the rule of inference known as "inference to the best explanation" is not necessarily to be committed either to realism or to anti-realism about science. Perhaps the clearest understanding of the practical workings and pitfalls of inference to the best explanation can be reached by examining an example.

Our example is taken from Jane Kelley's recent work in El Salvador. At the end of the 1979 field season at the Postclassic site of Cihuatan, a potentially significant archaeological feature was found at a corner of the temple platform in the Cerro de San Dieguito excavations. This feature, which its discoverers called an "oven," or "horno," was some 50 cm. in diameter at its base, stood about 42 cm. high, and looked rather like a decapitated beehive. The broken opening was approximately 30 cm. in diameter and would probably have been smaller in its original state. The oven was made of clay or *bajareque* fired from the heat once contained within it. Near its base on the north side was a ventilator, or lateral opening, some 10 cm. in diameter with a clay trough that extended outward 12 cm. from the oven. Inside the oven were more than 2 kilos of charcoal and partially charred wood, and resting on top of the charcoal were three stones. In the charcoal itself were portions of a blowpipe.

The top of the oven was approximately 20 cm. below the modern ground surface, and the bottom of the oven was on the same level as the temple platform base which was located only a few centimeters away. Indeed, the west side of the oven was found while clearing that corner of the platform base, and a new excavation unit was opened to expose this feature. There was no detected evidence of any stratigraphic difference between the fill over the temple and that covering the oven, nor was there any such evidence in the profile of the excavation unit in which the oven was located. Although there were no ceramics in the oven, the sherds in the surrounding fill were within the range of the Postclassic ceramics of Cihuatan, and there were several large sherds in the excavation unit at the level of the bottom of the oven. Some obsidian was found in the oven itself as well as in the surrounding fill. *Talpuja* (consolidated volcanic material suitable for facing buildings), presumably from the platform facing, was incorporated into the base of the oven, suggesting that the oven did, to some extent, postdate the temple.

The evidence that was both available and thought by the research team to be the most relevant at the time of excavation was the following:

1. A feature that appeared to be designed to create higher temperatures than can be achieved in open fires.
2. A blowpipe.
3. A quantity of charcoal topped by three rocks that could have formed a rest for a container holding the materials for which a high temperature was required.
4. A bottom ventilator, or opening, that could serve as an air intake (aided by the blowpipe).
5. Apparent homogeneity and continuity of fill over the oven and platform base.
6. The fact that all artifacts in proximity to the oven were within the range of Postclassic Cihuatan specimens.
7. The common horizontal level of the oven base and the platform base.
8. The fact that *talpuja* from the platform facing was used in the base of the oven—suggesting that the temple platform had been exposed when the oven was built.

On the basis of this evidence, the hypothesis was formulated that the oven had served a metallurgical function in the Postclassic period. Even at the time the hypothesis was first advanced, the archaeologists involved in the research found the presence of incompletely charred wood in the oven worrisome, for Cihuatan is in an area of tropical forest in which perishable materials such as wood do not ordinarily last very long. But the presence of the wood did not seem sufficient to warrant rejection of the metallurgy hypothesis in a Postclassic context. Another piece of evidence had been gathered, the significance of which was not immediately apparent; relatively more, indeed many more, botanical remains were recovered by flotation in the excavation unit containing the oven than elsewhere over the temple or in temple fill. Although later results were to show that this evidence was significant, it was not considered as being of sufficient weight to eliminate the primary working hypothesis at the outset.

As regards the oven's use, alternative theories were considered by the research team. For example, the idea that it had been a kiln was entertained, but the oven's size would have precluded the firing of any but the smallest pots, and so that possibility was rejected. The hypothesis that it had been a device for processing plant products was also suggested, as were other hypotheses, but that the oven had been related to metallurgy seemed, overall, the best account of its function. The strongest supporting evidence was, undoubtedly, the blowpipe—for numerous accounts of pre-Columbian metallurgy from both South America and Mesoamerica describe blowpipes, and many pictorial representations of metallurgy show metal workers using blowpipes.

The oven, of course, had failed to provide the researchers with slag, cast pieces, or, indeed, any evidence that would constitute an intuitively satisfying confirmation of the metallurgical thesis. For this reason, the archaeologists were forced to look to other lines of evidence for support of that part of their hypothesis directed at the oven's use. A review of the archaeological evidence from Cihuatan, apart from that taken from the oven excavation, revealed two items of interest. One was a stone "pendant" that looked very much as though its inspiration had been a copper bell; the other was the discovery of what appeared to be copper-bearing rocks in a ceramic vessel that had been deliberately buried. Against those pieces of evidence was the fact that there is no record of metal objects having been found at Cihuatan; but this observation was not conclusive, for little archaeological work has been done in the area. Only one Cihuatan burial has been excavated by archaeologists and, although there are recurring rumors that several tombs have been opened by nonarchaeologists, there is no information available as to their contents. Since metallurgy had been accepted both to the north and to the south of Cihuatan for earlier time periods, the fact that the evidence for metallurgy at Cihuatan was largely circumstantial was less problematic than it might otherwise have been.

The metallurgy hypothesis was then refined, with consideration given to the sort of metallurgical process in which the oven might have been used. The complete absence of metal or slag in the oven ruled out a version of the brazier technique in which ore is placed in direct contact with the charcoal and the blowpipe is employed to force air into the brazier from the top. Rather, the nature of the archaeological evidence suggested that a crucible had been set on the three rocks found on the top of the charcoal, and this in turn indicated that the process involved had not been the extraction of metal from ore, but the preparation of metal for casting.

One further aspect of the Cihuatan researchers' reasoning should be mentioned. Several large, ceramic figures of Xipe Totec have been found in El Salvador, and Xipe Totec figures from Cihuatan are known. Central Mexican influences are much debated for the Salvadoran Postclassic, and of course there are the enigmatic Pipil, the Nahuatl speakers of El Salvador who are assumed to have migrated from Central Mexico at a time, or times, not yet settled upon (but *see* Fowler 1981). Xipe Totec was associated with metallurgy in Central Mexico, so another—albeit tenuous—piece of evidence for the metallurgy hypothesis was identified.

Apparently, then, the metallurgy hypothesis had been arrived at, and subsequently supported, in a methodologically sound manner. In order first to form a theory about their discovery—the "oven"—and then to evaluate the strength of that hypothesis, the researchers followed the leads provided by the archaeological feature itself. Then they expanded their field of inquiry: looked for additional data, relied upon analogies from other sites, used distributional and comparative evidence, considered alternative hypotheses, paid particular attention to the stratigraphic situation of the oven itself, undertook a major review of pre-Columbian metallurgy, and so on. Through all of this work, no strongly disconfirming data were uncovered.

A week before the argument for the metallurgy hypothesis was to be presented to a professional audience, the results of a Carbon 14 dating showed the charcoal from the "oven" to have an age of 165+/- 65 B.P., suggesting a very recent date for this feature—

probably late eighteenth or early nineteenth century. A second Carbon 14 date was in the same range, and since there had been a large amount of relatively uncontaminated charcoal available for dating, there seemed no good reason to dispute the dates obtained. Thus, the researchers were driven to reject the belief that their discovery had constituted evidence of metallurgical activities in Postclassic Cihuatan.

Of course, scientists often continue to argue for a hypothesis in the face of strongly disconfirming evidence, and often such persistence is fully justifiable. Archaeologists have been known to reject Carbon 14 dates that turn out to be incompatible with preconceived ideas about chronological placement based on other evidence. Sometimes the rejection of Carbon 14 dates is justified and at other times it is questionable. In the case of the Cihuatan oven, as the evidence was reevaluated in the light of the Carbon 14 dates, it became clear that the wrong bits of evidence had been emphasized. What emerge as relevant in the new interpretive context are the presence of incompletely charred wood and the higher return from flotation in the excavation unit and in the fill from the oven itself, both of which, given the tropical environment as Cihuatan, suggest a fairly recent age for the "oven." The most plausible explanation of the oven before the Carbon 14 dates were available seemed to be the Postclassic metallurgy hypothesis; but the dates, if they were to be accepted, introduced new evidence that actually seemed to render the initial hypothesis highly implausible. What had appeared to be the best explanation of the available evidence was eliminated by new evidence, and it became necessary to seek a new explanation that would best account for both the old and the new evidence. Thus, in spite of the rather impressive evidence favoring the original hypothesis, it is clear that the dating of the material alters our view of what is the best explanation of the data in this case. This might be so even if we did not previously have before us all of the relevant plausible hypotheses; when the evidence against the previously favored hypothesis appears sufficiently compelling, this can sometimes cause us to reconsider and add to the class of possible hypotheses a new one that seems more reasonable in light of the new evidence.

The Cihuatan example illustrates certain other features of science that are of increasing interest, such as the relationship between inference and technology. Fifty years ago—i.e., long before the appearance of radiocarbon dating in archaeology—the Cihuatan researchers' original analysis would probably have been accepted. A new technique has since been developed, one now so central to the discipline that to disregard the dates it produces would be irresponsible. Further refinements in Carbon 14 technology will bring still other changes. For example, ten years hence it may be that dates will be available almost immediately, thus saving archaeologists the time they now sometimes must spend arguing for inferences, such as that advanced by the archaeologists working at Cihuatan, which are ultimately refuted by that evidence. The more sensitive the instruments at their disposal, the more sophisticated archaeological inferences can be. Parallel developments in the other techniques used in archaeology would have similar effects.

Another point that the Cihuatan example nicely illustrates concerns the dangers necessarily attending the selection of pieces of evidence upon which to focus. Archaeologists can never give careful consideration to all the data they uncover; they must, then, choose those which they think are the most relevant and work with them. Such a process must involve the exercise of judgment and may lead to incorrect conclusions. Sometimes, and unavoidably so, archaeologists start off down the wrong path and only come to realize their error when forcefully disconfirming evidence is discovered. Since not all hypotheses are as easily or immediately disconfirmed as that of the Cihuatan oven, archaeologists should pay particular attention to their starting point in framing ideas and hypotheses.

A deeper, although related lesson can also be learned. Even if archaeologists could in any particular case integrate all the available data into their deliberations, there would still be the possibility that the conclusions they reached were wrong. The reasoning involved in building a case for a new hypothesis is inductive: pieces of evidence are cited in favor of a proposition that, if accepted, will expand our body of knowledge. All of those pieces of evidence can be just what the scientist claims they are and yet the conclusion reached can still be false, and this because such a conclusion necessarily goes beyond the data. It is always possible to go wrong where inductive reasoning is involved. No matter how frequently two kinds of events have been correlated in the past, it remains possible that their correlation has been accidental and will not occur in the future. The point of reflecting on what counts as good reasoning is to minimize the likelihood of that happening. The principles that have emerged from such reflection are aimed at maximizing the reliability of results, facilitating the identification of best-confirmed hypotheses and the like.

The process of getting straight on these matters is a philosophical one. In general, what we expect from philosophy is clarification. Although—somewhat paradoxically—conflicts in the underlying theory nearly always emerge from philosophical reflection, there is much to be learned from such conflicts. Consider, for example, the lottery paradox which we have discussed in Chapters 5 and 6. Two entirely acceptable propositions—(1) that any particular ticket in a million-ticket fair lottery will not win and (2) that, because the lottery is a fair one, some ticket will win—taken together yield the result both that some ticket will win and that no ticket will win. What this leads us to see is that the widely held "firmness" concept of confirmation, upon which a hypothesis is viewed as confirmed when it is nearly certain in light of its supporting evidence, will not alone found a theory of acceptance. The paradox teaches us, that is, that high probability is not sufficient for acceptance of a hypothesis.

A very important part of the Cihuatan oven example concerns the use of analogy. The researchers found themselves having to look analogically both at similar features found in other contexts and at similar contexts to see whether the hypothesized use for the "oven" feature fit reasonably with other things known about the area. But analogical reasoning goes far beyond the confines of this particular example; indeed, analogy appears to us (*contra* Gould 1980 and Freeman 1968) to be one of the most important, indeed critical, tools that archaeologists—not to say all scientists—use. Having discussed analogical argument at some length in Chapter 6, we turn here to a few additional and summary points about this form of inference.

Analogy and realism revisited

The role of analogical reasoning in theorizing of all kinds can scarcely be overestimated, but in the context of philosophy of science, the importance of analogy did not come to the fore in theoretical discussions until logical positivism had lost some of its sway. Whereas logical positivists had viewed confirmation as occurring in the context of individual words (operationism) or sentences (verificationism), their successors thought it more appropriate and ultimately more profitable to evaluate whole theories. They took a more holistic approach to confirmation, and saw in the use of analogy a way of making the comparisons among theories which are necessary to scientific explanation.

Rom Harré is prominent among the philosophers who have focussed on the importance of analogical reasoning in science. That Harré's concept of the scientific enterprise differs dramatically from that of the positivists is revealed in statements such as the following where he claims that "Scientists, in much of their theoretical activity, are trying to form a *picture* of the mechanisms of nature which are responsible for the phenomena we observe" (1970:34–35, emphasis added). Scientists are, according to Harré, attempting to answer the question "Why is it that the patterns of phenomena are the way they are?". It is, he says, through the formulation of theory that this is accomplished: "A theory answers this question by supplying an account of the constitution and behavior of those things whose interactions with each other are responsible for the manifested patterns of behaviour" (Ibid.:35). "To achieve this," he continues:

> a theory must very often fill in gaps in our knowledge of the structures and constitutions of things. This it does—by conceiving of a model for the presently unknown mechanism of nature. Such a model is, in the first instance, no more than a putative analogue for the real mechanism. The model is itself modelled on things and materials and processes which we do understand. In a creative piece of theory construction, the relation between the model of the unknown mechanism and what it is modelled on is also a relation of analogy. Thus, at the heart of a theory are various modelling relations which are types of analogy (Ibid.:35).

Harré goes on to say that:

> Part of the great importance of models in science derives from their role as progenitors of hypothetical mechanisms. It often happens that the antecedents of an effect are well known, but the causal mechanism by which the antecedents bring about the effect is not. Consider, for instance, the long history of the use of catalysts, and of

antibiotics, before the way they worked was found out. Then, in the imagination, we make a model for the unknown mechanism. Whatever *is* in the black box, one might say, could be like this. (Ibid.:39)

Harré distinguishes a number of kinds of models, and discusses the relations among models, analogies, and metaphors. Models are to be viewed as more than just sources of picturesque terminology. They are rather "the source of genuine science-extending existential hypotheses" (Harré 1970:47). Models form the basis of theories, which are meant to explain certain phenomena.

In a similar vein, Mary Hesse develops a model of science which

> interprets scientific theory in terms of a network of concepts related by laws, in which only pragmatic and relative distinctions can be made between the 'observable' and the 'theoretical'—theoretical concepts are introduced by analogy with the observational concepts constituting the natural descriptive language. Scientific language is therefore seen as a dynamic system which constantly grows by metaphorical extension of natural language, and which also changes with changing theory and with reinterpretation of some of the concepts of the natural language itself (Hesse 1974:34–35).

For our purposes, what is interesting about these particular characterizations of the role of analogical reasoning is that an acceptance of scientific realism underlies them. Analogical arguments are here viewed as ways of making corrections between the known and the unknown in such a way as to result in genuine knowledge of the causal mechanisms at work in the world. What is more, this has been the "received" view about the place of analogical reasoning in the construction of scientific theories: we are said to be searching for "real" causes—an account of what is "really" in the black box—and not just for plausible stories.

Given the pervasiveness of the realism underlying this standard view of analogy, it is clear that we need to ask whether an emphasis on analogical argument requires this metaphysic, or whether that emphasis is also compatible with the anti-realism we have been putting forward as an alternative possibility. To answer this, let us look first at analogical inference in another context, and then take one last view of the nature of theories.

One of the areas in which analogical arguments figure most

centrally is the law. It is often observed that the doctrine of precedent—the principle in common-law jurisdictions which, at its broadest, requires that like cases be decided alike—is applied through analogical reasoning, for the identification of like cases is seen as involving the drawing of analogies. As Berman and Greiner say in *The Nature and Function of Law*:

> In a legal system which attaches primary importance to the authority of past judicial decisions (precedents), as in England [Canada] and the United States, analogical reasoning in adjudication characteristically takes the form of (a) the search for a fact-situation in a previously decided case comparable to the fact-situation of the case before the court, (b) extraction from the previously decided comparable case of the principle upon which that case was decided, and (c) application of that principle to the case at hand (1972:417).

Edward Levi describes the legal reasoning process as follows:

> The basic pattern . . . is reasoning by example. It is reasoning from case to case. It is a three-step process described by the doctrine of precedent in which a proposition descriptive of the first case is made into a rule of law and then applied to a next similar situation. The steps are these: similarity is seen between cases; next the rule of law inherent in the first case is announced; then the rule of law is made applicable to the second case (1949:1–2 footnote deleted).

We know, of course, that not every analogy is a good one. Between any two phenomena in science, fact-situations in legal cases, or whatever, there will always be both some similarities and some differences. Clearly, what we need for the drawing of well-founded conclusions on the basis of analogical reasoning are strongly relevant similarities. As Levi says of this issue in legal reasoning, "The problem for the law is: 'When will it be just to treat different cases as though they were the same?'" (Ibid.:3). Since relevance is always context dependent, it is difficult to say a great deal at a general level about when likenesses will warrant inferences by analogy. Even so, we can at least say that if an analogy between two legal cases has been pointed to and is a good one—i.e., more than just suggestive, then if the two are ultimately disposed of differently, one ought to be able to give convincing reasons for that difference in treatment. If one cannot, then it is hard to avoid the conclusion that a mistake has been made and an injustice done.

The identification of similarities and differences, and of reasoning on the basis of those similarities and differences, is crucial in archaeology, and, as in the law, if different conclusions arise from strongly supported and apparently analogous cases, we need to be given reasons why different conclusions were reached. For example, if strong analogies are identified between two subsistence regimes, then a convincing argument would have to be made to show why different conclusions were drawn about, say, division of labor. With respect to the legal situation, it is clear that analogies are used not necessarily to construct explanations of what has happened (though they *may* be used for this purpose) but also to determine solutions to particular cases on the basis of what has been thought to be a just outcome in similar past situations. Their use in this context is thus often primarily normative, and there is no essential tie to any external reality.

The recognition that the insistance on the task of describing an independent reality is by no means the only way of understanding the scientific enterprise or intellectual life in general has become more and more widespread in philosophy. Besides Van Fraassen's important book, which we have discussed at some length, recent work by Rorty (1979,1982), Putnam (1978,1979), and Goodman (1978,1984) has contributed a great deal to this side of the debate. All are quite radical in an important sense. Rorty, for example, offers a reinterpretation of 'rationality' and 'objectivity' according to which it is self-deceptive to try to explain these concepts in terms of conditions of accurate representation. He even claims that philosophers do not have "a special kind of knowledge about knowledge" (1979:393) and that there is no such thing as a special "philosophical method" or "philosophical point of view."

Goodman goes so far as to argue that there is no such thing as a single world of which we may have alternative descriptions, but rather that there are as many worlds as there are right versions of them, and that rightness is not necessarily a matter of truth in the sense of correspondence with some independent "real world." Thus, he says:

truth is no more a necessary than a sufficient consideration for a choice of a statement. Not only may the choice often be of a statement that is the more nearly right in other respects over one that is the more nearly true, but where truth is too finicky, too uneven, or does not fit comfortably with other principles, we may choose the nearest amenable and illuminating lie. Most scientific laws are of this sort: not assiduous reports of detailed data but sweeping Procrustean simplifications (1978:121).

For Goodman, science and art (including the traditional humanities) are equally modes of understanding, ways of knowing:

Science seeks relevant, significant, illuminating principles, often setting aside trivial or overcomplicated truths in favor of powerful unifying approximations. And art, like science, provides a grasp of new affinities and contrasts, cuts across worn categories to yield new organizations, new visions of the worlds we live in (1984:5).

Putnam, originally the arch realist, has recently changed his views quite significantly so that he is now in substantial agreement with Goodman on this sort of point. He argues, further, that moral knowledge, in the sense of practical knowledge of how to live, is as important as standard intellectual knowledge, but involves feeling and imagination—the sort of insight we gain through literature rather than science.

And much of the emptiness of current social science arises from the attempt to study social and psychological questions with an entirely false ideal of "objectivity" which misses even the connections of the social sciences with each other, in addition to missing the questions of greatest importance to moral reflection (1978: 93).

We are thus encouraged to try to see the search for knowledge and understanding, of which archaeology is a part, in a more integrative and less fragmented way than some recent theorizing would suggest; but this need not involve a search for *the* true representation of the reality that is our past.

On philosophizing in archaeology

At the beginning of this chapter, we noted that underlying the many considerations involved in employing philosophical concepts in archaeology was the fundamental issue of the relationship between theories and the objects or subjects of theorizing. Roughly speaking, two polar attitudes about this fundamental issue are detectable in archaeology, as in many sciences. On the one hand there are those who see theories as remote, nebulous, and irrelevant to the real business of the discipline. On the other there are those who see theories as all pervasive, inescapable, and influencing every aspect of archaeology, or whatever discipline is being considered. It is probably the case that disciplines characterized by poorly articulated theories and an abundance of nuts-and-bolts activities will have a great many practitioners who eschew theory; and this characterization may well apply to archaeology.

Overt criticism of philosophizing in archaeology comes both from archaeologists and from philosophers. Before mounting our final arguments about the positive aspects of philosophizing in archaeology, we would like to look briefly at two recent negative commentaries. Flannery's (1982) amusing and acidic remarks in "The Golden Marshalltown" hit a responsive chord in the hearts of many archaeologists. In this, the most recent of Flannery's parables, the character identified as the "philosopher of the 60's" comes off very badly, while the "Old Timer" is the hero of the vignette. Archaeological philosophers, by devoting themselves to "method and theory," have exploited a shortcut to career development through abstract, philosophically based criticism. A result is that archaeologists will now "believe anything" (1982:277). We take Flannery's main point here to be that a large gap has existed between philosophizing and the substantive aspects of the discipline, and this is a point a number of other authors have made— albeit in less readable prose. Flannery advocates a rejection of philosophy and a return to real archaeology. At the same time, he goes on to tell us through the persona of the "Old Timer" that archaeology is like a game of football—a game of strategy. The important innovation of recent years are due, he feels, to the game strategies of practiced coaches and players. Included here are what he calls "differing philosophies" (1982:271). As Wylie has commented: "Thus, far from offering an unequivocal condemnation of philosophical reflection and analysis, Flannery seems to share the view that it is indispensible so long as it is developed in analysis of archaeological practice and problems" (1985:486).

Lester Embree, a philosopher, was a participant in the symposium, "The Prospects for a Philosophy of Archaeology," which was referred to earlier in this chapter. In his paper, "Archaeology without a Capital P," he argues that philosophy is what philosophers do and archaeology is what archaeologists do. Archaeologists can use philosophy as a source of stimuli, but the bottom line is that archaeological philosophizing is not philosophy; "People without credentials are amateurs" (Embree 1983:2). Although we reject Embree's insistence on disciplinary credentials as the key to legitimacy as well as his sharp drawing of disciplinary domains, his comments raise some challenging questions. It is true that much of the "amateur" philosophizing in archaeology has been counterproductive, and it is equally true that a philosophy of archaeology will not be a transplanted philosophy of science.

Flannery and Embree are not alone in criticizing philosophy in archaeology. Schiffer (1981) has noted the increasingly nonproductive interaction between philosophers and archaeologists; philosophers for the most part have criticized archaeologists in the archaeological forum in condescending and nonconstructive ways. The effect has been to dampen archaeological enthusiasm for philosophical matters and to affirm traditional disciplinary boundaries. Schiffer has felt that "from time to time problems of a philosophical nature" (1981:901) may arise in archaeology and at such times, philosophical concerns are appropriate in archaeology. Schiffer seems to suggest that in future we should limit the use of philosophy or philosophical approaches to a small number of specific issues. He has suggested also that the interaction pattern between philosophers and archaeologists should be altered, with philosophers learning from archaeology rather than simply criticizing.

In spite of fundamental differences between Flannery and Schiffer, Alison Wylie (1982b and 1985) builds a case for some interesting similarities in the two views. Both authors identify as a basic, underlying problem the failure on the part of archaeological philosophers to keep in touch with the concrete problems of archaeology. Philosophical commentaries have also failed to provide relevant and constructive guidelines for productive research. The results have been disappointing. Whether the breakdown is considered to have occurred in the one-sided interaction between philosophers and archaeologists or within archaeology itself, both authors perceive a critical gap between words and deeds as well as opportunism on the part of both philosophers or archaeologists. Both authors recommend a sharp curtailment of what they take to be philosophical discussion, although their individual thoughts concerning how and where curtailment should occur are quite different.

Wylie, however, sees both authors as qualifying their positions even as they press them. Schiffer's examples seem to suggest that it would be desirable to involve philosophers in, say, ecological debates "if they could be made to take archaeology itself seriously" (Wylie 1982b:11); and Flannery's "Old Timer," who tells us that archaeology, like football, is a game of strategy, would presumably grant that useful innovations in football strategy have occurred in recent years. Wylie infers that "archaeology, too, is a game that will not be played effectively and, more to the point, whose effectiveness will not develop, except through *constant critical assessment and innovation at a strategic level* which he [Flannery] allows is philosophical" (Wylie,1982b:11, emphasis in the original; cf. Flannery 1982:271). What Flannery objects to is that the principles of the play or game have become an end in themselves, detached from the real business of doing archaeology.

As we interpret some of the negative criticisms of philosophizing in archaeology, our conclusions are that there is substantial agreement on past problems; yet even the most severe critics nonetheless acknowledge the importance of theorizing or employing game strategies. The whole point of philosophizing, of course, is to create links between theories and the concrete work of a discipline. Whether the links are made with or without paying conscious attention to philosophy, archaeology does deal with theories. As we have discussed more fully elsewhere, as Watson and Wylie (1983) and Gibbon (1983) argued in their respective contributions to "The Prospects for a Philosophy of Archaeology" symposium, and as Binford (1977:3) recognizes, "facts" are inevitably "theory-laden." Archaeologists always select "facts" with regard to a particular theory or idea since they can never consider all the potential data available, and interpretation of facts is controlled by theory. It is surely better to explicitly recognize the "theory-laden" nature of archaeological enterprises, in order to assess and control the theories used, than to ignore this factor.

Theories help us to put things into a framework, to take the next step. Theories provide structures for concrete data, and this is an end in itself, since the alternative is disciplinary chaos. Theories guide research. Any interpretation is affected by theory as well as by myriad other factors, including those usually called sociological.

With the challenge to the appropriateness of logical positivism as a charter for archaeological philosophizing, archaeologists have lost a coherent program as well as the illusion that it would solve their problems. This has given way to a situation that is, admittedly, less neat and tidy, less programmatic, but which is quite healthy in its shift to an exploratory/critical stance grounded in substantive archaeological issues. While universal laws would be the most satisfying, and deduction the most attractive form of logical explanation, alternatives that embody less certainty will be explored for their potential usefulness. We have argued that inference-to-the-best explanation may prove to be useful to archaeologists. Inductive strategies and induction as an explanatory form must be explored. More attention must be directed to confirmation.

Archaeologists must further internalize the well-known maxim that you don't get answers to unasked questions, and the question to a large extent determines what counts as explanation—a point Wesley Salmon (1983), drawing on van Fraassen's (1980) recent thoughts on explanation, has made very nicely. His (standard) example of the way in which a "why" question affects the explanatory answer centers on the famous bank robber Willie Sutton, who answered a journalist's question, "Why do you rob banks?," with, "Because that is where the money is." The point developed by Salmon concerns contrast classes. For the bank robber, the contrast class consisted of classes of places to rob, whereas the journalist was interested in why Sutton robbed banks rather than pursuing another occupation. An economist asking Sutton the same question might have been satisfied with an explanation that identified Sutton as being from a working-class background in an area of high unemployment, whereas a psychologist might have wanted to know why this one individual among his impoverished peers turned to crime. As Salmon noted, "ask a philosophical question and you may get a philosophical answer" (1983:10).

Given the importance of the questions, it follows that archaeologists might pay more attention to the discovery side of the discovery/justification fence, recognizing that there are few guidelines in traditional philosophy. Sociological and historical factors appear to exercise great sway over the formulation of questions and of research strategies, and, if for no other reason, should be better understood.

It seems that archaeology must rely on analogy in the formulation of hypotheses and theories to a greater degree than must other social sciences, such as, for example, sociology. In our view, even archaeologists who explicitly reject analogy (i.e., Gould 1980; Freeman 1968) are only placing the analogy they reject at a distance, where it is more difficult to control, rather than dispensing with it entirely. If there is general agreement that analogy is indeed central to archaeological formulations, philosophical arguments about relevance and strong vs. weak analogies can provide a useful point of departure for more rigorous handling of this concept.

The self-scrutiny initiated by the appeal to philosophy in the founding days of the New Archaeology, as David Clarke (1973) so succinctly argued, has resulted in an irreversible loss of innocence. Archaeological horizons have been expanded. The initial expansion had the temporary effect of detaching the new visionary perspective from the empirical base. Corrective measures are underway to weld the empirical base to theoretical concerns and to sort out what is useful to archaeology from the rich tapestry of philosophical writings. The emerging philosophy of archaeology must be firmly grounded in the context and content of archaeology. Critical appeal to philosophy and judicious selection of appropriate concepts will provide useful guidance. After all, there is no need to fall into pits whose existence has already been flagged. It is to be hoped that the result will not be a partisan Old or New Archaeology, but a better and stronger archaeology.

Notes

1. This case was also used in Hanen and Kelley 1983a and Kelley 1980. Mario Aliphat was in charge of the excavation and developed the metallurgy hypothesis.

/ References Cited

Achinstein, Peter
1968   *Concepts of Science.* Johns Hopkins University Press, Baltimore.
1970   The Problem of Theoretical Terms. In *Readings in the Philosophy of Science,* edited by Baruch A. Brody, pp. 234–250. Prentice-Hall, Englewood Cliffs, New Jersey. Originally published in *American Philosophical Quarterly* 2:193–203.
1971   *Law and Explanation.* Oxford University Press, London.

Achinstein, P., and S. F. Barker (editors)
1969   *The Legacy of Logical Positivism.* Johns Hopkins University Press, Baltimore.

Ackermann, R. J.
1965   Deductive Scientific Explanation. *Philosophy of Science* 32:155–167.

Adams, Richard E. W.
1973   The Collapse of Maya Civilization: A Review of Previous Theories. In *The Classic Maya Collapse,* edited by T. Patrick Culbert, pp. 21–34. A School of American Research Book. University of New Mexico Press, Albuquerque.

Adams, Richard E. W. (editor)
1977   *The Origins of Maya Civilization.* University of New Mexico Press, Albuquerque.

Adovasio, J. M., J. D. Dunn, J. Donahue, R. Stuckenrath, J. Gilday, and K. Lord
1978   Meadowcroft Rockshelter. In *Early Man in America from a Circum-Pacific Perspective,* pp. 140–180. Occasional Papers of the Department of Anthropology, University of Alberta No. 1. Archaeological Researchers International, Edmonton.

379

Agenbroad, Larry. D.
1984   Mammoth Trampling and Other Bone Modification Processes at the Hot Springs Mammoth Site, South Dakota. *Abstracts*, First International Conference on Bone Modification, pp. 1–2. Carson City, Nevada.

Alexander, Herbert L.
1978   The Legalistic Approach to Early Man Studies. In *Early Man in America from a Circum-Pacific Perspective*. Occasional Papers of the Department of Anthropology, University of Alberta No. 1, pp. 20–22. Archaeological Researchers International, Edmonton.

Allen, W. L., and J. B. Richardson
1971   The Reconstruction of Kinship from Archaeological Data: The Concepts, the Methods, and the Feasibility. *American Antiquity*. 36: 41–53.

Ascher, Robert
1970   Murder in Eden or the New Dogma of Man's Origin. *The Cornell Plantations* 26. Pp. 7–12.
1973   Analogy in Archaeological Inference. In *In Search of Man*, edited by E. L. Green, pp. 403–410. Little, Brown, Boston. Originally published 1961, *Southwestern Journal of Anthropology* 17:317–325.

Ayer, A. J.
1946   *Language, Truth and Logic*. Second Edition. Dover Publications, New York.

Ayer, A. J. (editor)
1959   *Logical Positivism*. The Free Press, Glencoe.

Bar-Hillel, Y.
1968   The Acceptance Syndrome. In *The Problem of Inductive Logic*, edited by I. Lakatos, pp. 150–161. North-Holland Publishing Company, Amsterdam, Holland.

Barnes, Barry
1972a   *Sociology of Science*. Penguin Books, Harmondsworth, England.
1972b   Sociological Explanation and Natural Science. *European Journal of Sociology* 13:373–391.
1973   The Comparison of Belief Systems: Anomaly versus Falsehood. In *Modes of Thought*, edited by R. Horton and R. Finnegan, pp. 182–198. Faber and Faber, London.
1974   *Scientific Knowledge and Sociological Theory*. Routledge and Kegan Paul, London.
1982   *T. S. Kuhn and Social Science*. Columbia University Press, New York.

Barrais, D. A., and Joan E. Freeman
1958   Late Woodland Pottery in Wisconsin as Seen from Aztalan. *The Wisconsin Archaeologist* 39 (1): 35–61.

Beebe, B.
1983　Evidence of Carnivore Activity in a Late Pleistocene/Holocene Archaeological Site (Bluefish Cave I), Yukon Territory. In *Carnivores, Human Scavengers and Predators: A Question of Bone Technology*. Proceedings of the 15th Chacmool Symposium, pp. 1–14. Archaeological Association, University of Calgary, Calgary.

Bennet, C. L.
1979　Radiocarbon Dating with Accelerators. *American Scientist* 67:450–457.

Berman, Harold J., and William R. Greiner
1972　*The Nature and Function of Law*. Foundation Press, Mineola.

Bernal, Ignacio
1980　*A History of Mexican Archaeology*. Thames and Hudson, London.

Bhaskar, R.
1978　*A Realist Theory of Science*. Harvester Press, Sussex.

Biddick, K. A., and J. Tomenchuk
1975　A New Method for Locating, Recording and Plotting Observed Features on Long Bones with Special Emphasis on Fractures. *Journal of Field Archaeology* 2:230–249.

Bielawski, E.
1982　Inuit Heritage Education: Arctic Archaeology in the Social Context of the Northwest Territories, Canada. In *Directions in Archaeology: A Question of Goals*. Proceedings of the 14th Annual Conference, pp. 353–362. Chacmool, The Archaeological Association of the University of Calgary, Calgary.

Binford, Lewis R.
1962　Archaeology as Anthropology. *American Antiquity* 28:217–225.
1964　A Consideration of Archaeological Research Design. *American Antiquity* 29:425–451.
1965　Archaeological Systematics and the Study of Culture Process. *American Antiquity* 31:203–210.
1967　Smudge Pits and Hide Smoking: the Use of Analogy in Archaeological Reasoning. *American Antiquity* 32:1–12.
1968　Archaeological Perspectives. In *New Perspectives in Archaeology*, edited by Sally R. Binford and Lewis R. Binford, pp. 5–32. Aldine Publishing Company, Chicago.
1972　*An Archaeological Perspective*. Seminar Press, New York.
1977　General Introduction. In *For Theory Building in Archaeology*, pp. 1–10. Academic Press, New York.
1978a　On Covering Law and Theories in Archaeology. *Current Anthropology* 19:631–632.
1978b　*Nunamiut Ethnoarchaeology*. Academic Press, New York.

1981   *Bones: Ancient Men and Modern Myths*. Academic Press, New York.

1984   Bones of Contention: A Reply to Glynn Isaac. *American Antiquity* 49:164–167.

Binford, Lewis R. (editor)
1977   *For Theory Building in Archaeology*. Academic Press, New York.

Black, Max
1962   *Models and Metaphors*. Cornell University Press, Ithaca.

Blakey, Michael L.
1983   Sociopolitical Bias and Ideological Production in Historical Archaeology. In *The Socio-Politics of Archaeology*, edited by Joan M. Gero, David M. Lacy, and Michael L. Blakey, pp. 5–16. Research Report Number 23. Department of Anthropology, University of Massachusetts, Amherst.

Bloor, D.
1971   Two Paradigms of Scientific Knowledge? *Science Studies* 1 (1):101–115.

1973   Are Philosophers Averse to Science? In *Meaning and Control*, edited by D. O. Edge and J. N. Wolf, pp. 1–17.Tabistock, London.

Boas, Franz
1920   The Classification of American Languages. *American Anthropologist* 22: 367–376.

1932   The Aims of Scientific Research. *Science* 76: 605–613.

Bonnichsen, Robson
1973   Some Operational Aspects of Human and Animal Bone Alteration. In *Mammalian OsteoArchaeology: North America*, edited by B. Miles Gilbert, pp. 9–24. Special Publication of the Missouri Archaeological Society, Columbia.

1978a   Clovis: Migration or *In Situ* Development? In *Abstracts of the Fifth Biennial Meeting*, pp. 133–135. American Quaternary Association, University of Alberta, Edmonton.

1978b   Critical Arguments for Pleistocene Artifacts from the Old Crow Basin, Yukon: A Preliminary Statement. In *Early Man in America from a Circum-Pacific Perspective.* pp. 102–118. Occasional Papers of the Department of Anthropology, University of Alberta No. 1, Archaeological Researchers International, Edmonton.

1983   The Broken Bone Controversy: Some Issues Important for the Study of Early Archaeological Sites. In *Carnivores, Human Scavengers and Predators: A Question of Bone Technology*, pp. 241–270. Proceedings of the 15th Chacmool Symposium. Archaeological Association of the University of Calgary, Calgary.

1984   Ginsberg's Broken Bones. *Abstracts*, First International Conference on Bone Modification, p. 4. Carson City, Nevada.

Bonnichsen, Robson, and Richard Will

1980   Cultural Modifications of Bone: The Experimental Approach to Faunal Analysis. In *Mammalian Osteology*, edited by B. Miles Gilbert, pp. 7–30. Modern Printing Company, Laramie.

Brain, C. K.

1969   The Contribution of Namib Desert Hottentots to an Understanding of Australopithecine Bone Accumulations. *Scientific Papers of the Namib Desert Research Station* 39: 13–22.

1984   The Evidence for Bone Modification by Early Hominids in Southern Africa. *Abstracts*, First International Conference on Bone Modification, pp. 5–6. Carson City, Nevada.

Braithwaite, R. B.

1953   *Scientific Explanation.* Cambridge University Press, Cambridge.

1970   Laws of Nature and Causality. In *Readings in the Philosophy of Science*, edited by B. Brody, pp. 55–63. Prentice-Hall, Englewood Cliffs, New Jersey.

Bruner, J.

1957   On Perceptual Readiness. *Psychological Review* 64:123–152.

Brunhouse, Robert L.

1973   *In Search of the Maya*. University of New Mexico Press, Albuquerque.

Bryan, Alan L.

1983   Bone Alteration Patterns as Clues for the Identification of Early Man Sites, or An Attempt to Demystify the Search for Early Americans. In *Carnivores, Human Scavengers and Predators: A Question of Bone Technology*, pp. 193–218. Proceedings of The 15th Chacmool Symposium, Archaeology Association, University of Calgary, Calgary.

Bryan, Alan L. (editor)

1978   Early Man in America from a Circum-Pacific Perspective. *Occasional Papers of the Department of Anthropology, University of Alberta* No. 1. Archaeological Researchers International, Edmonton.

Burian, Richard M.

1977   More Than a Marriage of Convenience: On the Inextricability of History and Philosophy of Science. *Philosophy of Science* 44:1–42.

Buss, I. O.

1961   Some Observations on Food Habits and Behavior of the African Elephant. *Journal of Wildlife Management* 25:131–148.

Butler, Robert

1965   The Structure and Function of the Old Cordilleran Concept. *American Anthropologist* 67:1120–1131.

Butler, William B.
1979   Archaeology and Prohibition. *Plains Anthropologist* 24:67–71.
Butterfield, H.
1957   *The Origins of Modern Science: 1300–1800*. Revised Edition. The Free Press, New York.
Butts, Robert E. and John W. Davis (editors)
1970   *The Methodological Heritage of Newton*. University of Toronto Press, Toronto.
Campbell, Norman R.
1920   *Physics: The Elements*. Cambridge University Press, Cambridge.
1953   *What is Science*. Reprinted. Dover Publications, New York. Originally published 1921, Methuen, London.
Carnap, Rudolf
1936 andTestability and Meaning. *Philosophy of Science* 3:420–468; 4:1–40.
1937
1962   *Logical Foundations of Probability*. Second Edition. First Edition 1950. University of Chicago Press, Chicago.
1963   Replies and Systematic Expositions. In *The Philosophy of Rudolf Carnap*, edited by P. A. Schilpp, pp. 859–1013. Open Court. La Salle, Illinois.
1970   Theories as Partially Interpreted Formal Systems. In *Readings in the Philosophy of Science*, edited by Baruch A. Brody, pp. 190–199. Prentice-Hall, Englewood Cliffs, New Jersey.
Century, J. R.
1977   Original Supply Planning System Advocates Reserve Optimism. *Oilweek* 28(13): 38–40.
1978   Conventional Petroleum Assessments: Facts and Fallacies. Paper presented to the Canadian Society of Petroleum Geologists, Calgary, Alberta.
Chamberlain, T. C.
1897   The Method of Multiple Working Hypotheses. *Journal of Geology* 39 (2):155–165. Reprinted in *Science* 148 (1965), and in *In Search of Man*, edited by E. L. Green, 1973, Little, Brown, Boston.
Chard, Chester
1969   *Man in Prehistory*. McGraw-Hill, New York.
Chesterton, G. K.
1951   The Honour of Israel Gow. *The Father Brown Omnibus*, Dodd, Mead and Company, New York.
Church, Alonzo
1949   Review of *Language, Truth and Logic*, by A. J. Ayer. *The Journal of Symbolic Logic* 14:52–53.
Cinq-Mars, Jacques
1979   Bluefish Cave 1: A Late Pleistocene Eastern Berengia Cave De-

posit in the Northern Yukon. *Canadian Journal of Archaeology*. 3:1–32.

Clarke, David L.

1970   Review of *Explanation in Archaeology: An Explicitly Scientific Approach* by P. J. Watson, S. A. LeBlanc, and C. L. Redman. *Antiquity* 46:237–239.

1972   Models and Paradigms in Contemporary Archaeology. In *Models in Archaeology*, edited by David L. Clarke, pp. 1–60. Methuen and Company, London.

1973   Archaeology: A Loss of Innocence. *Antiquity* 47:6–18.

Cohen, L. Jonathan

1977   *The Probable and the Provable*. Clarendon Press, Oxford.

Cohen, Mark Nathan

1977   *The Food Crisis in Prehistory*. Yale University Press, New Haven and London.

Cohen, R. S., and Ernest Nagel

1934   *An Introduction to Logic and Scientific Method*. Harcourt, Brace and Company, New York.

Collingwood, R. G.

1922   Are History and Science Different Kinds of Knowledge? *Mind* 31:443–451.

1939   *An Autobiography*. Oxford University Press, Oxford.

1940   *An Essay on Metaphysics*. The Clarendon Press, Oxford.

1946   *The Idea of History*. The Clarendon Press, Oxford.

Conkey, Margaret W.

1978   Participation in the Research Process: Getting Grants. Paper presented at the 77th Annual Meeting of the American Anthropological Association, Los Angeles.

1982   Archaeological Research, Gender Paradigms, and Invisible Behavior. Paper Presented at the 81st Annual Meeting of the American Anthropological Association, Washington, D.C.

Conkey, Margaret W., and Janet D. Spector

1984   Archaeology and the Study of Gender. In *Advances in Archaeological Method and Theory*, Vol. 7, edited by Michael Schiffer. pp. 1–38. Academic Press, New York.

Copleston, F. C.

1956   *Contemporary Philosophy*. Burns and Oates, London.

Cordell, Linda S.

1979   *Cultural Resources Overview of the Middle Rio Grande Valley, New Mexico*. U.S. Government Printing Office, Washington D.C.

Cordell, Linda S., and Fred Plog

1979   Escaping the Confines of Normative Thought: A Re-evaluation of Puebloan Prehistory. *American Antiquity* 44:405–429.

Culbert, T. Patrick
1977   Maya Development and Collapse: An Economic Perspective. In *Social Process in Maya Prehistory*, edited by Norman Hammond. pp. 509–530. Academic Press, New York.
Culbert, T. Patrick (editor)
1973   *The Classic Maya Collapse*. A School of American Research Book. University of New Mexico Press, Albuquerque.
Culin, Stewart
1903   American Indian Games. *American Anthropologist* 5:58–64.
Daniel, Glyn (editor)
1981   *Toward a History of Archaeology*. Thames and Hudson, London.
Darwin, F., Ed.
1887   *The Life and Letters of Charles Darwin*. J. Murray, London.
Darwin, F. and A. C. Seward (editors)
1903   *More Letters of Charles Darwin*, Vol. 1. J. Murray, London.
David, Nicholas
1971   The Fulani Compound and the Archaeologist. *World Archaeology* 3(2): 111–131.
Davidson, Donald
1963   Action, Reasons, and Causes. *The Journal of Philosophy* 60:685–700.
Davis, E. Mott
1979   The First Quarter Century of the Texas Archaeological Society. *Bulletin of the Texas Archaeological Society* 50:159–194.
DeBoer, W. R.
1982   Archaeology as a Myth-Making and Myth-Serving Profession. In *Directions in Archaeology: A Question of Goals*, pp. 363–367. Proceedings of the 14th Annual Conference, Archaeological Association of University of Calgary, Calgary.
DeBoer, W. R., and D. W. Lathrap
1979   The Making and Breaking of Shipibo-Conibo Ceramics. In *Ethnoarchaeology: Implications of Ethnography for Archaeology*, edited by Carol Kramer, pp. 102–138. Columbia University Press, New York.
Deetz, James
1965   The Dynamics of Stylistic Change in Arikara Ceramics. *Illinois Studies in Anthropology*, No. 4. University of Illinois Press, Urbana.
Dewey, John
1938   *Logic: The Theory of Inquiry*. Henry Holt and Company, New York.
Dincauze, Dena F.
1984   The Colonization of Archaeologists. In *Proceedings 1984*, edited

by William J. Mayer-Oakes and Alice W. Portnoy, pp. 49–53. American Society for Conservation Archaeology.

Dincauze, Dena F., and Michael Roberts
1978   The Case for Historic Preservation—Will it Survive with Double Standards? *ASCA Newsletter* 4(6):10–14.

Doran, J. E., and F. R. Hodson
1975   *Mathematics and Computers in Archaeology.* Harvard University Press, Cambridge.

Doyel, David E.
1982   Medicine Men, Ethnic Significance and Cultural Resource Management. *American Antiquity* 47:634–642.

Dray, W.
1957   *Laws and Explanation in History.* Oxford University Press, Oxford.
1959   "Explaining What" in History. In *Theories of History*, edited by P. Gardiner, pp. 403–408. The Free Press, New York.

Ducasse, C. J.
1925   Explanation, Mechanism and Teleology. *The Journal of Philosophy* 22:150–155.

Duhem, Pierre
1954   *The Aim and Structure of Physical Theory.* Princeton University Press, Princeton. Translated From 1914 Second Edition.

Dumond, Don E.
1977   Science as Archaeology: The Saints Go Marching In. *American Antiquity* 42:330–349.

Dunnell, Robert C.
1980   Evolutionary Theory and Archaeology. *Advances in Archaeological Method and Theory*, Vol. 3, edited by Michael B. Schiffer, pp. 35–99. Academic Press, New York.
1982   Science, Social Science and Common Sense: The Agonizing Dilemma of Modern Archaeology. *Journal of Anthropological Research* 38:1–25.

Dworkin, Ronald
1977   The Model of Rules I and II. In *Taking Rights Seriously*, pp. 14–80. Harvard University Press, Cambridge.

Dyson, Stephen L.
1985   Two Paths to the Past: A Comparative Study of the Last Fifty Years of *American Antiquity* and the *American Journal of Archaeology*. *American Antiquity* 50(2):452–463.

Eberle, R., D. Kaplan, and R. Montague
1961   Hempel and Oppenheim on Explanation. *Philosophy of Science* 28:418–428.

Embree, Lester
1983   Archaeology Without a Capital P. Paper presented at the 48th

Annual Meeting of the Society for American Archaeology, Pittsburgh.

Erasmus, Charles J.
1950   Patolli, Pachisi, and the Limitation of Possibilities. *Southwestern
         Journal of Anthropology* 6:369–387.

Erasmus, Charles J., and Waldemar R. Smith
1967   Cultural Anthropology in the United States Since 1900: A Quan-
         titative Analysis. *Southwestern Journal of Anthropology* 23:111–
         140.

Ewing, A.C.
1934   *Idealism: A Critical Survey.* Methuen and Company, London.
1973   Meaninglessness. In *A Modern Introduction to Philosophy,* edited
         by Paul Edwards and Arthur Pap, pp. 770–779. Third Edition.
         The Free Press, New York.

Fagan, Brian M.
1972   *In The Beginning: An Introduction to Archaeology.* Little, Brown and
         Company, Boston.

Feigl, Herbert
1945   Operationalism and Scientific Method. *Psychological Review*
         52:250–259 and 284–288.

Feyerabend, Paul K.
1975   *Against Method: Outline of an Anarchistic Theory of Knowledge.*
         Humanities Press, Atlantic Highlands, New York.

Figgins, Jesse D.
1927   The Antiquity of Man in America. *Natural History* 33:229–239.

Flannery, Kent V.
1967   Culture History and Cultural Process: A Debate in American
         Archaeology. *Scientific American* 217(2):119–122.
1972   The Cultural Evolution of Civilizations. *Annual Review of Ecology
         and Systematics* 3:399–426.
1973   Archaeology with a Capital "S." In *Research and Theory in Current
         Archaeology,* edited by Charles L. Redman, pp. 47–53. John
         Wiley and Sons, New York.
1976   *The Early MesoAmerican Village.* Academic Press, New York.
1982   The Golden Marshalltown: A Parable for the Archaeology of the
         1980's. *American Anthropologist* 84:265–278.

Ford, Richard I.
1973   Archaeology Serving Humanity. In *Research and Theory in Current
         Anthropology,* edited by Charles L. Redman, pp. 83–93. John
         Wiley and Sons, New York.

Forman, Sylvia Helen
1977   Occupational Status of Women in Anthropology Departments,
         1976–1977. *Anthropological Newsletter* 18(9):10–12.

1978   Occupational Status of Women in Anthropology Departments
            1977–1978. *Anthropological Newsletter* 19(8):18–22.

Fowler, Don D.
1984   Response to "Scholarship as Stewardship" by William J. Mayer-
            Oakes. *Proceedings 1984*, edited by William J. Mayer-Oakes and
            Alice W. Portnoy, pp. 45–48. American Society for Conservation
            Archaeology.

Fowler, William R., Jr.
1981   The Pipil-Nicarao of Central America. Unpublished Ph.D. dis-
            sertation, Department of Archaeology, University of Calgary,
            Calgary.

Francis, Peter D., and Eric C. Poplin (editors)
1982   Directions in Archaeology: A Question of Goals. *Proceedings of
            the 14th Annual Conference*, Archaeology Association University
            of Calgary, Calgary.

Freeman, L. G., Jr.
1968   A Theoretical Framework for Interpreting Archaeological Mate-
            rials. In *Man the Hunter*, edited by R. B. Lee and I. DeVore, pp.
            262–267. Aldine Publishing Company, Chicago.

Fritz, John M., and Fred T. Plog
1970   The Nature of Archaeological Explanation. *American Antiquity*
            35:405–412.

Frontiers: A Journal of Women's Studies, Vol. 2.

Gero, Joan M.
1983   Gender Bias in Archaeology: A Cross Cultural Perspective. In
            The Socio-Politics of Archaeology, edited by Joan M. Gero, David
            M. Lacey, and Michael L. Blakey, pp. 51–58. *Research Reports
            Number* 23. Department of Anthropology, University of Mas-
            sachusetts, Amherst.
1985   Socio-politics and the Woman-at-Home Ideology. *American An-
            tiquity* 50(2):342–350.

Gero, Joan M., David M. Lacy, and Michael L. Blakey (editors)
1983   The Socio-Politics of Archaeology. *Research Reports Number 23.*
            Department of Anthropology, University of Massachusetts,
            Amherst.

Gibbon, Guy
1983   Realism: A Better Metaphysics for Archaeology? Paper delivered
            at the 48th Annual Meeting of the Society for American Archae-
            ology, Pittsburgh.
1984   *Anthropological Archaeology.* Columbia University Press, New
            York.

Giere, Ronald N.
1973   History and Philosophy of Science: Intimate Relationships or

Marriage of Convenience? *British Journal for the Philosophy of Science* 24:282–297.

1976   Empirical Probability, Objective Statistical Methods and Scientific Inquiry. In *Foundations of Probability Theory, Statistical Inference, and Statistical Theories of Science*, Vol. 2, edited by C. A. Hooker and W. Harper, pp. 63–93. D. Reidel Publishing Company, Dordrecht, Holland.

1979   *Understanding Scientific Reasoning*. Holt, Rinehart and Winston, New York.

Gifford, Carol A., and Elizabeth A. Morris
1985   Digging for Credit: Early Archaeological Field Schools in the American Southwest. *American Antiquity* 50(2):395–411.

Gladwin, Thomas
1958   Canoe Travel in the Truk Area: Technology and its Psychological Correlates. *American Anthropologist* 60(5):893–899.

Glock, Albert E.
1985   Tradition and Change in Two Archaeologies. *American Antiquity* 50(2):464–477.

Gombrich, E. H.
1961   *Art and Illusion*. Second Edition. Princeton University Press, Princeton.

Goodman, Nelson
1965   *Fact, Fiction and Forecast*. Second Edition. The Bobbs-Merrill Company, Indianapolis.

1966   Comments. *Journal of Philosophy* 63:328–331.

1972a   The Way the World Is. In *Problems and Projects*, pp. 24–32. The Bobbs-Merrill Company, Indianapolis and New York.

1972b   Seven Strictures on Similarity. In *Problems and Projects*, pp. 437–446. The Bobbs-Merrill Company, Indianapolis and New York.

1973   *Fact, Fiction and Forecast*. Third Edition. The Bobbs-Merrill Company, Indianapolis.

1978   *Ways of Worldmaking*. Hackett Publishing Company, Indianapolis.

1984   *Of Minds and Other Matters*. Harvard University Press, Cambridge.

Gould, Richard A.
1980   *Living Archaeology*. Cambridge University Press, Cambridge.

Gould, Richard A., and Patty Jo Watson
1982   A Dialogue on the Meaning and Use of Ethnoarchaeological Reasoning. *Journal of Anthropological Archaeology* 1:355–381.

Griffith, Belver C., and Nicholas C. Mullins
1972   Coherent Social Groups in Scientific Change. *Science* 177:959–964.

Gruber, H. E., and P. H. Barrett
1974   *Darwin on Man.* E. P. Dutton and Company, New York.

Grunbaum, Adolf
1976a   Ad Hoc Auxiliary Hypotheses and Falsification. *The British Journal for the Philosophy of Science* 27:329–362.
1976b   Can a Theory Answer More Questions than One of its Rivals? *The British Journal for Philosophy of Science* 27:1–33.
1976c   Is the Method of Bold Conjectures and Attempted Refutations Justifiably the Method of Science? *The British Journal for the Philosophy of Science* 27:105–136.
1976d   Is Falsifiability the Touchstone of Scientific Rationality? Karl Popper versus Inductivism. In *Essays in Memory of Imre Lakatos.* Boston Studies in the Philosophy of Science, Vol. 39, edited by R.S. Cohen and M.W. Wartofsky, pp. 213–252. D. Reidel Publishing Company, Dordrecht, Holland.
1979   Is Freudian Psychoanalytic Theory Pseudo-Scientific by Karl Popper's Criterion of Demarcation? *American Philosophical Quarterly* 16(2):131–141.

Hacking, Ian
1965   *Logic of Statistical Inference.* Cambridge University Press, Cambridge.
1975   *The Emergence of Probability.* Cambridge University Press, Cambridge.

Hagstrom, W. O.
1965   *The Scientific Community.* Basic Books, New York.

Hanen, Marsha
1967   Goodman, Wallace and the Equivalence Condition. *Journal of Philosophy* 64:271–280.
1971   Confirmation and Adequacy Conditions. *Philosophy of Science* 38:361–368.
1975   Confirmation, Explanation and Acceptance. In *Analysis and Metaphysics,* edited by R. Lehrer, pp. 93–129. D. Reidel Publishing Company, Dordrecht, Holland.

Hanen, Marsha, and Jane H. Kelley
1983a   Inference to the Best Explanation in Archaeology. Paper presented at the 48th Annual Meeting of the Society for American Archaeology, Pittsburgh.
1983b   Social and Philosophical Frameworks for Archaeology. In The Socio-Politics of Archaeology, edited by Joan M. Gero, David M. Lacy, and Michael L. Blakey, pp. 107–117. *Research Report Num-*

ber 23, Department of Anthropology, University of Massachusetts, Amherst.

Hanen, Marsha, Margaret J. Osler, and Robert G. Weyant (editors)
1980   *Science, Pseudo-Science and Society*. Wilfred Laurier University Press, Waterloo.

Hanson, F. Allen
1974   Is Anthropology Scientific? Paper presented to the 73rd Annual Meeting of the American Anthropological Association, Mexico City.

Hanson, N. R.
1958   *Patterns of Discovery: An Inquiry into the Conceptual Foundations of Science*. Cambridge University Press, Cambridge.

Hardesty, D. L.
1980   The Use of General Ecological Principles in Archaeology. *Advances in Archaeological Method and Theory*, Vol. 3, edited by Michael B. Schiffer, pp. 158–187. Academic Press, New York.

Hardy, Kenneth R.
1974   Social Origins of American Scientists and Scholars. *Science* 185:497–506.

Harman, G.
1965   Inference to the Best Explanation. *Philosophical Review* 74:88–95.
1970   Induction. In *Induction, Acceptance and Rational Belief*, edited by M. Swain, pp. 83–100. D. Reidel Publishing Company, Dordrecht, Holland.

Harre, Romano
1970   *The Principles of Scientific Thinking*. University of Chicago Press, Chicago.
1979   *Social Being*. Basil Blackwell, Oxford.

Harre, R., and P. F. Secord
1972   *The Explanation of Social Behavior*. Basil Blackwell, Oxford.

Harrington, C. R., Robson Bonnichsen, and Richard E. Morlan
1975   Bones Say Man Lived in Yukon 27,000 Years Ago. *Canadian Geographical Journal* 91:42–48.

Harrison, Peter D., and D. L. Turner II (editors)
1978   *Prehistoric Maya Agriculture*. University of New Mexico Press, Albuquerque.

Hart, H. L. A.
1961   *The Concept of Law*. Oxford University Press, Oxford.

Haury, Emil
1958   Evidence at Point of Pines for a Prehistoric Migration from Northern Arizona. In *Migrations in New World Culture*. University of Arizona Social Science Bulletin, No. 27, pp. 1–6. University of Arizona Press, Tucson.

Haury, Emil W., E. B. Sayles, and William W. Wasley
1958   The Lehner Mammoth Site, Southeastern Arizona. *American Antiquity* 25:1–30.

Haynes, Vance
1978   The Clovis Culture. In *Abstracts of the Fifth Biennial Meeting*, pp. 129–132. American Quarternary Association, University of Alberta, Edmonton.

Hempel, Carl G.
1945   Studies in the Logic of Confirmation. *Mind* 54:1–26 and 97–121. Reprinted in *Aspects of Scientific Explanation*, by Carl G. Hempel, The Free Press, New York, 1965.
1962   Deductive-Nomological vs. Statistical Explanation. *Minnesota Studies in the Philosophy of Science*, Vol. III, edited by H. Feigl and G. Maxwell, pp. 98–169. University of Minnesota Press, Minneapolis.
1965   *Aspects of Scientific Explanation*. The Free Press, New York.
1966   *Philosophy of Natural Science*. Prentice-Hall, Englewood Cliffs, New Jersey.
1968   Operationism, Observation, and Theoretical Terms. In *Philosophy of Science*, edited by Arthur Danto and Sidney Morgenbesser, pp. 101–120. Meridian Books, New York.
1970   A Logical Appraisal of Operationism. In *Readings in the Philosophy of Science*, edited by Baruch A. Brody, pp. 200–210. Prentice-Hall, Englewood Cliffs, New Jersey.

Hempel, Carl G., and Paul Oppenheim
1948   Studies in the Logic of Explanation. *Philosophy of Science* 15:135–175. Reprinted with a Postscript in 1965 in *Aspects of Scientific Explanation*, by Carl G. Hempel, pp. 245–295. The Free Press, New York.

Hesse, Mary B.
1966   *Models and Analogies in Science*. University of Notre Dame Press, Notre Dame, Indiana.
1974   *The Structure of Scientific Inference*. MacMillan, London and Basingstoke.

Hester, Thomas R.
1981   CRM Publications: Dealing with Reality. *Journal of Field Archaeology* 8(4):493–496.

Hill, James N.
1970   Broken K Pueblo: Prehistoric Social Organization in the American Southwest. *Anthropological Papers of the University of Arizona*, Vol. 18. University of Arizona Press, Tucson.

Hole, Frank
1973 Questions of Theory in the Explanation of Culture Change in Prehistory. In *The Explanation of Culture Change: Models in Prehistory*, edited by Colin Renfrew, pp. 19–34. Gerald Duckworth and Company, London.

Hole, Frank, and Robert Heizer
1973 *An Introduction to Prehistoric Archaeology.* Third Edition. Holt, Rinehart and Winston, New York.

Holton, Gerald
1962 Models for Understanding the Growth and Excellence of Scientific Research. In *Excellence and Leadership to Democracy*, edited by S. R. Graubard and G. Holton, pp. 94–131. Columbia University Press, New York.

Hosler, Dorothy, Jeremy A. Sabloff, and Dale Runge
1977 Simulation Model Development: A Case Study of the Classic Maya Collapse. In *Social Process in Maya Prehistory*, edited by Norman Hammond, pp. 553–590. Academic Press, New York.

Hospers, John
1946 On Explanation. *The Journal of Philosophy* 43:337–356.

Hrdlicka, Ales
1907 Skeletal Remains Suggesting or Attributed to Early Man in North America. *Bureau of American Ethnology Bulletin* 33, pp. 21–28. Smithsonian Institution, Washington D.C.

1912 Early Man in South America. In collaboration with W. H. Holmes, Bailey Willis, Fred Eugene Wright, and Clarence N. Fenner. *Bureau of American Ethnology Bulletin* 52. Smithsonian Institution, Washington D.C.

1918 Recent Discoveries Attributed to Early Man in North America. *Bureau of American Ethnology Bulletin* 66. Smithsonian Institution, Washington, D.C.

Hsu, Francis L. K.
1979 The Cultural Problem of the Cultural Anthropologist. *American Anthropologist* 81:517–532.

Hughes, O. L., C. R. Harrington, J. A. Janssens, J. V. Matthews, Jr., R. E. Morlan, N. W. Rutter, and C. E. Schweger
1981 Upper Pleistocene Stratigraphy, Paleoecology and Archaeology of the Northern Yukon Interior, Eastern Berengia: I Bonnet Plume Basin. *Arctic* 34(4):329–365.

Hull, Clark L.
1943 *Principles of Behavior.* Appleton-Century-Crofts., New York.

Hull, David L.
1975 Review of *Philosophy of Natural Science*, by Carl Hempel, *The Structure of Scientific Revolutions*, Second Edition, by Thomas S.

Kuhn, and *Galileo: A Philosophical Study*, by Dudley Shapere. *Systematic Zoology* 24:395–401.

Hume, David
1955   *An Enquiry Concerning Human Understanding*. The Liberal Arts Press, New York.

Hurlbert, Beverly McElligott
1976   Status and Exchange in the Profession of Anthropology. *American Anthropologist* 78:272–284.

Ikawa-Smith, Fumiko
1975   Japanese Ancestors and Paleolithic Archaeology. *Asian Perspectives* 18:15–25.

Irving, W. N.
1978   Pleistocene Archaeology in Eastern Berengia. In *Early Man from a Circum-Pacific Perspective*. Occasional Papers of the Department of Anthropology, University of Alberta, No. 1, edited by Alan L. Bryan, pp. 96–101. Archaeological Researchers International, Edmonton.
1982   Old Crow Bone Studies 1982. Paper delivered at the 15th Annual Chacmool Symposium, Archaeological Association of the University of Calgary, Calgary.

Irving, W. N., and C. R. Harrington
1973   Upper Pleistocene Radiocarbon-Dated Artefacts from the Northern Yukon. *Science* 179:335–340.

Irving, W. N., A. V. Jopling, and I. Kritsch Armstrong
1984   Pleistocene Stratigraphy and Paralithic Technology in the Old Crow Basin: 1984. *Abstracts*, First International Conference on Bone Modification, pp. 18–19, Carson City, Nevada.

Isaac, Glynn
1978   Food Sharing and Human Evolution: Archaeological Evidence from the Plio-Pleistocene of East Africa. *Journal of Anthropological Research* 34:311–325.
1983a   Review of *Bones: Ancient Men and Modern Myths*, by L. R. Binford. *American Antiquity* 48:416–419.
1983b   Bones in Contention: Competing Explanations for the Juxtaposition of Early Pleistocene Artifacts and Faunal Remains. In *Animals in Archaeology*. Ed. by J. Clutton-Brock and C. Grigson. *British Archaeological Reports*. International Series 163:3–19.

Jacobson, Jerome
1979   Recent Developments in South Asian Prehistory and Protohistory. In *Annual Review of Anthropology*, Vol. 8, edited by B. J. Siegel, A. R. Beals, and S. A. Tylor, pp. 467–502. Annual Reviews, Palo Alto.

Jarvie, I. C.
1979   The Notion of a Social Science. In *Recent Approaches to the Social Sciences*, edited by H. K. Betz, pp. 76–88. University of Calgary, Calgary.

Jeffrey, R.
1956   Valuation and Acceptance of Scientific Hypotheses. *Philosophy of Science* 23:237–246.
1971   Statistical Explanation vs. Statistical Inference. In *Statistical Explanation and Statistical Relevance*, edited by Wesley Salmon, pp. 19–28. University of Pittsburgh Press, Pittsburgh.

Jennings, Jesse D.
1978   *Ancient Native Americans*. W. H. Freeman and Company, San Francisco.

Jevons, W. Stanley
1920   *The Principles of Science*. Reprinted. MacMillan Publishing and Company, London. Originally published 1874.

Joad, C. E. M.
1950   *A Critique of Logical Positivism*. University of Chicago Press, Chicago.

Johnson, LeRoy, Jr.
1972   Problems in "Avant-Garde" Archaeology. *American Anthropologist* 74:366–377.

Jopling, A. V., W. N. Irving, and B. F. Beebe
1981   Stratigraphic, Sedimentological and Faunal Evidence for the Occurrence of Pre-Sangamonian Artifacts in Northern Yukon. *Arctic* 34(1):3–33.

Kanner, Barbara
1977   Bibliography, Part II. In *A Widening Sphere*, edited by Martha Vicinus, pp. 199–270. Indiana University Press, Bloomington.

Kaplan, Bernice
1974   Statement. *Newsletter of the American Anthropological Association* 15:29.

Kaplan, D.
1961   Explanation Revisited. *Philosophy of Science* 28:429–436.

Keene, Arthur S.
1983   Biology, Behavior and Borrowing: A Critical Examination of Optimal Foraging Theory in Archaeology. In *Archaeological Hammers and Theories*, edited by James A. Moore and Arthur S. Keene. pp. 137–155. Academic Press, New York.

Keene, Arthur S., and William K. MacDonald
1980   Whatever Happened to the New Archaeology? An Examination of the Recent History of Archaeological Approaches in North

America and their Relationship to Anthropology. Paper presented at the 45th Annual Meeting of the Society for American Archaeology, Philadelphia.

Kelley, Jane H.

1979   To What Tune Shall We Dance. Paper delivered at the Canadian Archaeological Association Meetings, Vancouver.

1980   Evidence and Inference: A Cautionary Tale. Paper presented at the 45th Annual Meeting of the Society for American Archaeology, Philadelphia.

Kense, Francois J.

1982   Microwave Oven or Hearthpots: Redirection in Third World Archaeology. In *Directions in Archaeology: A Question of Goals.* Proceedings of the 14th Annual Conference, pp. 345–351. Archaeological Association, University of Calgary, Calgary.

Kim, J.

1963   Discussion, On the Logical Conditions of Deductive Explanation. *Philosophy of Science* 30:286–291.

King, Thomas F.

1984   Has Conservation Really Helped? *Proceedings 1984* edited by William J. Mayer-Oakes and Alice W. Portnoy, pp. 25–30. American Society for Conservation Archaeology.

Klejn, Leo S.

1977   A Panorama of Theoretical Archaeology. *Current Anthropology* 18(1):1–42.

Kluckhohn, Clyde

1940   The Conceptual Structure in Middle American Studies. In *The Maya and Their Neighbors,* edited by A. M. Tozzer, pp. 41–51. Appleton-Century-Croft Company, New York.

Knorosov, Yurii V.

1963   *Pis'mennost indeitsev maiia (Writing of the Maya Indians).* Academy of Sciences, Moscow-Leningrad, USSR.

Kroeber, Alfred L.

1916   Zuni Potsherds. *Anthropological Papers of the American Museum of Natural History,* Vol. 18, Part 1. American Museum of Natural History, New York.

1931   Historical Reconstruction of Culture Growths and Organic Evolution. *American Anthropologist* 33:149–156.

1948   *Anthropology.* Harcourt, Brace, New York.

Kuhn, Thomas J.

1957   *The Copernican Revolution: Planetary Astronomy in the Development of Western Thought.* Harvard University Press, Cambridge.

1970a   *The Structure of Scientific Revolutions.* Second Edition. University of Chicago Press, Chicago. First Edition, 1962.

1970b   The Logic of Discovery or Psychology of Research? In *Criticism and the Growth of Knowledge*, edited by Irme Lakatos and Alan Musgrave, pp. 1–23. Cambridge University Press, Cambridge.

1970c   Reflections on My Critics. In *Criticism and the Growth of Knowledge*, edited by Imre Lakatos and Alan Musgrave, pp. 231–278. Cambridge University Press, Cambridge.

Kyburg, H.

1963   A Further Note on Rationality and Consistency. *The Journal of Philosophy* 60:463–465.

1965   Comments. *Philosophy of Science* 32:147–151.

1968   The Rule of Detachment in Inductive Logic. In *The Problem of Inductive Logic*, Vol. 2, edited by Imre Lakatos, pp. 98–119. North-Holland Publishing Company, Amsterdam, Holland.

Lacy, David, and Robert J. Hasenstab

1983   The Development of Least Effort Strategies in CRM: Competition for Scarce Resources in Massachusetts. In The Socio-Politics of Archaeology, edited by Joan M. Gero, David M. Lacy, and Michael L. Blakey, pp. 31–50. *Research Report Number* 23. Department of Anthropology, University of Massachusetts, Amherst.

Lakatos, Imre

1970   Falsification and the Methodology of Scientific Research Programs. In *Criticism and the Growth of Knowledge*, edited by Imre Lakatos and Alan Musgrave, pp. 91–196. Cambridge University Press, Cambridge.

Lakotos, Imre, and Alan Musgrave (editors)

1970   *Criticism and the Growth of Knowledge*. Cambridge University Press, Cambridge.

Laughlin, William S.

1975   Ecosystem, Holocene History, and Siberian Origin. *Science* 189:507–515.

Laws, R. M.

1966   Age Criteria for the African Elephant, *Loxodonta a. africanus. East African Wildlife Journal* 4:1–37.

Leatherdale, W. H.

1974   *The Role of Analogy, Model and Metaphor in Science*. North-Holland Publishing Company, Amsterdam, Holland.

LeBlanc, Steven

1973   Two Points of Logic Concerning Data, Hypotheses, General Laws and Systems. In *Research and Theory in Current Archaeology*, edited by Charles L. Redman, pp. 199–214. John Wiley and Sons, New York.

LeMoine, Genevieve M., and A. Scott MacEachern (editors)

1983   *Carnivores, Human Scavengers and Predators: A Question of Bone*

Technology. Proceedings of the 15th Chacmool Symposium. Archaeological Association, University of Calgary, Calgary.

Leone, Mark P.

1972   Issues in Anthropological Archaeology. In *Contemporary Archaeology*, edited by Mark P. Leone, pp. 14–27. Southern Illinois Press, Carbondale and Edwardsville.

1973   Archaeology as the Science of Technology: Morman Town Plans and Fences. In *Research and Theory in Current Archaeology*, edited by Charles L. Redman, pp. 125–150. John Wiley and Sons, New York.

1978   Time in American Archaeology. In *Social Archaeology: Beyond Subsistence and Dating*, edited by Charles L. Redman, pp. 25–36. Academic Press, New York.

1982   Opinions About Recovering Mind. *American Antiquity* 47(4):742–760.

Lesser, A.

1935   Functionalism in Social Anthropology. *American Anthropologist* 37:386–393.

1939   Research Procedure and Laws of Culture. *Philosophy of Science* 6:345–355.

Levi, Edward H.

1949   *An Introduction to Legal Reasoning.* University of Chicago Press, Chicago.

Levi, Isaac

1967   *Gambling with Truth.* Alfred A. Knopf, New York.

Levi-Strauss, Claude

1963   Split Representation in the Art of Asia and America. In *Structural Anthropology*, translated by Claire Jacobson and Brooke Grundfest Schoepf, pp. 245–268. Basic Books, New York.

Levin, Michael E.

1973   On Explanation in Archaeology: A Rebuttal to Fritz and Plog. *American Antiquity* 38:387–395.

Lindsay, Alexander J., Jr.

1983   Explaining an Anasazi Migration to East Central Arizona. Paper read at the Society for American Archaeology Meetings, April 30, 1983, Pittsburgh.

Lipe, William

1978   Contracts, Bureaucrats, and Research: Some Emerging Problems of Conservation Archaeology in the United States. In *Archaeological Essays in Honor of Irving B. Rouse*, edited by R. C. Dunnell and E. S. Hall, pp. 121–147. Mouton, The Hague, Holland.

1984   Conservation for What? *Proceedings 1984*, edited by William J.

Mayer-Oakes and Alice W. Portnoy, pp. 1–11. American Society for Conservation Archaeology.

Lister, Florence C., and Robert H. Lister
1968   *Earl Morris and Southwestern Archaeology*. University of New Mexico Press, Albuquerque.

Longacre, William A.
1966   Changing Patterns of Social Integration: A Prehistoric Example from the American Southwest. *American Anthropologist* 68:94–103.
1968   Some Aspects of Prehistoric Society in East-Central Arizona. In *New Perspectives in Archaeology*, edited by S. R. Binfore and L. R. Binford, pp. 89–102. Aldine Publishing Company, Chicago.
1974   Kalinga Pottery Making: The Evolution of a Research Design. In *Frontiers of Anthropology*, edited by Murray J. Leaf, pp. 51–67. D. Van Nostrand Company, New York.
1981   CRM Publication: A Review Essay. *Journal of Field Archaeology* 8(4):487–491.

Lorenzo, Jose Luis
1978   Early Man Research in the American Hemisphere: Appraisal and Perspectives. In *Early Man in America from a Circum-Pacific Perspective*, edited by Alan Bryan, pp. 1–9. Occasional Papers of the Department of Anthropology, University of Alberta, No. 1. Archaeological Researchers International, Edmonton.

Losee, John
1972   *A Historical Introduction to the Philosophy of Science*. Oxford University Press, London.

MacCormac, Earl R.
1976   *Metaphor and Myth in Science and Religion*. Duke University Press, Durham.

MacDonald, William K. (editor)
1976   Digging for Gold: Papers on Archaeology for Profit. *Research Reports in Archaeology: Technical Reports* No. 5. Museum of Anthropology, University of Michigan, Ann Arbor.

MacNeish, Richard S.
1978   *The Science of Archaeology* Duxbury Press, Belmont, California.

Mandelbaum, Maurice
1967   *The Problem of Historical Knowledge*. Second Edition. Harper and Row, New York.

Martin, Paul S.
1971   The Revolution in Archaeology. *American Antiquity* 36:1–8.

Martin, Paul S., and Fred Plog
1973   *The Archaeology of Arizona*. Doubleday/Natural History Press, New York.

Masterman, Margaret
1970   The Nature of a Paradigm. In *Criticism and the Growth of Knowledge*, edited by Imre Lakatos and Alan Musgrave, pp. 59–89. Cambridge University Press, Cambridge.

Maull, Nancy
1976   Reconstructed Science as Philosophical Evidence. *PSA 1976. Proceedings of the 1976 Biennial Meeting of the Philosophy of Science Association*, Vol. 1, edited by F. Suppe and P. D. Asquith, pp. 119–130. Philosophy of Science Association, East Lansing, Michigan.

Maxwell, Grover
1970   The Ontological Status of Theoretical Entities. In *Readings in the Philosophy of Science*, edited by Baruch A. Brody, pp. 224–233. Prentice-Hall, Englewood Cliffs, New Jersey.

McGimsey, Charles R.
1972   *Public Archaeology*. Seminar Press, New York.
1981   Archaeology: A Profession in Transition. *Early Man* 3(3):28–32.

McGimsey, Charles R., and Hester A. Davis (editors)
1977   The Management of Archaeological Resources: The Airlie House Report. *Special Publication of the Society for American Archaeology.*

McMullin, Ernan
1975   History and Philosophy of Science: A Marriage of Convenience? *Boston Studies in the Philosophy of Science* 32:515–531.

Medawar, Peter B.
1969   *Induction and Intuition in Scientific Thought*. American Philosophical Society, Philadelphia.

Meltzer, David J.
1979   Paradigms and the Nature of Change in American Archaeology. *American Antiquity* 44:644–657.

Merton, Robert K.
1970   *Science, Technology, and Society in Seventeenth Century England*. Howard Fertig, New York.

Mill, John Stuart
1843   *A System of Logic*. John W. Parker, London.

Miller, Daniel
1980   Archaeology and Development. *Current Anthropology* 21(6):709–726.
1982   Explanation and Social Theory in Archaeological Practice. In *Theory and Explanation in Archaeology*, edited by Colin Renfrew, Michael J. Rowlands, and Barbara Abbott Segraves, pp. 83–95. Academic Press, New York.

Moctezuma, Eduardo Matos
1980   New Finds in the Great Temple. *National Geographic* 158(6):767–775.

Morgan, Charles G.
1970   Kim on Deductive Explanation. *Philosophy of Science* 37:434–439.
1973   Archaeology and Explanation. *World Archaeology* 4:259–276.
1974–   Explanation and Scientific Archaeology. *World Archaeology*
1975   6:133–137.

Morlan, Richard E.
1978   Early Man in the Northern Yukon Territories: Perspective as of 1977. In *Early Man in America from a Circum-Pacific Perspective*, edited by Alan Bryan, pp. 78–95. Occasional Paper, Department of the Department of Anthropology, University of Alberta, No. 1. Archaeological Researchers International, Edmonton.
1983   Spiral Fractures on Limb Bones: Which Ones Are Artificial? In *Carnivores, Human Scavengers and Predators: A Question of Bone Technology*, pp. 241–270. Proceedings of the 15th Chacmool Symposium, Archaeological Association, University of Calgary, Calgary.
1984   Problems of Interpreting Modified Bone from the Old Crow Basin, Yukon Territory. *Abstracts*, First International Conference on Bone Modification, p. 26. Carson City, Nevada.

Morlan, Richard E., and John V. Matthews, Jr.
1978   New Dates for Early Man. *Geos* Winter: 2–5.

Morley, Sylvanus G.
1946   *The Ancient Maya*. Stanford University Press, Stanford.

Moseley, Michael E.
1975   *The Maritime Foundations of Andean Civilization*. Cummings, Menlo Park, California.

Mundkur, Balaji
1976   The Cult of the Serpent in the Americas: Its Asian Background. *Current Anthropology* 17:429–455.

Murdock, George Peter
1949   *Social Structure*. MacMillan Publishing Company, New York.

Nagel, Ernest
1961   *The Structure of Science: Problems in the Logic of Scientific Explanation*. Harcourt, Brace and World, New York.

Newton-Smith, W. H.
1981   *The Rationality of Science*. Routledge and Paul Kegan, Boston, London, and Henley.

Oakley, Ann
1974   *The Sociology of Housework*. Martin Robertson, London.

Omer, I. A.
1970   On the D-N Model of Scientific Explanation. *Philosophy of Science* 37:417–433.

Orme, Byrony
1973   Archaeology and Ethnography. In *Explanation of Culture Change: Models in Prehistory*, edited by Colin Renfrew, pp. 481–492. Duckworth, London.

Osborn, Alan J.
1977   Strandloopers, Mermaids, and Other Fairy Tales: Ecological Determinants of Marine Resource Utilization—The Peruvian Case. In *For Theory Building in Archaeology*, edited by Lewis R. Binford, pp. 157–205. Academic Press, New York.

Passmore, John
1967   Logical Positivism. *The Encyclopedia of Philosophy*, Vol. 5, pp. 52–57. Editor in chief, Paul Edwards. MacMillan Publishing Company and The Free Press, New York.

Paynter, Robert
1983   Field or Factory? Concerning the Degradation of Archaeological Labor. In The Socio-Politics of Archaeology, edited by Joan M. Gero, David M. Lacy, and Michael L. Blakey, pp. 17–30. *Research Report Number* 23, Department of Anthropology, University of Massachusetts, Amherst.

Platt, J. R.
1964   Strong Inference. *Science* 146:347–353.

Plog, Fred T.
1974   *The Study of Prehistoric Change*. Academic Press, New York.
1975   Systems Theory in Archaeological Research. *Annual Review of Anthropology*, Vol. 4, pp. 207–224. Annual Reviews, Palo Alto.
1979   Alternative Models of Prehistoric Change. In *Transformations, Mathematical Approaches to Culture Change*, edited by Colin Renfrew and Kenneth L. Cooke, pp. 221–236. Academic Press, New York.
1982   Is a Little Philosophy (Science) a Dangerous Thing? In *Theory and Explanation in Archaeology*, edited by Colin Renfrew, Michael J. Rowlands, and Barbara Abbott Segraves, pp. 25–34. Academic Press, New York.

Plog, Stephen
1978   Social Interaction and Stylistic Similarity: A Reanalysis. In *Advances in Archaeological Method and Theory*, Vol. 1, edited by Michael E. Schiffer, pp. 143–182. Academic Press, New York.
1980   *Stylistic Variation in Prehistoric Ceramics*. Cambridge University Press, Cambridge.

Polach, H. A., and J. Golson
1966   Collection of Specimens for Radiocarbon Dating and Interpretation of Results. *Australian Institute of Aboriginal Studies Manual* No. 2. Australia National University, Canberra.

Popper, Karl
1935   *Logik der Forschung.* Springer, Vienna, Austria.
1950   *The Open Society and its Enemies.* Revised Edition. Princeton University Press, Princeton.
1961   *The Logic of Scientific Discovery.* Science Editions, New York.
1962   *Conjectures and Refutations: The Growth of Scientific Knowledge.* Basic Books, New York and Routledge and Kegan Paul, London.
1976   *Unended Quest: An Intellectual Autobiography.* Open Court Publishing Company, LaSalle, Illinois.

Price, Derek J. De Solla
1963   *Little Science, Big Science.* Columbia University Press, New York.

Price, William J., and Lawrence W. Bass
1969   Scientific Research and the Innovative Process. *Science* 164:802–806.

Putnam, H.
1962   What Theories are Not. In *Logic, Methodology and Philosophy of Science.* Proceedings of the 1960 International Congress, edited by E. Nagel, P. Suppes, and A. Tarski, pp. 240–251. Stanford University Press, Stanford.
1975   *Mathematics, Matter and Method,* Vol. 1. Cambridge University Press, Cambridge.
1978   *Meaning and the Moral Sciences.* Routledge and Kegan Paul, London.
1979   Reflections on Goodman's *Ways of Worldmaking. Journal of Philosophy* 76:603–618.

Quinton, Anthony
1967   Karl Raimund Popper. *The Encyclopedia of Philosophy,* Vol. 6, pp. 398–401. MacMillan Publishing Company and The Free Press, New York.

Raab, L. Mark
1984   ASCA and a Living Conservation Ethic. *Proceedings 1984,* edited by William J. Mayer-Oakes and Alice W. Portnoy, pp. 16–20. American Society for Conservation Archaeology.

Raab, L. Mark, and Albert C. Goodyear
1984   Middle Range Theory in Archaeology: A Critical Review of Origins and Applications. *American Antiquity* 49(2):255–268.

Ralph, Elizabeth K.
1971   Carbon-14 Dating. In *Dating Techniques for the Archaeologist,* ed-

ited by Henry N. Michael and Elizabeth K. Ralph, pp. 1–48. The M.I.T. Press, Cambridge.

Rathje, William L.
1973   Classic Maya Development and Denouement: A Research Design. In *The Classic Maya Collapse*, edited by T. Patrick Culbert. pp. 405–454. University of New Mexico Press, Albuquerque.

Raymond, J. Scott
1981   The Maritime Foundations of Andean Civilization: A Reexamination of the Evidence. *American Antiquity* 46(4):806–821.

Read, Dwight W., and Steven Le Blanc
1978   Descriptive Statements, Covering Laws, and Theories in Archaeology. *Current Anthropology* 19:307–335.

Redman, Charles L., and Patty Jo Watson
1970   Systemic, Intensive Surface Collection. *American Antiquity* 35:279–291.

Reeves, B. O. K.
1983   Pleistocene Archaeology in San Diego: Recent Research. Paper presented at the 11th International Congress for Anthropological and Ethnological Sciences, Vancouver.

Reichenbach, Hans
1938   *Experience and Prediction*. University of Chicago Press, Chicago.
1949   *The Theory of Probability*. University of California Press, Berkeley.

Reid, Thomas
1969   *Essays on the Active Powers of the Mind*. The M.I.T. Press, Cambridge, Massachusetts, and London, England.

Renfrew, Colin
1976   *Before Civilization: The Radiocarbon Revolution and Prehistoric Europe*. Penguin Books, Harmondsworth, England.
1983   Divided We Stand: Aspects of Archaeology and Information. *American Antiquity* 48(1):3–16.

Renfrew, Colin, Michael J. Rowlands, and Barbara Abbott Segraves (editors)
1982   *Theory and Explanation in Archaeology*. Academic Press, New York.

Richardson, James B., III
1981   Modeling the Development of Sedentary Maritime Economies on the Coast of Peru: A Preliminary Statement. *Annals of the Carnegie Museum* 50, pp. 139–150. Carnegie Museum of Natural History, Pittsburgh.

Roberts, Frank H. H.
1940   Developments in the Problem of the North American Paleo-Indian. In *Essays in Historical Anthropology in North America*.

Smithsonian Miscellaneous Collections, Vol. 100, pp. 51–116. Smithsonian Institution, Washington, D.C.

Roe, Anne

1953   A Psychological Study of Eminent Psychologists and Anthropologists, and a Comparison with Biological and Physical Scientists. *Psychological Monographs: General and Applied* 67:1–55.

1972   Patterns in Productivity of Scientists. *Science* 176:940–941.

Rogge, A. E.

1975   Processual Archaeology and the Philosophy of Science: Three Problems. Paper delivered at the Society for American Archaeology Annual Meeting, Dallas.

1976   A Look at Academic Anthropology: Through a Graph Darkly. *American Anthropologist* 78:829–843.

Rorty, Richard

1979   *Philosophy and the Mirror of Nature.* Princeton University Press, Princeton.

1982   *Consequences of Pragmatism.* University of Minnesota Press, Minneapolis.

Rowe, John H.

1965   The Renaissance Foundations of Anthropology. *American Anthropologist* 67(1):1–20.

1966   Diffusionism and Archaeology. *American Antiquity* 31:334–337.

Sabloff, Jeremy A., T. W. Beale, and A. M. Kurland, Jr.

1973   Recent Developments in Archaeology. *Annals of the American Academy of Political and Social Science* 408:103–118.

Sahlins, Marshall

1972   *Stone Age Economics.* Aldine-Atherton, Chicago and New York.

Salmon, Merrilee H.

1976   "Deductive" versus "Inductive" Archaeology. *American Antiquity* 44:376–380.

1978   What Can Systems Theory do for Archaeology? *American Antiquity* 43:174–183.

1982   *Philosophy and Archaeology.* Academic Press, New York.

Salmon, Merrilee H., and Wesley C. Salmon

1979   Alternative Models of Scientific Explanation. *American Anthropologist* 81:61–74.

Salmon, Wesley C.

1966a   Verifiability and Logic. In *Mind, Matter and Method: Essays in Philosophy of Science in Honor of Herbert Feigl,* edited by Paul K. Feyerabend and Grover Maxwell, pp. 354–376. University of Minnesota Press, Minneapolis.

1966b   *The Foundations of Scientific Inference.* University of Pittsburgh Press, Pittsburgh.

1971   *Statistical Explanation and Statistical Relevance.* With contributions by Richard C. Jeffrey and James G. Greeno. University of Pittsburgh Press, Pittsburgh.

1973   *Logic.* Second Edition. Prentice Hall, Englewood Cliffs, New Jersey.

1975a   Theoretical Explanation. In *Explanation*, edited by Stephen Korner, pp. 118–145. Basil Blackwell, Oxford.

1975b   Confirmation and Relevance. *Minnesota Studies in the Philosophy of Science*, Vol. 6, edited by G. Maxwell and R. Anderson, pp. 3–36. University of Minnesota Press, Minneapolis.

1983   The Formulation of Why Questions. Paper presented at the 48th Annual Meetings of the Society for American Archaeology, Pittsburgh.

Sanders, William T.

1962   Cultural Ecology of the Maya Lowlands, Part 1. *Estudios de Cultura Maya* 2:79–121.

1963   Cultural Ecology of the Maya Lowlands, Part 2. *Estudios de Cultura Maya* 3:203–241.

1973   The Cultural Ecology of the Lowland Maya: A Reevaluation. In *The Classic Maya Collapse*, edited by T. Patrick Culbert, pp. 325–365. A School of American Research Book. University of New Mexico Press, Albuquerque.

Sanders, William T., and Barbara J. Price

1968   *Mesoamerica: The Evolution of a Civilization.* Random House, New York.

Sanjek, Roger

1978   The Position of Women in Major Departments of Anthropology, 1967–76. *American Anthropologist* 80(4):894–904.

Sartorius, Rolf

1966   The Concept of Law. *Archives for Philosophy of Law and Social Philosophy* 52:161–193.

1975   *Individual Conduct and Social Norms.* Dickenson Publishing Company, Encino and Belmont, California.

Saul, Frank P.

1973   Disease in the Maya Area: The Pre-Columbian Evidence. In *The Classic Maya Collapse*, edited by T. Patrick Culbert, pp. 301–324. University of New Mexico Press, Albuquerque.

Saunders, Jeffrey J.

1977   Lehner Ranch Revisited. In *Paleoindian Lifeways*. The Museum Journal, Vol. 17, edited by Eileen Johnson, pp. 48–64. The West Texas Museum Association, Lubbock.

1978   *Mammuthus* in the Clovis Horizon of North America. *Abstracts of* the Fifth Biennial Meeting, American Quaternary Association, pp. 99–103. University of Alberta, Edmonton.

Scheffler, Israel
1963   The Anatomy of Inquiry. Alfred A. Knopf, New York.
1967   Science and Subjectivity. The Bobbs-Merrill Company, Indianapolis.
1970   The Fictionalist View of Scientific Theories. In Readings in the Philosophy of Science, edited by Baruch A. Brody, pp. 211– 222. Prentice-Hall, Englewood Cliffs, New Jersey.

Schiffer, Michael B.
1976   Behavioral Archaeology. Academic Press, New York.
1981   Some Issues in the Philosophy of Science. American Antiquity 46:899–908.

Schiffer, Michael B., and John H. House
1977   Cultural Resource Management and Archaeological Research: The Cache Project. Current Anthropology 18(1):43–68.

Schlick, Moritz
1949   Meaning and Verification. In Readings in Philosophical Analysis. edited by Herbert Feigl and Wilfrid Sellers, pp. 146–170. Appleton-Century-Crofts, New York.

Schneider, Harold K.
1977   Prehistoric Transpacific Contact and the Theory of Culture Change. American Anthropologist 79:9–25.

Schwartz, Douglas
1970   The Postmigration Culture: A Base for Archaeological Inference. In Reconstructing Prehistoric Pueblo Societies, edited by William A. Longacre, pp. 174–193. University of New Mexico Press, Albuquerque.

Scriven, Michael
1958   Definitions, Explanations, and Theories. Minnesota Studies in the Philosophy of Science, Vol. 2, edited by H. Feigl, M. Scriven, and G. Maxwell, pp. 99–195. University of Minnesota Press, Minneapolis.
1959a   Truisms as the Grounds for Historical Explanations. In Theories of History, edited by P. Gardiner, pp. 443–475. The Free Press, New York.
1959b   Explanation and Prediction in Evolutionary Theory. Science 130:477–482.
1962   Explanations, Predictions, and Laws. Minnesota Studies in the Philosophy, Vol. 3, edited by H. Feigl and G. Maxwell, pp. 170–230. University of Minnesota Press, Minneapolis.
1963   The Temporal Asymmetry between Explanations and Predic-

tions. *Philosophy of Science: The Delaware Seminar*, Vol. 1, pp. 97–105. John Wiley and Sons, New York.

Sellards, E. H.
1952   *Early Man in America: A Study in Prehistory*. University of Texas Press, Austin.

Sellars, Mary
1973   The Secret Notebook for the Practicing Archaeologist: With Preliminary Notes Toward an Ethno-science of Archaeology. *Plains Anthropologist* 18:140–148.

Sellars, Wilfrid
1970   The Language of Theories. In *Readings in the Philosophy of Science*, edited by Baruch A. Brody, pp. 343–353. Prentice-Hall, Englewood Cliffs, New Jersey.

Semenov, S. A.
1964   *Prehistoric Technology*. Translated by M. W. Thompson. Cosy, Adams, and Mackay, London.

Service, Elman R.
1969   Models for the Methodology of Mouthtalk. *Southwestern Journal of Anthropology* 25:68–80.

Shapere, D.
1969   Notes toward a Post-Positivistic Interpretation of Science. In *The Legacy of Logical Positivism*, edited by P. Achinstein and S. F. Barker, pp. 115–160. Johns Hopkins University Press, Baltimore.

Shepard, Anna O.
1936   The Technology of Pecos Pottery. In *The Pottery of Pecos*, Vol. 2. Papers of the Southwestern Expedition 7. Phillips Academy, Andover, and Yale University Press, New Haven.

Sikes, S. K.
1971   *The Natural History of the African Elephant*. Weidenfeld and Nicholson, London.

Skyrms, Brian
1975   *Choice and Chance*. Second Edition. Dickenson Publishing Company, Encino, California.

Smith, Marion A.
1955   The Limitations of Inference in Archaeology. *Archaeological Newsletter* 6:3–7.

Smith, Watson
1952   Excavations in the Big Hawk Valley. *Museum of Northern Arizona Bulletin* 24. Northern Arizona Society of Science and Art, Flagstaff.

Soffer, Olga
1983   Politics of the Paleolithic in the USSR: A Case of Paradigms Lost.

In The Socio-Politics of Archaeology, ed. by Joan M. Gero, David
M. Lacy, and Michael L. Blake, pp. 91–105. *Research Report
Number* 23. Department of Anthropology, University of Massachusetts, Amherst.

South, Stanley
1977   *Method and Theory in Historical Archaeology.* Academic Press, New
York.

Sneed, J.
1971   *The Logical Structure of Mathematical Physics* D. Reidel Publishing
Company, Dordrecht, Holland.

Spaulding, Albert C.
1953   Review of *Measurements of Some Prehistoric Design Developments in
the Southeastern United States*, by James A. Ford. *American Anthropologist* 55(4):588–591.
1968   Explanation in Archaeology. In *New Perspectives in Archaeology*,
edited by S. R. and L. R. Binford, pp. 33–40. Aldine Publishing
Company, Chicago.
1973   Archaeology in the Active Voice: The New Archaeology. In
Research and Theory in Current Archaeology, edited by Charles L.
Redman, pp. 337–354. John Wiley and Sons, New York.

Speth, John
1983   Review of *Advances in Archaeological Method and Theory*, Vol. 5,
edited by Michael J. Schiffer. *American Antiquity* 48(3):649–651.

Stanford, Dennis, Robson Bonnichsen, and Richard E. Morlan
1981   The Ginsberg Experiment: Modern and Prehistoric Evidence of a
Bone Flaking Technology. *Science* 212:438–439.

Stegmuller, W.
1976   *The Structure and Dynamics of Theories.* Springer-Verlag, New
York.

Stevens, S. S.
1935a   The Operational Basis of Psychology. *American Journal of Psychology* 47:323–330.
1935b   The Operational Definition of Psychological Concepts. *Psychological Review* 42:517–527.
1936   Psychology: The Propaedeutic Science. *Philosophy of Science* 3:90–103.
1945   Symposium on Operationism. *Psychological Review* 52:241–294.

Stewart, Joe D.
1979   The Formal Definitions of Decorative Traditions in the Jornada
Area: A Case Study of Lincoln Black-on-Red Design. In *Jornada
Mogollan Archaeology*, edited by Patrick H. Beckett and Regge N.
Wiseman, pp. 295–344. New Mexico State University, Las
Cruces.

Stickel, E. Gary
1979   More on Theory Building in Archaeology. *Current Anthropology*
        20(3):621–624.

Struever, Stuart
1968   Problems, Methods and Organization: A Disparity in the
        Growth of Archaeology. *Anthropological Archaeology in the Amer-*
        icas, pp. 131–133. The Anthropological Society of Washington,
        Washington, D.C.

Suppe, Frederick (editor)
1977   *The Structure of Scientific Theories*. University of Illinois Press,
        Urbana and Chicago.

Swallow, D. A.
1974   The Anthropologist as Subject. *Cambridge Anthropologist* 1:51–60.

Tainter, Joseph A., and David A. Gellio
1980   *Cultural Resources Overview: Mt. Taylor Area, New Mexico*. U.S.
        Government Printing Office, Washington D.C.

Tamplin, Morgan
1981   Development Archaeology: Educating or Exploiting the Third
        World? Paper delivered at the 14th Annual Conference of the
        Archaeology Association of the University of Calgary, Calgary.

Taylor, R.E.
1985   The Beginnings of Radiocarbon Dating in *American Antiquity:* A
        Historical Perspective. *American Antiquity* 50(2): 309–325.

Taylor, Walter W.
1948   *A Study of Archaeology*. Memoir 69 of the American Anthropolog-
        ical Association.
1954   Southwestern Archaeology: Its History and Theory. *American*
        Anthropologist 56 (4):561–575.

Teggart, J. F.
1916   Prolegomena to History: The Relation of History to Literature,
        Philosophy, and Science. *Publications in History* 4(3). University
        of California Press, Berkeley.
1918   *The Processes of History*. Yale University Press, New Haven.
1919   Anthropology and History. *Journal of Philosophy, Psychology and*
        Scientific Methods 16:691–696.
1925   *The Theory of History*. Yale University Press, New Haven.

Thagard, Paul R.
1978   The Best Explanation: Criteria for Theory Choice. *The Journal of*
        Philosophy 75:76–92.

Thomas, David
1979   *Naturalism and Social Science, A Post-Empiricist Philosophy of Social*
        Science. Cambridge University Press, Cambridge.

Thomas, David Hurst

1974   *Predicting the Past: An Introduction to Anthropological Archaeology.* Holt, Rinehart and Winston, New York.

Thompson, Raymond

1956   The Subjective Element in Archaeology. *Southwestern Journal of Anthropology* 12:327–332. Reprinted in 1971 in *Man's Imprint from the Past*, edited by James Deetz, pp. 148–154. Little, Brown and Company, Boston.

1958a   Preface. Migrations in New World Culture History. *University of Arizona Social Science Bulletin* no. 27, edited by Raymond H. Thompson, pp. v–vii. University of Arizona Press, Tucson.

1958b   Modern Yucatecan Maya Pottery Making. *Memoir of the Society for American Archaeology* No. 15.

Toll, H. Wolcott, Thomas C. Windes, and Peter J. McKenna

1980   Late Ceramic Patterns in Chaco Canyon: The Pragmatics of Modeling Ceramic Exchange. In *Models and Methods in Regional Exchange*. SAA Papers no. 1, pp. 95–117. Society for American Archaeology.

Tolstoy, Paul

1963   Cultural Parallels between Southeast Asia and Meso-America in the Manufacture of Bark Cloth. In *Transactions of the New York Academy of Sciences*, Series 2, 25, pp. 646–662.

1966   Method in Long Range Comparison. *Actas del 36 Congreso de Americanistas*, Vol. 1, pp. 69–89. Seville, Spain.

1972   Diffusion as Explanation and as Event. In *Early Chinese Art and Its Possible Influence in the Pacific Basin*, edited by N. Bernard, pp. 823–841. Intercultural Arts Press, New York.

Toulmin, Stephen

1953   *The Philosophy of Science: An Introduction.* Hutchinson, London.

1958   *The Uses of Argument.* Cambridge University Press, Cambridge.

1961   *Foresight and Understanding.* Hutchinson, London.

1970   Does the Distinction Between Normal and Revolutionary Science Hold Water? In *Criticism and the Growth of Knowledge*, edited by Irme Lakatos and Alan Musgrave, pp. 39–47. Cambridge University Press, Cambridge.

1972   *Human Understanding*, Vol. 1. Princeton University Press, Princeton.

Tribe, Laurence H.

1970–71 Trial by Mathematics: Precision and Ritual in the Legal Process. *Harvard Law Review* 84:1329–1393.

Trigger, Bruce G.

1980   Archaeology and the Image of the North American Indian. *American Antiquity* 45:662–676.

Tuggle, H. D., A. H. Townsend, and T. J. Riley
1972   Laws, Systems and Research Design: A Discussion of Explanation in Archaeology. *American Antiquity* 37:3–12.

Tylor, E. B.
1878   On the Game of Patolli in Ancient Mexico, and its Probable Asiatic Origin. *Journal of the Anthropological Institute of Great Britain and Ireland* Vol. 8, pp. 116–121.
1896   On American Lot-Games as Evidence of Asiatic Intercourse Before the Time of Columbus. *Internationales Archiv fur Ethnographie* 9: 55–67.

Vance, Carole
1975   Sexual Stratification in Academic Anthropology, 1974–75. *Anthropology Newsletter* 16(4):10–12.

van der Merwe, Nikolaas J.
1982   Carbon Isotopes, Photosynthesis and Archaeology. *American Scientist* 70:596–606.

van Fraassen, B. C.
1980   *The Scientific Image*. Clarendon Press, Oxford.

von Bertalanffy, Ludwig
1972   The History and Status of General Systems Theory. In *Trends in General Systems Theory*, edited by George J. Klir, pp. 21–41. Wiley-Interscience, New York.

von Mises, R.
1939   *Probability, Statistics and Truth*. William Hodge and Company, London.
1951   *Positivism*. Dover Publications, New York.

von Neumann, John, and Oskar Morgenstern
1944   *Theory and Games and Economic Behavior*. First Edition. Princeton University Press, Princeton.

Warren, A. Helene
1967   Petrographic Analysis of Pottery and Lithics. In *An Archaeological Survey of the Chuska Valley and the Chaco Plateau., New Mexico*, edited by Arthur H. Harris, James Schoenwetter, and A. H. Warren. *Museum of New Mexico. Research Records* 4:104–134.
1969   Tonque: One Pueblo's Glaze Pottery Industry Dominated Middle Rio Grande Commerce. *El Palacio* 76:36–42.
1970   Notes on the Manufacture and Trade of Rio Grande Glazes. *The Artifact* 8:1–7.

Wartofsky, M. W.
1976   The Relation between Philosophy of Science and History of Science. *Boston Studies in the Philosophy of Science* 39:717–737.

Washburn, Dorothy Kostes
1977   A Symmetry Analysis of Upper Gila Ceramic Design. *Papers of the Peabody Museum of Archaeology and Ethnology*, Vol. 68. Harvard University Press, Cambridge.

Wasserstrom, Richard A.
1961   *The Judicial Decision*. Stanford University Press, Stanford.

Watkins, John
1970   Against "Normal Science." In *Criticism and the Growth of Knowledge*, edited by Irme Lakatos and Alan Musgrave, pp. 25–37. Cambridge University Press, Cambridge.

Watson, J. B.
1924   *Psychology from the Standpoint of a Behaviourist*. J. B. Lippincott Company, Philadelphia.

Watson, Patty Jo
1979   The Idea of Ethnoarchaeology: Notes and Comments. In *Ethnoarchaeology: Implications of Ethnography in Archaeology*, edited by Carol Kramer, pp. 277–287. Columbia University Press, New York.

Watson, Patty Jo, and Alison Wylie
1983   The Prospects for Philosophical Analysis in Archaeology. Paper presented at the 48th Annual Meeting of the Society for American Archaeology, Pittsburgh.

Watson, Patty Jo, Stephen A. LeBlanc, and Charles L. Redman
1971   *Explanation in Archaeology: An Explicitly Scientific Approach*. Columbia University Press, New York and London.
1984   *Archaeological Explanation*. Columbia University Press, New York.

Watson, Richard A.
1976   Inference in Archaeology. *American Antiquity* 41:58–66.

Webster, David
1977   Warfare and the Evolution of Maya Civilization. In *The Origins of Maya Civilization*, edited by Richard E. W. Adams, pp. 335–372. University of New Mexico Press, Albuquerque.

Wendorf, Fred
1979   Changing Values in Archaeology. *American Antiquity* 44:641–643.

Westfall, Richard S.
1971   *Force in Newton's Physics: The Science of Dynamics in the Seventeenth Century*. MacDonald, London.

White, J. Peter
1977   Crude, Colourless and Unenterprising? Prehistorians and Their Views on the Stone Age of Sunda and Sahul. In *Sunda and Sahul*,

edited by J. Allen, J Golson and R. Jones, pp. 13–30. Academic Press, New York.

Whitehead, Alfred North
1953   *Science and the Modern World.* Cambridge University Press, Cambridge.

Wildesen, Leslie E.
1980   The Status of Women in Archaeology: Results of a Preliminary Survey. *Anthropology Newsletter* 21(5):5–7.
1984   The *Really* New Archaeology: A Decade of Progress. *Proceedings 1984,* edited by William J. Mayer-Oakes and Alice W. Portnoy, pp. 60–62. American Society for Conservation Archaeology.

Willey, Gordon R., (editor)
1974   *Archaeological Researches in Retrospect.* Winthrop Publishers, Cambridge.

Willey, Gordon R., and Jeremy A. Sabloff
1974   *A History of American Archaeology.* Thames and Hudson, London.

Willey, Gordon R., and Demitri B. Shimkin
1973   The Maya Collapse: A Summary View. In *The Classic Maya Collapse,* edited by T. Patrick Culbert, pp. 457–501. University of New Mexico Press, Albuquerque.

Williams, Barbara
1981   *Breakthrough: Women in Archaeology.* Walker and Company, New York.

Wilmsen, Edwin
1970   Review of *Man in Prehistory,* by Chester S. Chard. *American Anthropologist* 72(3):689–691.

Wilson, David
1975   *The New Archaeology.* Alfred A. Knopf, New York.

Winter, W. H.
1964   Elephant Behavior. *East African Wildlife Journal* 2:163–164.

Wittgenstein, Ludwig
1968   *Philosophical Investigations.* Third Edition. Translated by G.E.M. Anscombe. Basil Blackwell, Oxford.

Wobst, H. Martin, and Arthur S. Keene
1983   Archaeological Explanation as Political Economy. In The Socio-Politics of Archaeology, edited by Joan M. Gero, David M. Lacy, and Michael L. Blakey, pp. 79–89. *Research Report Number 23.* Department of Anthropology, University of Massachusetts, Amherst.

Woodall, J. Ned, and Phillip J. Perricone
1981   The Archaeologist as Cowboy: The Consequences of Professional Stereotype. *Journal of Field Archaeology* 8(4):506–508.

Woodbury, Richard B.

1973a   *Alfred V. Kidder*. Columbia University Press, New York.

1973b   Getting Round Archaeologists out of Square Holes. In *Research and Theory in Current Archaeology*, edited by Charles L. Redman, pp. 311–317. John Wiley and Sons, New York.

1985   Regional Archaeological Conferences. *American Antiquity* 50(2):434–444.

Woolley, Sir Leonard

1930   *Digging Up the Past*. Ernest Benn, London. Reprinted in Pelican Books, 1937 and 1960.

Wylie, Alison

1980   Analogical Inference in Archaeology. Paper presented at the 46th Annual Meeting of the Society for American Archaeology, Philadelphia.

1981   Positivism and the New Archaeology. Ph.D. dissertation, Department of Philosophy, State University of New York, Binghamton.

1982a   An Analogy by Any Other Name is Just as Analogical: A Commentary on the Gould-Watson Debate. *Journal for Anthropological Archaeology* 1:382–401.

1982b   The Prospects for Philosophical Analysis in Archaeology. Proposal for a Symposium at the 48th Annual Meeting of the Society for American Archaeology, Pittsburgh.

1985a   The Reaction Against Analogy. In *Advances in Archaeological Method and Theory*, Vol. 8, edited by Michael B. Schiffer, pp. 63–111. Academic Press, New York.

1985b   Between Philosophy and Archaeology. *American Antiquity* 50(2):478–490.

/ Permission Acknowledgments

Permission to quote from the following copyrighted materials is gratefully acknowledged:

Achinstein, Peter. Concepts of Science. Batimore: The Johns Hopkins
University Press. 1968. Quotes from pp. 102. 102–103 by permission of
the Johns Hopkins University Press. *Achinstein, Peter.* Law and Explanation. London: Oxford University Press. 1971. Quotes from pp. 97 and 110
by permission of Oxford University Press. *Binford, Lewis R.* General
Introduction. *In* for Theory Building in Archaeology, Lewis R. Binford,
ed. Orlando: Academic Press. Quotes from pp. 1, 2, 4–5, 5–6, 7 by
permission of Academic Press and the author. *Braithwaite, R. B.* Scientific
Explanation. Cambridge University Press. 1953. Quotes from pp. 319,
320, 321, 323 by permission of Cambridge University Press. *Collingwood,
R. G.* The Idea of History. Oxford: The Clarendon Press. Quotes from
pp. 214–215 by permission of Oxford University Press. *Erasmus, Charles J.*
Patolli, Pachisi, and the Limitation of Possibilities. Southwestern Journal
of Anthropology 6:369–387. 1950. Quotes from pp. 369–370, 372, 374,
382–387 by permission of the Journal of Anthropological Research.
Dumond, Don E. Science as Archaeology: The Saints Go Marching In.
American Antiquity, 42:330–349. 1977. Quotes from pp. 345–347 re-
produced by permission of the Society for American Archaeology and the
author from Science in Archaeology: The Saints Go Marching In by
Don E. Dumond. Vol. 42, 1977. *Feyerabend, Paul K.* Against Method:
Outline of an Anarchistic Theory of Knowledge. Atlantic Highlands,
New York: Humanities Press. Quotes from pp. 23, 27 by permission of
Verso/NLB. *Giere, Ronald N.* Understanding Scientific Reasoning. New
York: Holt, Rinehart and Winston. 1979. Quotes from p. 152 from UN-
DERSTANDING SCIENTIFIC REASONING by Ronald N. Giere. Copy-

right © 1979 by Holt Rinehart and Winston. Reprinted by permission of CBS College Publishing. *Grunbaum, Adolf.* Is Falsifiability the Touchstone of Scientific Rationality? Karl Popper versus inductivism. *In* Essays in Memory of Imre Lakatos. Boston Studies in the Philosophy of Science 39. R. S. Cohen and M. W. Wartofsky, eds. Dordrecht: D. Reidel Publishing Company. 1976. Quotes from pp. 218, 219 by permission of D. Reidel Publishing Company. *Grunbaum, Adolf.* Is the Method of Bold Conjectures and Attempted Refutations Justifiably the Method of Science. The British Journal for the Philosophy of Science 27:105–136. Quote from p. 107 by permission of the British Journal for the Philosophy of Science. *Harre, Romano.* The Principles of Scientific Thinking. Chicago: University of Chicago Press. 1970. Quotes from pp. 3, 4, 34–35, 39, 47 by permission of the University of Chicago Press from The Principles of Scientific Thinking by Romano Harre. Copyright 1970 University of Chicago. All rights reserved. *Harre, Romano, and P. F. Secord.* The Explanation of Social Behavior. Oxford: Basil Blackwell. Quotes from pp. 3, 6, 7, 8, 12, 40 by permission from Basil Blackwell and Mott Limited. *Haury, Emil.* Evidence at Point of Pines for a Prehistoric Migration from Northern Arizona. *In* Migrations in New World Culture. University of Arizona Social Science Bulletin 27:1–6. Tucson: University of Arizona. Quotes from pp. 1, 6 by permission of University of Arizona Press. *Hempel, Carl G.* Aspects of Scientific Explanation. New York: The Free Press. 1965. Quotes from pp. 6, 13, 55, 60, 62, 268–269, 360–361, 366, 367–368, 369, 374–375, 383, 400–401, 425, 426, 428, 429, 438–439, 445, 447, 452, 454, 457 by permission of The Free Press, a Division of Macmillan, Inc., from ASPECTS OF SCIENTIFIC EXPLANATION by Carl G. Hempel. Copyright © 1965 by The Free Press. *Hesse, Mary B.* Models and Analogy in Science. 1966. Notre Dame, Indiana: University of Notre Dame Press. Quotes from pp. 2–3, 5 by permission of University of Notre Dame Press. MODELS AND ANALOGIES by Mary B. Hesse © 1966 by University of Notre Dame Press. *Jeffrey, R.* Statistical Explanation vs. Statistical Relevance. *In* Statistical Explanation and Statistical Relevance by Wesley Salmon. Pittsburgh: The University of Pittsburgh Press. 1971. Quotes from pp. 24, 27 by permission of the University of Pittsburgh Press. *Johnson, Leroy.* Problems in 'Avant Garde' Archaeology. American Anthropologist 74–366–377. 1973. Quotes from pp. 367, 368, 369, 370 reproduced by permission of the American Anthropological Association from AMERICAN ANTHROPOLOGIST 74 (4), August 1972. Not for further sale or reproduction. *Kuhn, Thomas J.* The Structure of Scientific Revolutions. Chicago: University of Chicago Press. 1970. Quotes from pp. 1–2, 5, 19–21, 24, 35, 56, 69, 89, 136, 153, 157, 175, 177–178, 180–182, 184–185, 187, 207, 237–238 by permission of the University of Chicago Press. The Structure of Scientific Revolutions copyright 1962, 1970 by the University of Chicago. All rights reserved.

Kuhn, Thomas J. Reflections on my Critics. *In* Criticism and the Growth of Knowledge. I. Lakatos and A. Musgrave, eds. Cambridge: Cambridge University Press. pp. 231–279. 1970. Quotes from pp. 247, 250, 253, 254–255, 261, 262, 264, 265, 266–267 by permission of Cambridge University Press. *Kuhn, Thomas J.* The Logic of Discovery or Psychology of Research. *In* Criticism and the Growth of Knowledge. I. Lakatos and A. Musgrave, eds. Cambridge: Cambridge University Press. Quotes from pp. 1, 2, 5, 6, 14, 15, 19, 22 by permission of Cambridge University Press. *Levin, Michael E.* On Explanation in Archaeology: A Rebuttal to Fritz and Plog. American Antiquity 38:387–395. 1973. Quotes from pp. 388, 391–394 reproduced by permission of the Society for American Archaeology and the author from On Explanation in Archaeology; A Reubttal to Fritz and Plog by Michael Levin. Vol. 38, 1973. *MacCormac, Earl R.* Metaphor and Myth in Science and Religion. Durham: Duke University Press. Copyright © 1976. Quotes from pp. 27, 34, 36, 37 by permission of Duke University Press. *Metaphor and Myth in Science and Religion* by Earl R. MacCormac Copyright © 1976 by Duke University Press. *Newton-Smith, W. H.* The Rationality of Science. Boston, London and Henley: Routledge and Paul Kegan. 1981. Quotes from pp. 237, 238, 250 by permission of Watson, Little Limited. *Rathje, William L.* Classic Maya Development and Denouement: A Research Design. *In* The Classic Maya Collapse. T. Patrick Culbert, ed. pp. 405–454. Albuquerque: University of New Mexico Press. Quotes from pp. 405, 406 by permission of the University of New Mexico Press. *Salmon, Wesley C.* The Foundations of Scientific Inference. 1966. Pittsburgh, The University of Pittsburgh Press. Quotes from pp. 64, 68, 79, 82, 83, 129 reprinted from THE FOUNDATIONS OF SCIENTIFIC INFERENCE by Wesley C. Salmon by permission of © 1967 by the University of Pittsburgh Press. © 1971 University of Pittsburgh Press. *Salmon, Wesley C.* Statistical Explanation and Statistical Relevance. 1971. Pittsburgh: University of Pittsburgh Press. Quotes from pp. 33, 36, 57, 60, 72, 77 reprinted from STATISTICAL EXPLANATION AND STATISTICAL RELEVANCE by Wesley C. Salmon by permission of the University of Pittsburgh Press. © 1971 by University of Pittsburgh Press. *Skryms, Brian.* Choice and Chance: An Introduction to Inductive Logic. Second Edition. 1975. Encino, California: Dickenson Publishing Company. Quotes from pp. 7, 8–9, 13–15, 40 of CHOICE AND CHANCE: AN INTRODUCTION TO INDUCTIVE LOGIC, Second Edition by Brian Skryms. © 1975 by Dickensen Publishing Company. Reprinted by permission of Wadsworth Inc. *Suppe, Frederick.* The Structure of Scientific Theories. 1977. Urbana and Chicago: The University of Illinois Press. Quotes from pp. 618, 636, 640, 649, 670, 672, 684, 702 by permission of the University of Illinois Press. *Thagard, Paul R.* The Best Explanation: Criteria for Theory Choice. The Journal of Philosophy LXXV/2:76–92 (February

1978). Quotes from pp. 77, 86, 87, 87–88, 89, 90 by permission of The Journal of Philosophy. *Van Fraasen, B. C.* The Scientific Image. 1980. Oxford, Clarendon Press. Quotes from pp. vii, 5, 18, 59, 64, 77, 88, 124, 128 by permission of Oxford University Press. *Watkins, John.* Against 'Normal Science'. *In* Criticism and the Growth of Knowledge. I. Lakatos and A. Musgrave, eds. 1970. Cambridge: Cambridge University Press. Quotes from pp. 27, 28, 31, 37 by permission of Cambridge University Press. *Willey, Gordon R., and Dimitri Shimkin.* The Maya Collapse: A Summary View. *In* The Classic Maya Collapse. T. Patrick Culbert, ed. 1973. Albuquerque: University of New Mexico Press. Quotes from pp. 489–490, 497 by permission of University of New Mexico Press.

/ Index

academic training, ix

acceptance: and confirmation, 246–56; criteria, 37, 39, 41, 237, 253, 255, 277, 290–98, 336–37

access to archaeological resources, difficulty of, 139

accretion and scientific progress, 83

Achinstein, Peter, 200–201; quoted, 53, 201, 220, 326

activism, political, 128, 145

Adams, Richard E.W., 319; quoted, 318, 321

adaptiveness and cultural change, 286–90

adequacy: concept, 22; conditions, 169, 194, 234–35, 238–39, 242, 253, 255, 352

affirming the consequent, 222

Against Method, 68

Agenbroad, Larry, 315

agencies, governmental, 143–49

Alaska Science Conference, 310

Alexander, Herbert L., 293–98

ambiguity, 49, 50

American Anthropological Association, 126, 127, 133, 279

American Anthropologist, 5

American Antiquity, 5, 153

American Bottoms Project, 148

analogical reasoning, 326–38

analogy, 12, 43, 49, 256–69, 290–98, 334–36; belongs to discovery, 263; historical, 263; proper use of, 332; testing, 269; use in archaeology, 4, 257, 263–66, 268–69, 338–43, 368–73, 378

analyses, philosophical, 5–6

anarchism, 162; theoretical, 95–96

animal behavior studies, 315, 342

anthropology, 67, 107; hiring practices, 127; model of explanation in, 201; recruitment base, 126; university programs, 126–27; use of subjective data, 206

Anthropology Newsletter, 126, 158

antiauthoritarianism, 161

antiquity of man, in New World, 113–14. *See also* Early Man debate antirealism, 215–16, 219, 270–71, 362–68, 370, 372–73

archaeological data: alternative accounts, 119; collection, 166, 317–26; difficulty of retrieval, 147, 277–78; generalization from, 220–24; lack of, 37, 232, 277–78, 306; new types, 312; ordering, 119; problems with, 317–26; representativeness, 37; tampering, 37

Archaeological Explanation, 358, 359–60

archaeological record, 37, 262

archaeological research: geographical patterns, 139, 314; grants, 135; group, 157; individual, 155; location selection, 123–24; multi-disciplinary, 309–17; problem choice, 103, 138, 150, 338

Archaeological Symposium of the University of Calgary on Bone Technology, 315

archaeological technicians, 150–51

archaeologists: career patterns, 129, 143; interaction with philosophers, 374–76; memoirs, 158; North American, 125–28; philosophical sophistication, 354–60; professional advancement, 38; technological sophistication, 150

archaeology: antitraditional, 109; borrows from other disciplines, 121; colonial, 137; "colonization" of, 44, 139; controversial theses in, 290–303, 309–17; cultural influences on, 123–43; decentralization, 143–44;

development, 25–26; as elitist occupation, 126; ethnographic, 108; expeditionary, 130; global spherical model, 141–43; goals, 100, 108, 146, 228; graduate programs, 132; historical, 108; and history, 232; institutional affiliations, 143–49; international, 137–43; and Kuhn-Popper debate, 94–97; laws in, 109, 174, 220–24; methodology, viii, 1, 4, 110, 276–79, 301, 344–47, 354; national, 137–38; national centers, 138–39; nature of, 113; philosophical concepts in, 23–28; recruitment base, 126; revolution in, 104; as science, 30–31; sociology of, 20; southwestern, 119; spherical model, 111–20, 290–98, 300–301, 308; theoretical foundations, 67, 256; traditional, 1, 104, 166; university programs, 131–33. *See also* New Archaeology

argument: by analogy, 44; from authority, 5, 55; *modus tollens*, 35

Ascher, Robert, 113; quoted, 265–66, 268

assumptions, assessing, 39, 278

attrition, role in paradigm cycle, 109

Aveleyra Arroyo de Anda, Luis, 117

Ayer, A.J., quoted, 13–14

Bacon, Francis, 53
Barnes, Barry, quoted, 106
Barrais, D.A., 10
Bass, Lawrence W., 151

Beebe, B., 315
Behavioral Archaeology, 109
behaviorism, 25, 201, 202–4, 205
belief, 160–61
Berman, Harold J., quoted, 371
bias: cultural, 123, 142–43, 290; disciplinary, ix, 39
Big Science vs. Little Science, 156, 159, 317
Binford, Lewis R., 3, 11, 23, 109, 116, 117, 315, 325, 329; quoted, 31, 33, 52, 263, 289, 355, 356–57
biology, 58, 304–5
Black, Max, 262
Blakey, Michael L., 123
Bonampak murals discovery, 181
bone technology, 309–17
Bonnichsen, Robson, 313, 331–38; quoted, 312, 331
borrowing, 2, 121–23, 151, 308, 350, 351
boundaries, disciplinary, 349–54
boundary conditions, 265–66
Boy Scout Syndrome, 129–30
Braithwaite, R.B., 170–72, 180; quoted, 247
burden of proof, 293, 295
butchering experiments, 314
Butler, Robert, 5, 62
Butterfield, Herbert, quoted, 66

Campbell, N.R., 260–61
Canada, archaeology in, 140
Canadian Archaeological Association, 144
Carnap, Rudolf, 7, 209, 233–34
case studies, 43, 109; archaeological, 275–79; of theory construction, 99
causation: concept, 249; and sci-

entific laws, 170; temporal asymmetry of, 197
ceramic analysis, 119–20
certainty, 47
Chaco phenomenon, 38
change: adaptive, 286–90; gradual, 115, 292; processes, 115, 289; revolutionary, 110; scientific, 75, 91–92
Chesterton, G.K., 360–61
China, archaeology in, 140
Cihuatan site, 362–68
citation counts, 136
Clarke, David, 108, 109, 378; quoted, 27, 113
Classic Maya Collapse, The, 323
classification, 182–87
Clovis culture, 326–43
cognitive significance, 10; archaeological problems of, 16; criteria of, 17, 96
Cohen, Jonathan, 243
Cohen, Mark, 181
Cohen, R.S., 4
collaboration, international, 140, 141
Collingwood, R.G., 190–92
Colombia, archaeology in, 140
commonality of interests, 157, 278
common ancestry, 302
communication: abbreviated, 112; among archaeologists, 152–59, 317; formal, 156–57; informal, 152–56, 326; patterns, 148
complexity: of hypotheses, 39; of philosophical theories, 351–52
complex phenomena, interpreting, 322
conclusions, false, 367–68
conditional probabilities, 306

DeBoer, Warren, 123, 293; quoted, 12
decentralization, 148
decision making, 295
deduction, 30, 33, 34, 39, 52–59, 377
deductive archaeology, 58
deductive-nomological model of explanation, 5, 17, 23, 26, 30, 96, 167–73, 223, 325, 352; alternatives to, 195–209; in archaeology, 276, 352; criticized, 175–95, 206–9, 233
deductive reasoning, 4, 14, 44–51
deductivism, 52–59, 79. *See also* hypothetico-deductive model
definitions, operational, 9
demarcation criterion, Popperian, 86
derivative law, 176
detectivism, 5
Dewey, John, 4
dichotomies, xi, 115, 308
diffusion, transoceanic, 298–309
diffusionism, 139, 279
Digging for Gold, 144
Dincauze, Dena, quoted, 144–45
"Directions in Archaeology: A Question of Goals," 144
disciplinary credentials, 375
disciplinary matrix, 64–65, 69–70, 92, 101
discovery-justification distinction, 34, 40–44, 165, 260, 356
discovery process, 30, 34, 40, 61, 82; archaeological, 43, 94, 377; influences on, 100
dispute resolution, 294
distinctive features, 332–33
dogma, scientific, 19, 95

dogmatism, 294
dominant theory, 63–64, 67, 71, 85, 92, 103
donor-recipient relationship, 120
Doyle, David E., 128
Dray, William, 182–95, 207
Duhem, Pierre, 7, 260; quoted, 261
Duhem-Campbell controversy, 260–61
Dumbarton Oaks Mini-Conferences, 155, 157
Dumond, Don, quoted, 67, 73, 239–41
Dunnell, Robert C., quoted, 52

Early Man debate, 290–98, 309–17. *See also* antiquity of man
ecology, 122, 359
economics, 7
economic transformation, 286–90
Edinburgh School, 20
Einstein, Albert, 7
elimination rule, 253
eliminative approach to confirmation, 36, 231
elitism, 132
Embree, Lester, 374–75
emic approach, in archaeology, 138
empirical adequacy, claims of, 215–16, 217, 270
empirical hypotheses, 228–29, 241
empiricism, 7, 21, 77, 202
Equivalence Condition of Adequacy, 238–39
Erasmuc, Charles, 303–9
Erkenntnis, 8
ethnoarchaeology, 39, 43, 202
ethnocentrism, 123

ethnographic analogies, 231, 262–63

Euro-centrism, 123, 139, 290

evaluation: of archaeological results, 116; of arguments, 51; of conflicting claims, 97, 298–309, 317–43; criteria for, 299–300; of explanations, 200; of test results, 39; of theories, 217

evidence: accumulation of, 334; nature of, 276, 298–303; selection of, 367; standards of, 113–14

evolutionary model of explanation, 213

evolutionary progress, 84

evolutionary theory, 108, 123

evolutionism. *See* cultural materialism

experimental controls, 116

experimentation, 35, 216–17, 309–17

explaining the evidence, 37

explanation, 2, 3, 21; in archaeology, 25, 26, 174, 219–24; as argument, 197–208; asymmetries of, 219; case history, 279–86; competing, 218; by concept, 182–87; conflated with testing, 54; deductive, 5, 24, 199; dispositional, 180–82; functional, 179, 300, 307; genetic, 179–89; ideographic, 283; inductive model, 26; models of, 359; partial, 196–97; pragmatic aspect, 187–88, 192, 193, 200, 207–8, 217; rational, 189–90, 192; recent models, 209–19; related to confirmation, 36, 222, 283; scientific, 22, 29, 40, 85; statistical models, 173–75, 233; symmetrical, 160–61; teleological, 171–72; theory, 251–52; as three-term relation, 219

explanation, D-N model. *See* deductive-nomological model of explanation

explanation, I-S model. *See* inductive-statistical model of explanation

Explanation in Archaeology, 27, 45

Explanation of Social Behaviour, The, 201

explanatory power, 56, 248, 361

explanatory relevance requirement, 169, 179

export of archaeological material, 140

Fact, Fiction and Forecast, 239

fact-law distinction, 296–97

facts, 2

fallacy of affirming the consequent, 35, 39

falsifiability, 17, 31, 35, 77, 78, 79–80, 83, 85, 86, 95

falsifiability criterion, 91

falsification, 265, 281–82

faunal collections, 314

Feigl, Herbert, 7

feminist scholarship, xi

Feyerabend, Paul K., 20, 68, 161, 210–11; quoted, 95, 96

field schools, 155

fieldwork, 129, 135, 155–56

firmness concept of confirmation, 233–36, 237, 242, 368

First International Conference on Bone Modification, 317

Flannery, Kent V., 23, 109, 117, 323, 374; quoted, 228
"follows from," 237
Folsom discovery, 114–15, 292
Food Crisis in Prehistory, The, 181–82
For Theory Building in Archaeology, 355–56
Ford, J., 294
Ford, Richard I., 124
foreign research, constraints on, 140–41
formalizability, assumption of, 17, 237, 351–52
Forman, Sylvia Helen, 134
formation processes, 312
Fowler, Don, quoted, 149
Freeman, Joan E., 10
Fritz, John M., 41, 46–47, 183; quoted, 54
Fulani Compound, 39
fundamental laws, 176
funding, research, 142

generalization, 45, 78; archaeological, 286–90; in history, 207; incorrect, 220–21; lawlike vs. accidental, 230–31; reason-giving, 184, 190, 192
general theory, 289–90, 356
genetic fallacy, 355
Gero, Joan M., 135–36
Giere, Ronald, 283; quoted, 328–29
Gifford, Carol, quoted, 155–56
Gladwin, Thomas, 123
Glock, Albert E., quoted, 124
Gödel, Kurt, 7
Goodman, Nelson, 239; quoted, 43, 230–31, 335, 372–73

Gould, Richard A., 265
Greiner, William R., quoted, 371
Griffith, Belver C., 153
Grünbaum, Adolf, quoted, 78, 80, 81
Guide to Departments of Anthropology, 132, 133

H-D method. See hypothetico-deductive model
Hacking, Ian, 353
Hanson, N.R., 42; quoted, 218
Hardy, Kenneth, 124–25
Harman, Gilbert, 329; quoted, 248–49
Harré, Rom, 201–6, 208; quoted, 18–19, 203, 204, 369–70
Harrington, C.R., 310–17
Harvard University, 133
Hasenstab, Robert J., quoted, 145–46
Haury, Emil, 186, 279–86
Haynes, Vance, 331–43
Helmholtz, Hermann, 7
Hempel, Carl, 2; controversial nature of his work, 29; criticism of his models of explanation, 180–95; influence on New Archaeology, 29; and logical positivism, 7, 18; models of explanation, 5, 165–66, 351–52; quoted, 8, 32, 34, 35–36, 49, 168–69, 172–73, 182–83, 187–88, 194, 195–96, 198, 229, 230, 259–60, 262, 266–67; study of confirmation, 226–27, 234; view of science, 83; work on discovery process, 41, 261; work on hypothesis testing, 32–36
Hempel-Dray controversy, 182–95

Hempel-Oppenheim model. *See* deductive-nomological model of explanation
Hesse, Mary, quoted, 261, 370
Hester, Thomas R., 148
Hill, James, 155
hiring practices, 134
historicists, 90
history, *sui generis* view, 190
history of archaeology, 158
history of science, 43, 67–69, 73, 83, 89–90, 91, 160, 212, 297
Holden, W.C., 129
Hole, Frank, 276
Holmes, W.H.H., 113
Hrdlicka, Ales, 113–14
Hull, D.L., quoted, 211
human behavior, explaining, 203–6
Hume, David, 7, 170
Hurlbert, Beverly McElligott, 132
hypercoherence, 323–24
hypotheses: alternative, 316; auxiliary, 79–80, 335; choice of, 81, 215–16, 317–26, 330; competing, 326–28, 361; confirmable, 248; formulation of, 4, 33, 34, 39, 40, 279–86, 338; higher order, 317–26; logically independent of data, 39; multiple, 39; multivariate, 324–26
hypothetico-deductive model, 23, 30, 31, 33, 42, 51, 55, 56, 77, 80, 85, 199, 220, 225, 358. *See also* deductivism

idealism, 7
identification of objects, 47
impartiality requirement, 295
importance, concept of, 295–96

incommensurability of theories, 69, 72–73, 76, 85, 87, 92
increase in firmness concept of confirmation, 233–37, 239, 242, 253
individualization, 191–92
induction, 4, 30, 44; combined with deduction, 39; distinguished from inductivism, 52–59; by elimination, 247; by enumeration, 55; role in hypothetico-deductive method, 33–34; role in traditional archaeology, 33, 44
inductive reasoning, 26, 44–51; in archaeology, 51, 225, 360–61
inductive-statistical (I-S) model of explanation, 173–75
inductivism, 52–59, 77, 79, 80
inductivism, narrow. *See* intuition
inference to the best explanation, 215, 252, 269–70, 276–77, 279–86, 301–2, 318–26, 329–38, 344, 360–68, 377
inference to the best of competing explanations, 249, 302, 329
inference: archaeological, 23, 30, 42, 94, 345, 367; eliminative, 325; establishing support for, 34; incompatible, 49; inductive, 81, 248, 254; inductive vs. deductive, 44–51, 174; relation to technology, 366–67; rules, 20, 41; warranted, 329–30
influence, individual, 117, 136
innovation, 121–22
instantiation condition, 78, 80
intellectual heritage, European, 123–24
intellectual territoriality, ix

interdisciplinarity, 104, 121, 159
interpretation, 86, 87
intuition, 19, 31, 54, 231
investigator bias, 127
invisible colleges, 105, 153–57
Irving, W.N., 310–17

Jacobson, Jerome, 137, 139;
    quoted, 138
Jarvie, I.C., quoted, 6
Jeffrey, R., quoted, 198–99, 207
Johnson, LeRoy, quoted, 220–21,
    228
joint authorship, 157
Journal of Unified Science, 8
journals, archaeological, 156
justification, 22, 30, 82, 159; in ar-
    chaeology, 94; distinguished
    from discovery, 34, 40–44, 165;
    procedures, 100

Kaplan, Bernice, 127
Keene, Arthur S., quoted, 136
Kelley, David, 155
Kelley, Jane, 362–68
Kidder, Alfred F., 158
King, Thomas F., quoted, 144
Klejn, Leo S., 11
Kluckhohn, Clyde, 4
Krieger, Alex, 117
Kroeber, Alfred L., 303–9
Kuhn, Thomas, 2; compared with
    Popper, 82–94; conservatism,
    115; criticized, 69–76, 103–4; in-
    fluence, 61–62; quoted, 64, 66–
    67, 68, 69, 71, 72, 74, 75, 82, 84,
    85, 86, 88, 102, 118, 122, 153–
    54, 330; relativism, 68, 84–85,
    106; sociology of science, 64,

101–4; view of science, 20, 42,
    63–69, 112, 119, 211–12
Kyburg, H., 177; quoted, 235

Lacy, David, quoted, 145–46
Lakatos, Imre, 212–13
language of science, 9, 261, 262
Lathraps, D.W., quoted, 12
Laughlin, William S., 140
Lavoisier, Antoine, 66
law, discipline of, 294–98, 345,
    371–72
lawlikeness, 167, 176, 227–33
lawlike statements, 230; in archae-
    ology, 231–33, 283; confirma-
    tion of, 252
laws, scientific, 11, 28, 226, 227–
    33; in archaeology, 227–30,
    286–90, 377; and causation, 170;
    distinguished from generaliza-
    tions, 227; importance in expla-
    nation, 200–1, 207, 226, 356;
    statistical vs. universal, 228–29
Leatherdale, W.H., quoted, 257
LeBlanc, Steven, 39, 359; quoted,
    31, 354
legalism, in archaeology, 293–94,
    345
legislation, archaeological, 143
Leone, Mark P., 107, 124; quoted,
    108, 143
Lesser, A., 4
Levi, Edward, quoted, 257–58,
    371
Levin, Michael, 46–47, 183;
    quoted, 41–42, 47, 184–85, 226
Levi Strauss, Claude, 302
limitation of possibilities princi-
    ple, 306–7
Lindsay, Alexander, Jr., 284

Lipe, William, quoted, 147
literature, submerged, 146–47, 148
local knowledge, 285
logic: of confirmation, 205, 282–83; of explanation, 205, of justification, 40, 52; role in archaeology, 47; of science, 160, 172, 199; systems of, 51
logical empiricism. *See* logical positivism
logical positivism, 2, 85, 159, 166–67, 202–4, 352, 368–69; development, 7–8; and philosophy of science, 6–17; problems of, 18–19; shift away from, 99–101, 357, 377
Logik der Forschung, 169
Longacre, William, 155; pottery-residence example, 220–24
Lorenzo, Jose Luis, quoted, 292–93
Losee, John, quoted, 53–54
lottery paradox, 197–98, 235, 236, 368
Lounsbury, Floyd, 155

MacCormac, Earl R., quoted, 257–59
McGimsey, Charles R., 37
Mach, Ernst, 7
Maldonado-Koerdell, Manuel, 117
Mandelbaum, Maurice, 4
Man in Prehistory, 107
Martin, Paul S., 115, 155; quoted, 52
Massachusetts Historical Commission, 146
Masterman, Margaret, 107
Mathews, Peter, 155

Matthews, John V., Jr., quoted, 311
maturity of disciplines, 121
maximal specificity requirement, 195
Maya Collapse, 317–26
Mayan glyph group, 155
Meadowcroft Shelter report, 291–92
meaningfulness, criteria of, 8, 16
Meltzer, David J., 70; quoted, 110
Merton, Robert, 289
metallurgy, pre-Columbian, 362–68
metaphor, in science, 258–59, 262, 370
metaphysics, traditional, 8
method, abandonment of, 95
methodology, role in science, 96–97
Mexico, archaeology in, 139
middle-range theory, 289–90, 356
migration-diffusion controversy, 337–38
migration hypotheses, 279–86, 292, 331
Mill, John Stuart, 53
minority groups, in anthropology, 126–27
mobility, professional, 133
models: in science, 260–61, 370; vs. practice, 344
monitoring process, between science and technology, 151
monographs, archaeological, 157
Morgan, Charles, quoted, 221, 223
Morlan, Richard, 311, 329
Morley, Sylvanius G., 321
Morris, Earl, 158

Morris, Elizabeth, quoted, 155–56
Mullins, Nicholas C., 153
multiplication rule for probabilities, 234, 235, 237, 243–44
museums, 143
mythology, national, 124

Nagel, Ernest, 2, 4, 176
National Geological Survey, 311
nationalism, 124
National Museum of Man, 311
National Science Foundation, 135, 139
neo-positivism, 211
Neurath, Otto, 7
New Archaeology, 7, 40, 107, 154, 239, 285, 350; philosophical basis, viii, 29, 44, 54, 159, 165, 174; problems of, 109, 202, 220, 351–54; rejects inductivism, 78, 225; as revolution, 1, 23, 67, 73, 105, 115; and technology, 151
"newish" archaeology, 355
new philosophy of science, 89–90
Newton-Smith, W.H., 20; quoted, 160, 161
normal science, 63–65, 74–75, 90–94, 103, 113; applied to archaeology, 65, 70–71, 92–93, 105, 109, 114
"normative," as technical term, 21–22
normative principle of action, 189
norms, scientific, 83
Northern Yukon Research Project, 311
novelty, in archaeology, 38
numerical taxonomy, 10

objectivity, scientific, xi, 2, 30, 94, 100, 106, 124, 213–14, 217; in social sciences, 116, 162
objectivity-subjectivity debate, 75, 116
observability, 11, 202–3, 215, 217
observation, 10, 32, 79–80, 85, 271, 356–57
Old Crow Flats, 309–17
operationism, 8–10, 369
opinion-changing tactics, 115, 290–98
Oppenheim, Paul, 165, 167, 227, 228; quoted, 229
origins research, 136, 317
Orme, Byrony, 262
Osborn, Alan, quoted, 231–32

palaeontology, 74
paradigm, 64, 69, 92, 105–6, 109; in anthropology, 107–8; in archaeology, 62, 70–71, 106–11; competing, 108; shared, 113; shifts, 99, 106, 114
parallel development, 307–8
partial knowledge, 231
patolli-pachisi case, 303–9, 333
Patterns of Discovery, 42
Paynter, Robert, 145
"peaceful Maya" concept, 181
Pecos Conferences, 157
peer review, 147
Perricone, Phillip J., quoted, 130
personality types, in science, 124
petrographic analysis, 120
philosophical concepts, 2, 3; archaeologists' misconceptions of, 30, 350
philosophizing, archaeological, 353–54, 373–79

philosophy, 24; of archaeology, viii, 378; of history, 4; of law, 243; of physics, 6; of psychology, 7, 25; scientific, 7
Philosophy and Archaeology, 358–59
Philosophy of Natural Science, 41
philosophy of science, 1–2, 83, 346; archaeological interest in, 3–6, 29, 122, 159, 165–66, 349–50; debates in, 2; and logical positivism, 6–17; recent developments, 3, 18–21, 99; and science, 21–23, 61
physical sciences, 6, 205
physics, 6, 8
plausibility, 37, 56, 241, 246, 254, 316, 326–38
Plog, Fred T., 41, 46–47, 155, 183, 286–90; quoted, 54, 286
Poincare, Jules, 7
Point of Pines site, 279–86
political conditions, international, 140
Popper, Karl, 2, 5, 7, 61–62; advocate of deductivism, 53, 77; compared with Kuhn, 82–94; criticism of, 79–82; emphasis on falsification, 31, 35, 77; model of causal explanation, 169–70; quoted, 41, 54, 88, 268; rejects inductivism, 33, 80–81, 225; view of corroboration, 253–54; view of science, 16–17, 61–62, 76–79
possibility, logical vs. physical, 13
post-empiricism, 166
post-positivism, 166
predictions, scientific, 11, 55–57, 81, 170, 194, 227; of individual events, 251; vs. retrodictions, 77

prehistory, 124, 232
prescriptions for good science, 165
preservation archaeologists, 131
Price, Derek J. De Solla, 153, 156
Price, William J., 151
"prime mover" accounts, 320–21
probability, 56, 79, 196, 199, 208; and plausibility, 328
probability calculus, 237–38, 242–46, 306; applied to archaeology, 244, 303–9; interpretations of, 243
problem formulation, 33
problem orientation: anthropological, 127; archaeological, 40, 94, 120, 278, 338
process theorists, 228
processualists. *See* process theorists
production process, in archaeology, 145
professional associations, archaeological, 100, 111
professionalism, 129, 144–45
progress, scientific, 48, 83–85, 91, 95, 101, 102
"Prospects for a Philosophy of Archaeology, The," 357, 374
psychoanalytic theory, 7
psychologists, 125
psychology, 8, 206
psychology of science, 124
Public Archaeology, 37
public funding of archaeology, 130
publicity, 130
Putnam, Hilary, quoted, 270–71, 373
puzzle solving, 63, 92, 101

puzzling facts, 187

Quinton, Anthony, quoted, 17

Raab, Mark, quoted, 147
radical subjectivism, 68, 97
Rathje, William L., 318–26
rationalism, 166–67
rationalists, 160
rationality, in explanation, 211, 212–14
rational vs. nonrational, 20
Read, Dwight, quoted, 354
realism, 85, 166–67
reasoning: forms of, 2; scientific, 106
reasons, as causes, 203–4, 208
reburial of prehistoric remains, 141
Received View of Theories, 209, 214
recruitment: in archaeology, 129–31; patterns, 127; role in paradigm expansion, 109
Redman, Charles L., 359
regional typologies, 136
regional variation, 326–38
Reichenbach, Hans, 7, 13
reintegration of paradigms, 110
relativism, 68, 75–76, 84–85, 87, 94, 161
relevance, 208, 252–53, 266, 269, 296
relevance concept of confirmation, 233–36, 239
relevant evidence, in archaeology, 286–90
Renfrew, Colin, quoted, 147
repeatability of results, 116
repetition of experiments, 78

Report of the 1984 Survey of Anthropology Ph.D's, 126, 134
representativeness of archaeological data, 37
research design, 40, 41, 94, 212, 278, 338
revolution, scientific, 72–73, 105
revolutionary science, 65–67, 74–75, 83, 86, 90–94, 103
rigor, methodological, 360
Riley, T.J., 5
risk, and hypothesis testing, 81
Roe, Anne, 125
Rogge, A.E., 157; quoted, 42, 58, 149, 154
role expectations, in archaeology, 129
Rorty, Richard, quoted, 372
routinization, 38
Russell, Bertrand, 7

S-R model of explanation. *See* statistical-relevance model of explanation
Sabloff, Jeremy A., 114
Sahlins, Marshall, quoted, 288
Salmon, Merrilee, 5, 358–59
Salmon, Wesley, 5, 23, 48, 49, 119, 208; quoted, 37, 177, 179–80, 196–98, 241, 242, 246–47, 266, 377; and statistical-relevance model, 195–99
sampling, 38, 305–6, 332
Sanders, William, quoted, 319–20
San Isabel Ixtapan mammoth discovery, 117
Sanjek, Roger, quoted, 134
Saul, Frank, 321
Saunders, Jeffrey, 338–43

scale, of archaeological projects, 148, 286–90, 337
Scheffler, Israel, quoted, 75
Schele, Linda, 155
Schiffer, Michael B., 24, 155; quoted, 109, 375
Schlick, Moritz, 7, 12–13
Schneider, Harold K., quoted, 153
School of American Research, 157, 321, 326
Schwartz, Douglas, 285
science: Big vs. Little, 156, 159, 317; as biological phenomenon, 271; difficulty of defining, 94–95; distinguished from history, 190–91; goals of, 204, 248; good vs. bad, 96; identifying marks of, 6; nonlogical aspects of, 101; practice of, 21, 43, 55, 82–83, 90, 213–14, 357; as process, 191; theories of, 94
sciences, differences among, 20
scientific community, 64, 153; behavior, 101; role, 87
scientific enterprise, 2, 19, 20, 26, 43, 369
Scientific Explanation, 170–71
Scientific Image, The, 214
scientific investigation, 204
scientific knowledge, 211–13
scientific method, 1, 20, 21, 51, 52–54, 58; applied to archaeology, 38, 297
scientific papers, role of, 156–57
scientific procedure, 80
scientific realism, 26, 202, 204, 212, 213–14, 269–72, 362–68, 368–73; archaeological interest in, 202, 205, 214, 216; vs. positivism, 204–6

scientific reputation, 116–18
scientific respectability, 15
scientist, problem of defining, 84–85
Scriven, Micahel, 5, 193–95, 196, 206–7
Secord, P.F., 201–6, 208
segmentation, in archaeology, 149
selection criteria, 76, 79, 91
selective concept of confirmation, 36, 238–40, 252, 254, 302, 327–30
Sellards, E.H., 117
Semmelweis, Ignaz, 32, 35
Shapere, Dudley, 90, 213–14
Shepard, Anna, 119
Shimkin, Demitri B., quoted, 323, 324–25
similarities, relevant, 38, 55, 183, 269, 371
similarity, 186
simplicity, 56, 334, 341, 361
Skyrms, Brian, quoted, 44, 48, 226, 306
Smith, Marion A., 11
social differentiation in archaeology, 136
social evolutionism, 188
socialization processes, 103
social psychology, 25
social sciences, 219–20; development, 6, 71; goals, 161; object of study in, 203; search for laws in, 220, 232–33
Society for American Archaeology, 100, 127, 128, 147, 149, 153, 284, 310, 357
Society for Empirical Philosophy, 18

Society of Professional Archaeology in the United States, 144
sociology, 6, 206; of archaeology, 99–101, 124–25, 158
sociology of science, 3, 20, 62, 124; applied to archaeology, 104–11; Kuhnian, 101–4; role of, 159–63
"Socio-Politics of Archaeology, The" 100
Soffer, Olga, quoted, 141
South, Stanley, quoted, 70
South Asia, archaeology in, 139
Southern Methodist University, 133
Southwestern Anthropological Research Group, 154
Spaulding, Albert C., quoted, 294
Special Consequence Condition of Adequacy, 234, 237, 255
specialization, ix, 102–3, 104, 159; disadvantages, 154
specificity, 266–68
Speth, John, quoted, 147
stalemates, in archaeology, 333
standardized mediocrity, 146
statistical explanation, cases of, 249–50; in social sciences, 251–52
statistical laws, 173
statistical-relevance model of explanation, 195–99, 358
statistical tests, 244–46
status, professional, 117, 354
stone tool data, 286–90
"strong program" in sociology of science, 160–63
structuralism, 108
Structure of Scientific Revolutions, The, 63, 64, 74

Struever, Stuart, 150
stylistic analysis, 120
subjectivity in science, 30. See also radical subjectivism
Suppe, Frederick, quoted, 209–10, 212, 213–14
suspension of judgment, 333
symmetry of explanation and prediction, 197
systems approach, 322–24
systems theory, 5, 108, 121, 201, 359

Taylor, Walter, 4
technical assistance, obtaining, 151
technological determinism, 288
technology of archaeology, 150–52
Teggart, J.F., 4
Tepexpan Man find, 117
testability requirement, 96, 169
test implications of hypotheses, 34–35
testing, 64, 66; complexity of, 79, 324; conflated with explanation, 54; importance of, 4, 29, 361; indirect methods of, 32–33, 324–25; methodology of, 279–86; Popperian view of, 77; role in normal science, 85–86; role in scientific procedures, 30, 55; second level of, 346; standards of, 113–14; use in archaeology, 31–40, 94, 344
textbooks: archaeological, 4, 42, 107, 109, 315–16; scientific, 105
Thagard, Paul, 330; quoted, 276, 334–36
theoretical terms, 204

theories, scientific, 43; choice,
101, 217, 335, 341; competing,
87, 92, 101; evaluation of, 15;
formulation of, 369; nature of,
3; rejection of, 83
theorizing, in archaeology, 358
Theory of Games and Economic Behaviour, 7
theory-practice bifurcation, 24,
349–54
thingization, 141
Thomas, David, 71
Thompson, Raymond, 4, 46–47,
51, 58, 116; quoted, 285
threshold cases, 116
Tolstoy, Paul, 298–303
total evidence requirement, 50
Toulmin, Stephen, 213; quoted,
74, 93
Townsend, A.H., 5
training of archaeologists, 103,
131–33, 138
Trigger, Bruce G., 139
triggering mechanisms, 319
true-false dichotomy, 14
truth claims, 215–16
truth condition, for scientific
laws, 229
truth in science, 67–68, 84, 100,
161, 228, 361, 372–73
Tuggle, H.D., 5
Tylor, E.B., 303–9
typological analysis, 65, 92–93,
119–20

Ultimate Agreement, 271
understanding, 200; and explanation, 355
unified science, 8
uniformity, principle of, 263–64

Union of Soviet Socialist Republics, archaeology in, 141
unique events, 38
U.S. Forest Service, 135
universality, 228, 230
universities: as institutional base
for archaeology, 143; structure
of, ix. *See also* archaeology,
graduate programs; archaeology, university programs; *names
of individual universities*
University of Arizona, 132
University of Calgary, 128
University of Khartoum, 142
University of South Carolina, 132

validation by witness, 116–17
values, 87–89
Vance, Carole, 133–34
van Fraassen, Bas, 21, 214–19,
269–72
verifiability, 11–12
verifiability criterion. *See* Verification Criterion of Meaning
verification, 78. *See also* confirmation
Verification Criterion of Meaning,
10–11, 13; criticized, 14–16
verificationism, 10–11, 83, 369
Vienna Circle, 7–8, 10, 16, 18, 83
von der Merve, Nikolaas, 152

Waismann, Friedrich, 7
Wasserstrom, Richard, 34
Watkins, John, quoted, 71–72, 86,
89
Watson, John B., 25
Watson, Patty Jo, 265, 359;
quoted, 43, 264
Watson, Richard, quoted, 264

weak verifiability requirement, 13
Weeteluktuk, Daniel, 128
Wendorf, Fred, 149
West Africa, archaeology in, 138–39
Whewell, William, 53
White, Peter, 286–90
Whitehead, Alfred North, 4
Wildesen, Leslie E., 135
Willey, Gordon R., 114; quoted, 323, 324–25
Wilmsen, Edwin, quoted, 107
Wilson, David, quoted, 112
witnessing. *See* validation by witness
Wittgenstein, Ludwig, 7, 10
Wobst, H. Martin, quoted, 136

women, professional status of, 133–36
women's groups, in archaeology, 128
Woodall, J. Ned, quoted, 130
Woodbury, Richard B., 132; quoted, 152
working conferences, 157, 326, 346
Wormington, Marie, 117
Wylie, Alison, 265, 375–76; quoted, 26, 269, 374, 376

Yukon Refugium, 311
Yoffee, Norman, 326

Zubrow, Ezra, 145